Pressure Vessels

Pressure Vessels

The ASME Code Simplified

J. Phillip Ellenberger, P.E.
Robert Chuse
Bryce E. Carson, Sr.

Eighth Edition

McGraw-Hill

New York   Chicago   San Francisco   Lisbon   London   Madrid
Mexico City   Milan   New Delhi   San Juan   Seoul
Singapore   Sydney   Toronto

The **McGraw·Hill** Companies

CIP Data is on file with the Library of Congress

Copyright © 2004, 1993, 1984, 1977, 1960, 1958, 1956, 1954 by The McGraw-Hill Companies, Inc. All rights reserved. Printed in the United States of America. Except as permitted under the United States Copyright Act of 1976, no part of this publication may be reproduced or distributed in any form or by any means, or stored in a data base or retrieval system, without the prior written permission of the publisher.

8 9 10 QVR/QVR 1 9 8 7 6 5 4 3 2

ISBN 0-07-143673-1

The sponsoring editor for this book was Kenneth P. McCombs, the editing supervisor was Stephen M. Smith, and the production supervisor was Sherri Souffrance. It was set in Century Schoolbook by Paul Scozzari of McGraw-Hill Professional's Hightstown, N.J., composition unit. The art director for the cover was Anthony Landi.

McGraw-Hill books are available at special quantity discounts to use as premiums and sales promotions, or for use in corporate training programs. For more information, please write to the Director of Special Sales, McGraw-Hill Professional, Two Penn Plaza, New York, NY 10121-2298. Or contact your local bookstore.

 This book is printed on recycled, acid-free paper containing a minimum of 50% recycled, de-inked fiber.

Information contained in this work has been obtained by The McGraw-Hill Companies, Inc. ("McGraw-Hill") from sources believed to be reliable. However, neither McGraw-Hill nor its authors guarantee the accuracy or completeness of any information published herein and neither McGraw-Hill nor its authors shall be responsible for any errors, omissions, or damages arising out of use of this information. This work is published with the understanding that McGraw-Hill and its authors are supplying information but are not attempting to render engineering or other professional services. If such services are required, the assistance of an appropriate professional should be sought.

Contents

Preface   ix

Chapter 1. Origin, Development, and Jurisdiction of the ASME Code   1

    History of the ASME Code   1
    Additions to the Code   3
    ASME Boiler and Pressure Vessel Committee   4
    Procedure for Obtaining the Code Symbol and Certificate   5
    The National Board of Boiler and Pressure Vessel Inspectors   7
    Shop Reviews by the National Board   8
    National Board Requirements   8
    Code Case Interpretations   9
    Canadian Pressure Vessel Requirements   13
    Department of Transportation Regulations   15
    Welded Repair or Alteration Procedure   16
    Managerial Factors in Handling Code Work   20
    Work-Flow Schedules   28
    Mill Orders   29
    Warehouse Orders   30
    Shop Organization   30

Chapter 2. Descriptive Guide to the ASME Code Section VIII, Division 1, Pressure Vessels   33

    Requirements for Establishing Design Thickness Based on Degree of Radiography (Code Pars. UW-11 and UW-12)   33
    Use of Internal-Pressure Cylindrical Shell Thickness Tables   42
    Use of Simplified External-Pressure Cylindrical Shell Thickness Charts   44
    Use of Internal-Pressure–Thickness Tables for Ellipsoidal and Torispherical Heads   46
    Elements of Joint Design for Heads   58
    Use of Pressure–Thickness Charts for Flat Heads and Bolted Flat Cover Plates   72
    Various Other Designs for Heads   77

Allowable Pressure, Pitch, and Thickness	79
Simplified Calculations for Reinforcement of Openings	83
Cryogenic and Low-Temperature Vessels	89
Impact-Test Exemptions	90
Service Restrictions and Welded Joint Category	97
Noncircular-Cross-Section-Type Pressure Vessels	99
Jacketed Vessels	101
Pressure Vessels Stamped with the UM Symbol	102
Stamping Requirements	103

Chapter 3. Design for Safety   105

Basic Causes of Pressure Vessel Accidents	105
Corrosion Failures	105
Stress Failures	105
Other Failures	106
Design Precautions	107
Inspection Openings	109
Quick-Opening Closures	111
Pressure Relief Devices	113

Chapter 4. Guide to Quality Control Systems for ASME Code Vessels: A Practical Guide to Writing and Implementing a Quality Control Manual   115

Chapter 5. Inspection and Quality Control of ASME Code Vessels   141

Shop Inspection	141
Tests	144
ASME Manufacturer's Data Reports	148
Field Assembly Inspection	149
Responsibility for Field Assembly	149
Inspection Procedure	150
Radiography	151
Conditions Requiring Radiography	151
Interpretation of Welding Radiographs	153
Sectioning	159
Hydrostatic Testing	161
Test Gage Requirements	162
Seven Simplified Steps to Efficient Weld Inspection	163

Chapter 6. Welding, Welding Procedure, and Operator Qualification   167

Examination of Procedure and Operator	167
Record of Qualifications	168
Welding Procedures and Qualification Simplified	168
Welding Procedure Specification Forms	170
Variables [Code Pars. QW-400, QW-250, and QW-350 and Tables QW-415 (WPS) and QW-416 (WPQ)]	173

Welder Qualifications	176
Welding Details and Symbols	180
Joint Preparation and Fit-Up	184
Effects of Welding Heat	189
Preheating	193
Postweld Heat Treatment	194
Conditions Requiring Postweld Heat Treatment	196

Chapter 7. Nondestructive Examination   199

Leak Testing	205
Acoustic Emission Testing	208

Chapter 8. ASME Code Section VIII, Division 2, Alternative Rules, and Division 3, Alternative Rules for High Pressure Vessels   211

Chapter 9. Power Boilers: Guide to the ASME Code Section I, Power Boilers   217

Chapter 10. Nuclear Vessels and Required Quality Assurance Systems   225

Power Plant Cycles	225
Pressurized Water Reactor Plants	226
Boiling Water Reactor Plants	227
Quality Assurance	231

Chapter 11. Department of Transportation Requirements for Cargo Tanks   241

History and Overview of 49 CFR §107 to §180	241
§178.345	242
Specification Cargo Tank MC-331 (§178.337)	244
Certification of Cargo Tanks (§178.340-10)	247
Maintenance of Specification Cargo Tanks—General Information (§180)	248
Glossary of Terms	251
Qualification and Maintenance of Specification Cargo Tanks (§180 to §180.417)	252
Purpose, Applicability, and General Requirements (§180.1 to §180.3)	252
Definitions (§180.403)	253
Qualification of Cargo Tanks (§180.405)	253
Upgrading Existing Specification MC-306, MC-307, and MC-312 Cargo Tanks [§180.405(c) to (g)]	254
Requirements for Tests and Inspections of Specification Cargo Tanks (§180.407)	256
Minimum Qualifications for Inspectors and Testers (§180.409)	264
Acceptable Results of Tests and Inspections (§180.411)	266
Corroded or Abraded Areas	266
Repair, Modification, Stretching, or Rebarreling of a Cargo Tank (§180.413)	267
Repair and Modifications	267

Test and Inspection Markings (§180.415)   268
Reporting and Record Retention Requirements (§180.417)   268

Chapter 12. ISO 9000 Quality Assurance Requirements   271

Introduction to ISO 9000 Standards   271
Assessment of the Quality Assurance Program   274
Notified Bodies and the CE Mark   277

Appendix A. Cylinder Volume Tables and Diagrams   279

Appendix B. Circumferences of Cylinders   285

Appendix C. Decimal Equivalents and Theoretical Weights of Steel Plates   297

Appendix D. Pipe Wall Thicknesses   299

Appendix E. Dry Saturated Steam Temperatures   301

Appendix F. Metric (SI) Units; English and Metric Conversions   305

Bibliography   311
Index   313

Preface

This eighth edition of *Pressure Vessels: The ASME Code Simplified* has been brought up to date by J. Phillip Ellenberger, who has some 40 years of experience in the field, with expanded details and some new sections that outline recent changes to the *Boiler and Pressure Vessel Code* of the American Society of Mechanical Engineers (ASME). Much of the material here is based on the content of the seventh edition, which was written by Robert Chuse and Bryce E. Carson, Sr., who together represented more than 60 years of experience in design, fabrication, and inspection of ASME Code pressure vessels. The wealth of real-world knowledge possessed by the authors makes this book a reference tool that is accurate, practical, and easily understood.

Since the publication of the seventh edition, various processes have been changed by the use of computers. However, what has not changed is the need for both theoretical and practical information to make a quality product.

The Code has been accepted for many years as the standard for the construction of safe boilers and pressure vessels. It is progressive and viable. Important changes and additions are made when required. For example, Code Par. UG-20, which deals with minimum design metal temperatures and impact-test requirements, continues to evolve as we learn more about how to alleviate problems from embrittlement. After many years of effort and study, the allowable stress values have been increased. This increase represents the continued improvement in material reliability and the ability to maintain adequate margins.

Note that the National Board of Boiler and Pressure Vessel Inspectors has assumed responsibility for the *Synopsis of Boiler and Pressure Vessel Laws, Rules and Regulations.*

This book is geared to meet the needs of engineering, manufacturing, repair, and testing companies that have to comply with Code requirements, and to make the Code work better, more efficient, and more profitable.

The U.S. Department of Transportation continues to modify the rules for cargo tanks. These changes are explained in rewritten Chap. 11.

As Europe unites economically, the adoption of international standards becomes essential. Just as overseas manufacturers must now conform to ASME standards for vessels to be exported to the United States, the ISO 9000 quality assurance system and the provisions of the Pressure Equipment Directive must be understood and implemented in order to export pressure vessels to Europe in the near future. These quality regulations are described and clarified in Chap. 12.

Welding is one of the most important functions in building pressure vessels. This book is a practical reference and text that can be used in schools and industry. It explains the theory of the weld, welding metallurgy, effects of welding heat, types of welding tests, welding procedures, and qualification of procedures so they can be written as required by the Code and understood clearly. Examples show how tests are prepared, completed, and correctly evaluated.

The various charts, tables, and forms on design, fabrication, and inspection of Code vessels are created to facilitate the reader's application of Code requirements. For example, estimators, engineers, and inspectors will find the thickness charts for cylindrical shells and dished heads for internal pressure useful because they provide information on required thicknesses at a glance. And there is a method given for using these charts with the new allowable stresses. All references to specific parts of the Code—subsections, parts, paragraphs, figures, tables, and appendixes—have been prefixed with the word *Code* (for example, Code Fig. 1-4) in order to distinguish them from references to the various figures, tables, and appendixes in this text.

It is the authors' hope that this book will clarify the reader's understanding of the Code and that it will encourage people in the industry to take all the necessary precautions in ordering, designing, fabricating, and inspecting Code vessels.

The data have been selected to answer questions most frequently asked by pressure vessel manufacturers and repair concerns. This information is advisory only, gained from the authors' long experience in the pressure vessel industry. There is no obligation on the part of anyone to adhere to the recommendations made.

It must be remembered that there is no alternative to reading and understanding the ASME Code. This book is not meant to replace the reading of the Code, but to clarify it.

Acknowledgments

This book reflects an industry-wide effort based on contributions of many persons and companies. Special appreciation goes to the following:

American Society of Mechanical Engineers, New York, New York

American Society for Nondestructive Testing, Columbus, Ohio

American Welding Society, Miami, Florida

Hartford Steam Boiler Inspection and Insurance Company, Hartford, Connecticut

L. W. Welding and Tank Repair Co., Uxbridge, Massachusetts

National Board of Boiler and Pressure Vessel Inspectors, Columbus, Ohio

National Tank Truck Carriers, Inc., Alexandria, Virginia

TUV America, Inc., Danvers, Massachusetts

Laura Baum, Katy, Texas

Karen Carson, East Stroudsburg, Pennsylvania

Eve Chuse, Leonia, New Jersey

Albert Egreczky, Basking Ridge, New Jersey

Donna Fanelli, Flemington, New Jersey

Greg Hunt, Buffalo, New York

Ralph Rockwood, Chicago, Illinois

Brent Wiersma, Uxbridge, Massachusetts

C. H. Willer, Miami, Florida

James Wolcott, Odessa, New York

Pressure Vessels

Chapter

1

Origin, Development, and Jurisdiction of the ASME Code

History of the ASME Code

On March 20, 1905, a disastrous boiler explosion occurred in a shoe factory in Brockton, Massachusetts, killing 58 persons, injuring 117 others, and causing a quarter of a million dollars in property damage. For years prior to 1905, boiler explosions had been regarded as either an inevitable evil or "an act of God" (see Figs. 1.1 and 1.2). But this catastrophic accident had the effect of making the people of Massachusetts see the necessity and desirability of legislating rules and regulations for the construction of steam boilers in order to secure their maximum safety. After much debate and discussion, the state enacted the first legal code of rules for the construction of steam boilers in 1907. In 1908, the state of Ohio passed similar legislation, the Ohio Board of Boiler Rules adopting, with a few changes, the rules of the Massachusetts Board.

Therefore, other states and cities in which explosions had taken place began to realize that accidents could be prevented by the proper design, construction, and inspection of boilers and pressure vessels and began to formulate rules and regulations for this purpose. As regulations differed from state to state and often conflicted with one another, manufacturers began to find it difficult to construct vessels for use in one state that would be accepted in another. Because of this lack of uniformity, both manufacturers and users made an appeal in 1911 to the Council of the American Society of Mechanical Engineers to correct the situation. The Council answered the appeal by appointing a committee "to formulate standard specifications for the construction of steam boilers and other pressure vessels and for their care in service."

Figure 1.1 The Brockton, Massachusetts, shoe factory. (*Courtesy of The Hartford Steam Boiler Inspection and Insurance Company.*)

The first committee consisted of seven members, all experts in their respective fields: one boiler insurance engineer, one material manufacturer, two boiler manufacturers, two professors of engineering, and one consulting engineer. The committee was assisted by an advisory committee of 18 engineers representing various phases of design, construction, installation, and operation of boilers.

Following a thorough study of the Massachusetts and Ohio rules and other useful data, the committee made its preliminary report in 1913 and sent 2000 copies of it to professors of mechanical engineering, engineering departments of boiler insurance companies, chief inspectors of boiler inspection departments of states and cities, manufacturers of steam boilers, editors of engineering journals, and others interested in the construction and operation of steam boilers, with a request for suggestions of changes or additions to the proposed regulations.

After three years of countless meetings and public hearings, a final draft of the first *ASME Rules for Construction of Stationary Boilers and For Allowable Working Pressures,* known as the 1914 edition, was adopted in the spring of 1915.

Figure 1.2 Shoe factory after the boiler explosion of March 20, 1905, which led to the adoption of many state boiler codes and the *ASME Boiler and Pressure Vessel Code*. (*Courtesy of The Hartford Steam Boiler Inspection and Insurance Company.*)

Additions to the Code

Since 1914, many changes have been made and new sections added to the Code as the need arose. The present sections are listed in the following order:

Section I. Power Boilers

Section II. Materials

    Part A: Ferrous Material Specifications

    Part B: Nonferrous Material Specifications

    Part C: Specifications for Welding, Rods, Electrodes, and Filler Metals

    Part D: Properties

Section III. Rules for Construction of Nuclear Components

    Subsection NCA: General Requirements for Divisions 1 and 2

      Division 1

        Subsection NB: Class 1 Components

        Subsection NC: Class 2 Components

        Subsection ND: Class 3 Components

        Subsection NE: Class MC Components

        Subsection NF: Supports

Subsection NG: Core Support Structures

Subsection NH: Class 1 Components in Elevated Temperature Service

Appendices

Division 2. Code for Concrete Containments

Division 3. Containment Systems for Storage and Transport Packaging of Spent Nuclear Fuel and High Level Radioactive Materials and Waste

Section IV. Heating Boilers

Section V. Nondestructive Examination

Section VI. Recommended Rules for Care and Operation of Heating Boilers

Section VII. Recommended Guidelines for the Care of Power Boilers

Section VIII. Pressure Vessels

Division 1

Division 2. Alternative Rules

Division 3. Alternative Rules for Construction of High Pressure Vessels

Section IX. Welding and Brazing Qualifications

Section X. Fiber-Reinforced Plastic Pressure Vessels

Section XI. Rules for Inservice Inspection of Nuclear Power Plant Components

ASME Boiler and Pressure Vessel Committee

The increase in the size of the Code reflects the progress of industry in this country. To keep up with this spontaneous growth, constant revisions have been required. The ASME Code has been kept up to date by the Boiler and Pressure Vessel Committee (currently consisting of more than 800 volunteer engineers and other technical professionals) which considers the needs of the users, manufacturers, and inspectors of boilers and pressure vessels. In the formulation of its rules for the establishment of design and operating pressures, the Committee considers materials, construction, methods of fabrication, inspection, certification, and safety devices. The ASME works closely with the American National Standards Institute (ANSI) to assure that the resulting documents meet the ANSI criteria for publication as American National Standards.

The members of the Committee do not represent particular organizations or companies but have recognized background and experience by which they are placed in categories, which include manufacturers, users of the products for which the codes are written, insurance inspection, regulatory, and general. The Committee meets on a regular basis to consider requests for interpretations and revisions and additions to Code rules as dictated by advances in technology. Approved revisions and additions are published semiannually as addenda to the Code.

To illustrate, boilers were operating in 1914 at a maximum pressure of 275 psi and temperature of 600°F. Today, boilers are designed for pressures as high as 5000 psi and temperatures of 1100°F, and pressure vessels for pressures of 3000 psi and over and for temperatures ranging from −350°F to more than 1000°F.

Each new material, design, fabrication method, and protective device brought new problems to the Boiler Code Committee, requiring the expert technical advice of many subcommittees in order to expedite proper additions to and revisions of the Code. As a result of the splendid work done by these committees, the *ASME Boiler and Pressure Vessel Code* has been developed; it is a set of standards that assures every state of the safe design and construction of all boiler and pressure vessels used within its borders and is used around the world as a basis for enhancing public health, safety, and welfare. Many foreign manufacturers are accredited under the provisions of the *ASME Boiler and Pressure Vessel Code*.

Procedure for Obtaining the Code Symbol and Certificate

Users of pressure vessels prefer to order ASME Code vessels because they know that such vessels will be designed, fabricated, and inspected to an approved quality control system in compliance with a safe standard. Pressure vessel manufacturers want the Code symbol and Certificate of Authorization so that they will be able to bid for Code work, thereby broadening their business opportunities. They also believe that authorization to build Code vessels will enhance the reputation of their shop.

If a company is interested in building Code vessels according to the ASME Section VIII, Division 1, Pressure Vessels Code, it should acquaint itself with Code Pars. U-2 and UG-92, which outline the manufacturer's responsibilities and define the requirements for an inspector. This third party in the manufacturer's plant, by virtue of being authorized by the state to do Code inspection, is the legal representation which permits the manufacturer to fabricate under state laws (the ASME Code).

Manufacturers who want to construct Code vessels covered by Section VIII, Division 1, obligate themselves with respect to quality and documentation (see Code Appendix 10, Quality Control Systems). A survey will be required for the initial issuance of an ASME Certificate of Authorization and for each renewal. The evaluation is performed jointly by the Authorized Inspection Agency and the jurisdictional authority concerned which has adopted, and also administers, the applicable boiler and pressure vessel legislation. When the jurisdictional authority does not make the survey, or the jurisdiction is the inspection agency, the National Board of Boiler and Pressure Vessel Inspectors will be asked to participate in the survey.

After the survey has been jointly made to establish that a quality control system is actually in practice, the National Board representative, if involved, and the jurisdiction authority will discuss their findings with the manufacturer. If the manufacturer's system does not meet Code requirements, the company will be asked to make the necessary corrections. The survey team will then forward its report to the ASME with the recommendation that the manufacturer receive the Certificate of Authorization or, if failure to meet standards exists, that the certificate not be issued.

All Code shops must follow the above procedures in order to obtain the Code symbol. If your shop wants the Code Certificate of Authorization, you should write to the Secretary of the Boiler and Pressure Vessel Code Committee, ASME, 3 Park Avenue, New York, NY 10016-5990, stating your desire to build such vessels. To help the secretary and the Subcommittee on Code Symbol Stamps evaluate your shop, the application must describe the type and size of vessel the shop is capable of building and the type and size of your equipment, especially fabricating equipment. If possible, you should arrange with some Authorized Inspection Agency, such as your state, city, or insurance company, to undertake the inspection service after your shop has received ASME certification. You can then inform the ASME that specific inspection arrangements have already been made.

The secretary will send a statement of your request to the Subcommittee on Code Symbol Stamps. A request will also be made to the chief inspector of the particular state, city, or other inspection agency governing your company and/or the National Board of Boiler and Pressure Vessel Inspectors to make a survey that will determine whether or not your company has a quality control system that is capable of designing and fabricating pressure vessels by ASME Code rules.

The secretary receives a report of this survey and will then forward the report to the Subcommittee on Code Symbol Stamps. Should the subcommittee judge favorably, your company will be issued a Code symbol and Certificate of Authorization. A manufacturer with the ability, integrity, and quality control system to design and fabricate good pressure vessels will have no difficulty in obtaining Code authorization.

The National Board of Boiler and Pressure Vessel Inspectors

The National Board of Boiler and Pressure Vessel Inspectors, an independent, nonprofit organization whose members are the jurisdictional officials responsible for enforcing and administrating the *ASME Boiler and Pressure Vessel Code,* was first organized in 1919. Since then, it has served for the uniform administration and enforcement of the rules of the *ASME Boiler and Pressure Vessel Code.* In drafting the first Code, the Boiler Code Committee realized that it had no authority to write rules to govern administration, and that compliance could be made mandatory only by the legislative bodies of states and cities.

Although some authorities thought that uniformity could be achieved by a code of uniform rules, they proved to be mistaken. Various states and cities adopted the ASME Code with the provision that boilers be inspected during construction by an inspector qualified under their own regulations; the boiler was then to be stamped with the individual local stamping to indicate its conformity with these regulations. This requirement created an unwieldy situation, for boilers constructed in strict accordance with the ASME Code still had to be stamped with the local stamping, thereby causing needless delay and expense in delivery of the vessel.

It was evident that some arrangement had to be made to overcome such difficulties. Therefore, boiler manufacturers met with the chief inspectors of the states and cities that had adopted the ASME Code and formed the National Board of Boiler and Pressure Vessel Inspectors for the purpose of presenting the ASME Code to governing bodies of all states and cities. Their aim was not only to promote safety and uniformity in the construction, installation, and inspection of boilers and pressure vessels but also to establish reciprocity between political subdivisions of the United States. Such ideals could best be carried out by a central organization under whose auspices chief inspectors, or other officials charged with the enforcement of inspection regulations, could meet and discuss their problems. The efforts of this first group of administrators succeeded in extending National Board membership to all Canadian provinces and most of the states and cities of the United States. It is now possible for an authorized shop to build a boiler or pressure vessel that will be accepted anywhere in the United States or Canada after it has been inspected by an Authorized Inspector holding a National Board Commission.

The ASME Code requires that Inspectors must meet certain minimum requirements of education and experience and must pass a written examination before they can be commissioned to perform Code inspections. One of the many functions of the National Board is the commissioning of Authorized Inspectors.

Shop Reviews by the National Board

Before the issuance of an ASME Certificate of Authorization to build Code vessels, the manufacturer must have and demonstrate a quality control system. This system has to include a written description explaining in detail the quality-controlled manufacturing process.

Before the issuance of renewal of a Certificate of Authorization, the manufacturer's facilities and organization are subject to a joint review by the inspection agency and the legal jurisdiction concerned. For those areas where there is no jurisdiction, or where a jurisdiction does not review a manufacturer's facility, and/or the jurisdiction is the inspection agency, the function may be carried out by a representative of the National Board of Boiler and Pressure Vessel Inspectors. When the review of the manufacturer's facilities and organization has been made jointly by the inspection agency, and/or the legal jurisdiction, or a representative of the National Board of Boiler and Pressure Vessel Inspectors, a written report will be sent to the American Society of Mechanical Engineers. The National Board of Boiler and Pressure Vessel Inspectors also participates in all nuclear surveys.

The National Board also issues Certificates of Authorization for the use of the National Board R stamp, to be applied to ASME Code and National Board–stamped boilers and/or pressure vessels on which repairs are to be made (see the section Welded Repair or Alteration Procedure in this chapter).

Another important service the National Board performs is the certification of safety valves and safety relief valves. The *ASME Boiler and Pressure Vessel Code* contains specific requirements governing the design and capacity certification of safety valves installed on Code-stamped vessels. The certification tests are conducted at a testing laboratory approved by the ASME Boiler and Pressure Vessel Committee.

In addition to safety valve certification, the National Board has its own testing laboratory for its experimental work and for certifying ASME safety and relief valves. The facilities of the laboratory are also available to manufacturers and other organizations for research, development, or other test work.

National Board Requirements

All Canadian provinces and most states require boilers and pressure vessels to be inspected during fabrication by an Inspector holding a National Board Commission and then to be stamped with a National Board standard number. Qualified and authorized boiler and pressure vessel manufacturers must be registered with the National Board of Boiler and Pressure Vessel Inspectors, 1055 Crupper Ave., Columbus, OH 43229. In addition, two data sheets on each vessel must be filed with the National Board, one copy of which is retained by the Board and the other sent to the administrative authority of the state, city, or province in which the vessel is to be used (see Code Par. UG-120).

The National Board of Boiler and Pressure Vessel Inspectors is constantly working to assure greater safety of life and property by promoting and securing uniform enforcement of boiler and pressure vessel laws and uniform approval of designs and structural details of these vessels, including the accessories that affect their safe operation; by furthering the establishment of a uniform Code; by espousing one standard of qualifications and examinations for inspectors who are to enforce the requirements of this Code; and by seeing that all relevant data of the Code are made available to members.

In prior years the Uniform Boiler and Pressure Vessel Laws Society issued a *Synopsis of Boiler and Pressure Vessel Laws, Rules, and Regulations*, the publishing of which has been assumed by the National Board. The book gives a summary of applicable laws of states, provinces, cities, and counties. It has proved valuable to users and manufacturers of boilers and pressure vessels. Figures 1.3 and 1.4 show political subdivisions of the United States and Canada.

Code Case Interpretations

As the Code does not cover all details of design, construction, and materials, pressure vessel manufacturers sometimes have difficulty in interpreting it when trying to meet specific customer requirements. In such cases the Authorized Inspector should be consulted. If the Inspector is unable to give an acceptable answer or has any doubts or questions about the proper interpretation of the intent of the Code, the question can be referred to the Inspector Supervisor's office. If the Inspector and his or her supervisor are not able to provide a ruling, the manufacturer may then request the assistance of the Boiler and Pressure Vessel Committee, which meets regularly to consider inquiries of this nature. In referring questions to the Boiler and Pressure Vessel Committee, it is necessary to submit complete details, sketches of the construction involved, and references to the applicable paragraphs of the Code; it is also useful to include opinions expressed by others. (See Code Appendix 16 for guidance in preparing inquiries to the ASME.)

Inquiries should be submitted by letter to the Secretary of the Boiler and Pressure Vessel Committee, ASME, 3 Park Avenue, New York, NY 10016-5990. (It should be noted that with the advent of the Internet, much of this work can be accomplished through the Web. The ASME can be found at www.asme.org.) The secretary will distribute copies of the inquiry to the committee members for study. At the next committee meeting, interpretations will be formulated.

Figure 1.3 Boiler and pressure vessel laws in the United States and Canada. (*Reprinted with permission of the National Board of Boiler and Pressure Vessel Inspectors, www.nationalboard.org, from* National Board Synopsis of Boiler and Pressure Vessel Laws, Rules and Regulations, © 2003.)

These interpretations (issued twice each year) may then become addenda to the Code. If so, they will first be printed in the Addenda supplement of the Code for subsequent inclusion in the latest edition of the Code (printed every three years). Addendas are issued once each year.

The addenda incorporate as many case interpretations as possible in order to keep the open file of cases to a minimum. For their own good,

BOILER AND PRESSURE VESSEL CODES

KEY:
- I – Power Boilers
- II – Materials
- III(1) – Nuclear Facility Components
- III(2) – Nuclear Facility (Code for Concrete Containments)
- IV – Heating Boilers
- V – Nondestructive Examination
- VI – Care and Operating of Heating Boilers
- VII – Care of Power Boilers
- VIII(1) – Pressure Vessels
- VIII(2) – Pressure Vessels (Alternate Rules)
- VIII(3) – Pressure Vessels (Alternate Rules for High Pressure)
- IX – Welding and Brazing Qualifications
- X – Fiber-Reinforced Plastic Pressure Vessels
- XI – Inservice Inspection, Nuclear

Y – Required by Law   N – Not Required by Law

UNITED STATES

Jurisdiction	I	II	III(1)	III(2)	IV	V	VI	VII	VIII(1)	VIII(2)	VIII(3)	IX	X	XI	Code Cases
Alabama	N	N	N	N	N	N	N	N	N	N	N	N	N	N	N
Alaska	Y	Y	N	N	Y	Y	Y	Y	Y	Y	Y	Y	Y	N	Y
Arizona	Y	Y	N	N	Y	Y	N	N	N	N	N	Y	N	N	N
Arkansas	Y	Y	Y	Y	Y	Y	Y	Y	Y	Y	Y	Y	Y	Y	Y
California	Y	Y	Y	Y	Y	Y	Y	Y	Y	Y	N	Y	Y	Y	N
Colorado	Y	Y	N	N	Y	N	N	N	Y	Y	N	N	Y	N	N
Connecticut	Y	Y	Y	N	Y	Y	Y	Y	N	N	N	Y	N	N	Y
Delaware	Y	Y	Y	Y	Y	Y	Y	Y	Y	Y	Y	Y	Y	Y	Y
Florida	Y	Y	N	N	Y	Y	Y	Y	N	N	N	Y	N	N	N
Georgia	Y	N	Y	Y	Y	N	N	N	Y	Y	Y	N	Y	Y	Y
Hawaii	Y	Y	Y	Y	Y	Y	Y	Y	Y	Y	N	Y	Y	Y	Y
Idaho	Y	Y	Y	Y	Y	Y	Y	Y	Y	Y	Y	Y	Y	Y	Y
Illinois	Y	Y	N	N	Y	Y	Y	Y	Y	Y	Y	Y	Y	N	N
Indiana	Y	Y	Y	Y	Y	Y	N	N	Y	Y	N	Y	N	N	N
Iowa	Y	Y	Y	Y	Y	Y	N	Y	Y	Y	Y	Y	Y	N	N
Kansas	Y	Y	N	N	Y	Y	Y	Y	Y	Y	Y	Y	Y	N	N
Kentucky	Y	Y	Y	Y	Y	Y	N	N	Y	Y	N	Y	N	N	Y
Louisiana	Y	N	N	N	Y	N	Y	Y	N	N	N	Y	N	N	Y
Maine	Y	Y	N	N	Y	Y	Y	Y	Y	Y	Y	Y	Y	N	N
Maryland	Y	Y	Y	Y	Y	Y	Y	Y	Y	Y	Y	Y	Y	Y	Y
Massachusetts	Y	Y	Y	N	Y	Y	N	N	Y	N	N	Y	Y	Y	Y
Michigan	Y	Y	Y	N	Y	Y	N	N	Y	N	N	Y	N	Y	Y
Minnesota	Y	Y	Y	Y	Y	Y	Y	Y	Y	Y	Y	Y	Y	Y	Y
Mississippi	Y	N	N	N	Y	N	N	N	Y	N	N	N	N	N	N
Missouri	Y	Y	N	N	Y	Y	Y	Y	Y	Y	N	Y	Y	N	Y
Montana	Y	Y	N	N	Y	Y	Y	Y	N	N	N	Y	N	N	N
Nebraska	Y	Y	N	N	Y	Y	Y	Y	Y	Y	Y	Y	Y	N	Y
Nevada	Y	Y	N	N	Y	Y	Y	Y	Y	Y	Y	Y	Y	N	N
New Hampshire	Y	Y	Y	Y	Y	Y	Y	Y	Y	Y	N	Y	Y	Y	N
New Jersey	Y	Y	Y	Y	Y	Y	Y	Y	Y	Y	N	Y	Y	Y	Y
New Mexico	Y	N	N	N	Y	N	N	N	N	N	N	Y	N	N	N
New York	Y	Y	N	N	Y	Y	N	N	Y	N	N	N	N	N	Y
North Carolina	Y	Y	Y	Y	Y	Y	N	N	Y	Y	Y	Y	Y	Y	N
North Dakota	Y	Y	N	N	Y	Y	N	N	Y	Y	Y	Y	Y	N	N
Ohio	Y	Y	Y	Y	Y	Y	Y	Y	Y	Y	N	Y	Y	Y	Y
Oklahoma	Y	Y	N	N	Y	Y	Y	Y	Y	Y	Y	Y	N	N	N
Oregon	Y	Y	Y	Y	Y	Y	Y	Y	Y	Y	Y	Y	Y	Y	Y
Pennsylvania	Y	Y	Y	Y	Y	N	N	N	Y	Y	Y	Y	N	Y	Y
Rhode Island	Y	Y	Y	Y	Y	Y	Y	Y	Y	Y	N	Y	Y	Y	Y
South Carolina	Y	Y	N	N	N	Y	N	N	N	N	N	Y	N	N	N
South Dakota	Y	N	N	N	Y	Y	N	N	N	N	N	Y	N	N	N
Tennessee	Y	Y	Y	Y	Y	Y	Y	Y	Y	Y	Y	Y	Y	Y	Y
Texas	Y	N	Y	Y	Y	N	N	N	Y	Y	N	N	N	Y	Y

Figure 1.4 Tabulation of boiler and pressure vessel codes and stamping requirements in the United States and Canada. (*Reprinted with permission of the National Board of Boiler and Pressure Vessel Inspectors, www.nationalboard.org, from* National Board Synopsis of Boiler and Pressure Vessel Laws, Rules and Regulations, © 2003.)

UNITED STATES continued...

Jurisdiction	I	II	III (1)	III (2)	IV	V	VI	VII	VIII (1)	VIII (2)	VIII (3)	IX	X	XI	Code Cases
Utah	Y	Y	N	N	Y	Y	Y	Y	Y	Y	Y	Y	Y	Y	Y
Vermont	Y	Y	N	N	Y	Y	N	N	Y	Y	N	Y	Y	N	N
Virginia	Y	Y	Y	Y	Y	Y	Y	Y	Y	Y	N	Y	Y	Y	Y
Washington	Y	N	Y	Y	Y	N	N	N	Y	Y	Y	N	Y	N	N
West Virginia	Y	Y	N	N	Y	Y	Y	Y	Y	Y	Y	Y	N	N	Y
Wisconsin	Y	Y	Y	Y	Y	Y	N	N	Y	Y	N	Y	Y	Y	Y
Wyoming	N	N	N	N	N	N	N	N	Y	Y	Y	N	N	N	N

CANADA

Jurisdiction	I	II	III (1)	III (2)	IV	V	VI	VII	VIII (1)	VIII (2)	VIII (3)	IX	X	XI	Code Cases
Alberta	Y	Y	Y	Y	Y	Y	Y	Y	Y	Y	Y	Y	Y	Y	Y
British Columbia	Y	Y	Y	Y	Y	Y	Y	Y	Y	Y	Y	Y	Y	Y	Y
Manitoba	Y	Y	Y	Y	Y	Y	Y	Y	Y	Y	N	Y	Y	Y	Y
New Brunswick	Y	Y	Y	Y	Y	Y	Y	Y	Y	Y	N	Y	Y	Y	N
Newfoundland and Labrador	Y	Y	N	N	Y	Y	Y	Y	Y	Y	N	Y	Y	N	N
Northwest Territories	Y	Y	N	N	Y	Y	Y	Y	Y	Y	N	Y	N	N	N
Nova Scotia	Y	Y	N	N	Y	Y	Y	Y	Y	Y	Y	Y	Y	N	Y
Nunavut Territory	Y	Y	Y	Y	Y	Y	Y	Y	Y	Y	N	Y	Y	Y	N
Ontario	Y	Y	Y	Y	Y	Y	Y	Y	Y	Y	Y	Y	Y	Y	Y
Prince Edward Island	Y	Y	Y	Y	Y	Y	Y	Y	Y	Y	N	Y	Y	N	Y
Québec	N	N	N	N	N	N	N	N	N	N	N	Y	N	N	N
Saskatchewan	Y	Y	Y	Y	Y	Y	N	N	Y	Y	N	Y	Y	Y	N
Yukon Territory	Y	Y	Y	Y	Y	Y	Y	Y	Y	Y	N	Y	Y	N	N

CITIES, COUNTIES AND TERRITORIES OF THE UNITED STATES

Jurisdiction	I	II	III (1)	III (2)	IV	V	VI	VII	VIII (1)	VIII (2)	VIII (3)	IX	X	XI	Code Cases
Albuquerque, New Mexico	Y	Y	N	N	Y	N	Y	Y	N	N	N	Y	N	N	Y
Buffalo, New York	Y	N	Y	Y	Y	N	N	Y	Y	N	N	N	N	N	Y
Chicago, Illinois	Y	Y	Y	Y	Y	Y	Y	Y	Y	Y	N	Y	Y	Y	Y
Denver, Colorado	Y	Y	Y	Y	Y	Y	Y	Y	Y	Y	N	Y	Y	Y	N
Detroit, Michigan	Y	Y	Y	Y	Y	Y	Y	Y	Y	Y	N	Y	Y	N	N
Los Angeles, California	Y	Y	Y	Y	Y	N	N	Y	Y	Y	N	Y	N	N	Y
Miami-Dade County, Florida	Y	N	N	N	Y	N	N	N	Y	Y	N	N	N	N	Y
Milwaukee, Wisconsin	N	N	N	N	N	N	N	N	N	N	N	N	N	N	N
New Orleans, Louisiana	Y	Y	Y	Y	Y	Y	Y	Y	Y	Y	N	Y	Y	Y	N
Puerto Rico	Y	Y	Y	Y	Y	Y	Y	Y	Y	Y	N	Y	Y	Y	Y
Seattle, Washington	Y	Y	Y	Y	Y	Y	Y	Y	Y	Y	N	Y	Y	Y	Y
Spokane, Washington	Y	Y	N	N	Y	Y	Y	Y	Y	Y	N	Y	Y	N	Y
St. Louis, Missouri	Y	Y	N	N	Y	Y	Y	Y	Y	Y	Y	Y	Y	N	N
Washington, D.C.	Y	Y	Y	Y	Y	Y	Y	Y	Y	Y	N	Y	Y	Y	Y

Figure 1.4 (*Continued*)

STAMPING REQUIREMENTS

UNITED STATES

Jurisdiction	National Board	ASME
Alabama	None	None
Alaska	R, VR	All, except HV, NV, UV, V
Arizona	R, VR	A, E, H, HLW, HV, M, PP, S, V
Arkansas	NR, R, VR	All
California	None	All
Colorado	NR, R, VR	All
Connecticut	R, VR	A, E, H, HLW, HV, M, N, NA, NPT, NV, PP, S, V
Delaware	NR, R, VR	All
Florida	R, VR	A, E, H, HLW, HV, M, PP, S, V
Georgia	NR, R, VR	A, E, H, HLW, HV, M, N, NA, NPT, NV, PP, RP, S, U, U2, U3, UM UV, V
Hawaii	R, VR	A, E, H, HLW, HV, M, PP, S, U, UM UV, V
Idaho	NR, R, VR	All
Illinois	R, VR	A, E, H, HLW, HV, M, N,* NA,* NPT,* NV,* PP, RP, S, U, U2, UM, UV, V
Indiana	NR, R	A, E, H, HLW, M, N, NA, NPT, PP, S, U, U2, UM UV, V
Iowa	R, VR	A, E, H, HLW, HV, M, N, NA, NPT, PP, S, U, U2, UV, V
Kansas	R, VR	A, E, H, HLW, HV, M, PP, S, U, UM, V
Kentucky	R, VR	A, E, H, HLW, HV, PP, S, U, UV, V
Louisiana	R, VR	A, E, H, HLW, HV, M, S
Maine	R, VR	A, E, H, HLW, HV, M, PP, RP, S, U, U2, U3, UM, V
Maryland	R	All
Massachusetts	NR, R, VR	H, M, N, NA, NPT, NV, PP, S, U, UM
Michigan	None	A, E, H, HLW, HV, M, N, NA, NPT, NV, PP, S, V
Minnesota	NR, R, VR	All
Mississippi	VR	A, E, H, HLW, HV, M, S, U, UV, V
Missouri	R, VR	All
Montana	R, VR	A, E, H, HV, S
Nebraska	R, VR	A, E, H, HLW, HV, M, PP, S, U, U2, U3, V
Nevada	NR, R, VR	A, E, H, HLW, HV, M, NV, PP, S, U, U2, UM, UV, V
New Hampshire	NR, R, VR	A, H, N, NA, NPT, PP, S, U, UV
New Jersey	R	All
New Mexico	None	A, E, H, HLW, HV, PP, S, V
New York	R, VR	A, E, H, M, PP, S, U, V
North Carolina	R, VR	All
North Dakota	R, VR	A, E, H, HLW, HV, M, PP, RP, S, U, U2, U3, UD, UV, UV3, V
Ohio	NR, R, VR	All
Oklahoma	R, VR	A, E, H, HLW, HV, M, PP, S, U, U2, UM, UV, V
Oregon	R, VR	A, H, HLW, N, NPT, PP, S, U, U2, U3, UM, V
Pennsylvania	NR, R, VR	A, E, H, HLW, HV, M, N, NA, NPT, NV, PP, S, U, U2, U3, UM, UV, V
Rhode Island	NR, R, VR	All
South Carolina	None	None
South Dakota	NR, R, VR	A, E, H, HLW, HV, M, PP, S, V
Tennessee	NR, R, VR	All
Texas	R, VR	A, E, H, HLW, HV, M, N, NA, NPT, NV, PP, S, U, U2, UV, V
Utah	NR, R, VR	A, E, H, HLW, HV, M, N, NA, NPT, NV, PP, RP, S, U, U2, U3, UV, V
Vermont	R, VR	A, E, H, HLW, HV, M, PP, S, U, U2, UM, UV, V
Virginia	None	All
Washington	NR, R, VR	All
West Virginia	R, VR	A, E, PP, S
Wisconsin	NR, R, VR	All
Wyoming	None	U, U2, U3, UV

* Comes under the jurisdiction of the Illinois Department of Nuclear Safety.

Figure 1.4 (*Continued*)

all Code manufacturers should subscribe to the Code Cases offered by the American Society of Mechanical Engineers. They will then automatically receive all cases issued by the Boiler and Pressure Vessel Committee, some of which may prove very useful.

Canadian Pressure Vessel Requirements

United States manufacturers of ASME Code vessels often receive orders for vessels to be installed in Canada. All provinces of Canada

CANADA

Jurisdiction	National Board	ASME
Alberta	None	All when manufactured outside Canada
British Columbia	None	All
Manitoba	R, VR	S, U, H when manufactured outside Canada
New Brunswick	NR, R, VR	H, HLW, N, NV, PP, S, U
Newfoundland and Labrador	None	All
Northwest Territories	None	A, E, H, HLW, HV, M, S, U, U2, UM, UV, V
Nova Scotia	R, VR	A, M, PP, S, U, U2, UM, UV, V
Nunavut Territory	None	A, E, H, HLW, HV, M, S, U, U2, UM, UV, V
Ontario	None	All
Prince Edward Island	NR, VR	H, HLW, HV, M, N, NA, NPT, NV, S, U, U2, UV, V
Québec	None	All when manufactured outside Canada
Saskatchewan	None	E, H, HLW, HV, M, S, U, U2, UM, UV, V
Yukon Territory	None	All

CITIES, COUNTIES AND TERRITORIES OF THE UNITED STATES

Jurisdiction	National Board	ASME
Albuquerque, New Mexico	R, VR	A, E, H, HLW, HV, M, S, U, UM, UV, V
Buffalo, New York	R, VR	H, S, U
Chicago, Illinois	All	All
Denver, Colorado	NR, R, VR	All
Detroit, Michigan	None	All
Los Angeles, California	NR, R, VR	A, H, HLW, HV, M, PP, S, U, UM, UV, V
Miami-Dade County, Florida	None	None
Milwaukee, Wisconsin	R	H, S, U
New Orleans, Louisiana	NR, R, VR	All
Puerto Rico	NR, R, VR	All
Seattle, Washington	NR, R, VR	All
Spokane, Washington	R, VR	A, E, H, HLW, M, PP, S, U, U2, UV, V
St. Louis, Missouri	R	H, S, U, U2
Washington, D. C.	VR	A, E, H, PP, U

Figure 1.4 (*Continued*)

have adopted the ASME Code and require vessels to be shop-inspected by an Inspector holding a National Board Commission and to be stamped with a provincial registration number in addition to the ASME symbol and National Board stamping.

But before construction begins, one important requirement has to be met in all provinces of Canada. Each manufacturer must submit blueprints and specification sheets in triplicate of all designs for approval and registration by the chief inspector of the province in which the vessel is to be used.

After the chief inspector receives the drawings, they are checked by an engineer to determine their compliance with the Code and also with provincial regulations. If the design does not meet the requirements, a report is sent to the manufacturer, explaining why it cannot be approved and requesting that the necessary changes be made and the corrected drawings returned. If the design complies in full, one copy of the drawing is returned to the manufacturer with the stamped approval and a registration number that must be stamped on the vessel in addition to the ASME symbol and National Board stamping.

Once a design has been approved and registered, any number of vessels of that design can be built and used in the province where it is approved. If a vessel of the same design is to be sent to another Canadian province, the manufacturer's letter to the chief inspector of that pro-vince should state the registration number of the original approval. The second province will then use the same registration number with the addition of its own province number. For example, a design first registered in Ontario might have been given the number 764. This number would be followed by a decimal point and then the number 5, which signifies the Province of Ontario. The registration would thus be 764.5. If this same design were used in the Province of Manitoba, it would be given the registration number 764.54, with the figure 4 denoting Manitoba.

The Canadian provinces also require their own manufacturer's affidavit form, with the registration number and the shop inspector's signature on the data sheets. Finally, when a vessel is delivered to a purchaser in Canada, an affidavit of manufacture bearing the registration number and the signature of the authorized shop inspector must be sent to the chief inspector of the province for which it is intended.

Department of Transportation Regulations

A new Chapter 11 in this book explains clearly the new federal regulations 49 CFR Parts ¶107, ¶178, and ¶180, formerly known as dockets HM-183 and HM-183A. This new regulation mandates and requires that all over-the-road specification cargo tank trucks be manufactured by a holder of an ASME Code symbol U stamp. Most of these cargo tank trucks will not bear the official ASME Code stamp; however, certain specification cargo tanks will be manufactured and continue to be stamped ASME if all the requirements of the Code are met. Additionally and most importantly, all specification cargo tanks will now be required to be repaired by the holder of a National Board R stamp.

These new regulations will require numerous companies to obtain the ASME U Code symbol stamp or the National Board R symbol stamp. Quality control programs will be required when no such programs have existed in the past. Training of personnel will now be involved in areas of nondestructive examination, testing, procedure qualification, and numerous other areas involving quality. This will cause additional expense to the facilities involved. New maintenance actions involving periodic tests and inspections are also mandated and must be carried out.

Welded Repair or Alteration Procedure

Repairs or alterations of ASME Code vessels may be made in any shop that manufactures Code vessels or in the field by any welding contractor qualified to make repairs on such vessels. Recognizing the need for rules for repair and alterations to boilers and pressure vessels, the National Board of Boiler and Pressure Vessel Inspectors in their inspection code, Chapter III, gives rules that many states have incorporated into their own boiler and pressure vessel laws. Many companies have been using these rules as a standard for repairs to pressure vessels.

The National Board also has six Certificates of Authorization for use of the National Board R stamp. Any repair organization may obtain Certificates of Authorization for use of the National Board R stamp to be applied to ASME Code and National Board–stamped boilers and/or pressure vessels on which repairs have been made.

Before an R stamp is issued, the firm making repairs must submit to a review of its repair methods. It must also have an acceptable written quality control System covering repair design, materials to be used, welding procedure specifications, welders' qualifications, repair methods, and examination. The review will be made by a National Board member jurisdiction where the repair company is located. If there is no National Board member jurisdiction, or, at the request of such jurisdiction, the review will be made by a representative of the National Board.

Manufacturers and assemblers who hold ASME Certificates of Authorization and Code stamps, with the exception of H (cast iron), V, UV, and UM stamps, may obtain the National Board R stamp without review provided their quality control system covers repairs.

Whenever the R stamp is to be applied to National Board–stamped boilers and pressure vessels, all repairs are subject to acceptance by an Inspector who holds a valid National Board Commission and a valid Certificate of Competency issued by a state or jurisdiction of the United States or Province of Canada.

The Authorized Inspector will be guided by the *National Board Inspection Code** as well as by any existing local rules governing repair of vessels. When the work is to be done on location, repairs to a vessel require greater skill than was required when the vessel was first shop-constructed. Utmost consideration must be given to field repairs, which may necessitate cutting, shaping, and welding with portable equipment. The practicability of moving the vessel from the site to a properly equipped repair shop should be considered. Before repairs are made, the method of repair must be approved by an Authorized Inspector. The

---

*The Code is published by the National Board of Boiler and Pressure Vessel Inspectors, 1055 Crupper Ave., Columbus, OH 43229.

Inspector will examine the vessel, identify the material to be welded, and compare it with the material to be used in repair. A check is then made to ensure that the welding contractor or shop has a qualified welding procedure for the material being welded and that the welder who does the work is properly qualified to weld that material. If the repair or alteration requires design calculations, the Inspector will review the calculations to assure that the original design requirements are met. For repair of Section VIII, Division 2 pressure vessels, the design calculations must be reviewed and certified by a Registered Professional Engineer experienced in pressure vessel design.

When the repairs or alterations and the necessary tests have been completed, one or more of the following reports are filled out (see Fig. 1.5): Form R-1, Report of Welded Repair; Form R-2, Report of Alteration; and Form R-3, Report of Parts Fabricated by Welding. (Form R-4, Report Supplementary Sheet, should be used for any information for which there is not enough room on one of the other three forms.) Some states and insurance companies have their own forms which must be signed by the contractor or manufacturer making the repair and by the Authorized Inspector.

When an alteration is made to a vessel, it shall comply with the section of the ASME Code to which the original vessel was constructed to conform. The required inspection shall be made by an Authorized Inspector holding a National Board Commission. No alteration shall be initiated without the approval of the Authorized Inspector.

When alterations are completed, Form R-2 shall be signed and certified, indicating that the alteration has met the applicable requirements. This R-2 report shall record the required data and shall clearly indicate what changes or alterations have been made to the original construction. On the R-2 report, under Remarks, the name of the original manufacturer, manufacturer's serial number, and the National Board number assigned to the vessel by the original manufacturer shall be stated. When the vessel has been altered, an original and one copy of the certified repair or alteration report shall be filed with the National Board of Boiler and Pressure Vessel Inspectors. A copy shall go to the jurisdictional authority, inspection agency, and the customer for either repairs or alterations when so requested by the applicable persons.

When the stamping plate is made for an altered vessel, it shall contain all the information required by the National Board Inspection Code (see Fig. 1.6). Note the difference in the stamping of the nameplate of an altered vessel from the stamping of the nameplate of a repaired vessel (see Fig. 1.7).

The words *Altered* or *Rerated* in letters $5/16$ in high shall be stamped on the plate in lieu of the Code symbol, National Board R symbol, and National Board number. The date shown on the nameplate shall be the

FORM R-1 REPORT OF WELDED REPAIR
in accordance with provisions of the National Board Inspection Code

1. Work performed by ____(1)_____ ____(2)____
   _____
   (name of repair organization)   (Form R No.)
   _____ ____(53)____
   (address)   (P.O. No. Job No. etc.)

2. Owner ____(3)_____
   (name)
   _____
   (address)

3. Location of installation ____(4)_____
   (name)
   _____
   (address)

4. Unit identification ____(5)____ Name of original manufacturer ____(6)____
   (boiler, pressure vessel)

5. Identifying nos.: ____(7)____ ____(8)____ ____(8)____ ____(8)____ ____(9)____
   (mfg serial no.) (National Board No.) (jurisdiction no.) (other) (year built)

6. NBIC Edition/Addenda: ____(10)_____ ____(10)____
   (edition)   (addenda)

   Original Code of Construction for Item: ____(11)____ ____(11)____
   (name/section/division) (edition/addenda)

   Construction Code Used for Repair Performed: ____(11)____ ____(11)____
   (name/section/division) (edition/addenda)

7. Description of work: ____(12)____
   (use supplemental sheet, Form R-4, if necessary)
   _____
   _____
   _____ Pressure Test, if applied ____(13)____ psi

8. Replacement Parts. Attached are Manufacturer's Partial Data Reports or Form R-3s properly completed for the following items of this report:
   ____(14)____
   _____
   (name of part, item number, data report type, mfr's. name and identifying stamp)

9. Remarks: ____(15)____
   _____
   _____

CERTIFICATE OF COMPLIANCE

I, ____(16)____, certify that to the best of my knowledge and belief the statements in this report are correct and that all material, construction, and workmanship on this Repair conforms to the National Board Inspection Code.
National Board "R" Certificate of Authorization No. ____(17)____ expires on ____(18)____,
Date ____(19)____, ____(20)____ Signed ____(21)____
                    (name of repair organization)   (authorized representative)

CERTIFICATE OF INSPECTION

I, ____(22)____, holding a valid Commission issued by The National Board of Boiler and Pressure Vessel Inspectors and certificate of competency issued by the jurisdiction of ____(23)____ and employed by ____(24)____ of ____(25)____ have inspected the work described in this report on ____(26)____, and state that to the best of my knowledge and belief this work complies with the applicable requirements of the National Board Inspection Code.
By signing this certificate, neither the undersigned nor my employer makes any warranty, expressed or implied, concerning the work described in this report. Furthermore, neither the undersigned nor my employer shall be liable in any manner for any personal injury, property damage or loss of any kind arising from or connected with this inspection.
Date ____(19)____, ____ Signed ____(27)____ Commissions ____(28)____
                              (inspector)   (National Board (incl endorsements), and jurisdiction, and no.)

Figure 1.5 Form R-1, Report of Welded Repair; Form R-2, Report of Alteration; Form R-3, Report of Parts Fabricated by Welding; Form R-4, Report Supplementary Sheet; and Guide for Completing National Board Form R Reports. (*Reprinted with permission of the National Board of Boiler and Pressure Vessel Inspectors, www.nationalboard.org, from 1998 National Board Inspection Code, © 1998.*)

FORM R-2 REPORT OF ALTERATION
in accordance with provisions of the National Board Inspection Code

1. Work performed by ___(1)_____ ___(2)___
                            (name of alteration organization)   (Form R No.)

   _____(address)_____ ___(53)___
                                          (P.O. No. Job No. etc.)

2. Owner ___(3)_____
               (name)

   _____(address)_____

3. Location of installation ___(4)_____
                            (name)

   _____(address)_____

4. Unit identification ___(5)___ Name of original manufacturer ___(6)___
                       (boiler, pressure vessel)

5. Identifying nos.: ___(7)___ ___(8)___ ___(8)___ ___(8)___ ___(9)___
               (mfg serial no.) (National Board No.) (jurisdiction no.) (other) (year built)

6. NBIC Edition/Addenda: ___(10)___ ___(10)___
                           (edition)   (addenda)

   Original Code of Construction for Item: ___(11)___ ___(11)___
                                       (name/section/division) (edition/addenda)

   Construction Code Used for Alteration Performed: ___(11)___ ___(11)___
                                                    (name/section/division) (edition/addenda)

7. Description of work: ___(12)___
                            (use supplemental sheet, Form R-4, if necessary)

   _____
   _____
   _____
   _____
   _____
   _____
   _____
   _____ Pressure Test, if applied ___(13)___ psi

8. Replacement Parts. Attached are Manufacturer's Partial Data Reports or Form R-3s properly completed for the following items of this report:
   ___(14)___
   _____
   _____
   _____
   _____
   _____
   _____
   _____
   _____
               (name of part, item number, data report type, mfr's. name and identifying stamp)

9. Remarks: ___(15)_____
   _____
   _____
   _____
   _____

Figure 1.5 (*Continued*)

Form R-2 (back)
_____ (2)
(Form R No.)

DESIGN CERTIFICATION

I, _____(16)_____, certify that to the best of my knowledge and belief the statements in this report are correct and that the Design Change described in this report conforms to the National Board Inspection Code.
National Board "R" Certificate of Authorization No._____(17)_____expires on__(18)__,_____
Date__(19)__, _____ _____(20)_____ Signed_____(21)_____
                        (name of design organization)   (authorized representative)

CERTIFICATE OF DESIGN CHANGE REVIEW

I, _____(22)_____, holding a valid Commission issued by The National Board of Boiler and Pressure Vessel Inspectors and certificate of competency issued by the jurisdiction of _____(23)_____ and employed by _____(24)_____ of _____(25)_____ have reviewed the design change as described in this report and state that to the best of my knowldge and belief such change complies with the applicable requirements of the National Board Inspection Code.
By signing this certificate, neither the undersigned nor my employer makes any warranty, expressed or implied, concerning the work described in this report. Furthermore, neither the undersigned nor my employer shall be liable in any manner for any personal injury, property damage or loss of any kind arising from or connected with this inspection.
Date__(19)__, _____ Signed_____(27)_____ Commissions _____(28)_____
                                    (Inspector)   (National Board (incl endorsements), and jurisdiction, and no.)

CONSTRUCTION CERTIFICATION

I, _____(16)_____, certify that to the best of my knowledge and belief the statements in this report are correct and that all material, construction, and workmanship on this Alteration conforms to the National Board Inspection Code.
National Board "R" Certificate of Authorization No._____(17)_____expires on__(18)__,_____
Date__(19)__, _____ _____(20)_____ Signed_____(21)_____
                        (name of alteration organization)   (authorized representative)

CERTIFICATE OF INSPECTION

I, _____(22)_____, holding a valid Commission issued by The National Board of Boiler and Pressure Vessel Inspectors and certificate of competency issued by the jurisdiction of __(23)_____ and employed by _____(24)_____ of __(25)_____ have inspected the work dedcribed in this report on __(26)__, _____and state that to the best of my knowledge and belief this work complies with the applicable requirements of the National Board Inspection Code.
By signing this certificate, neither the undersigned nor my employer makes any warranty, expressed or implied, concerning the work described in this report. Furthermore, neither the undersigned nor my employer shall be liable in any manner for any personal injury, property damage or loss of any kind arising from or connected with this inspection.
Date__(19)__, _____ Signed_____(27)_____ Commissions _____(28)_____
                                    (Inspector)   (National Board (incl endorsements), and jurisdiction, and no.)

Figure 1.5 (*Continued*)

date of alteration. This nameplate shall be permanently attached to the vessel adjacent to the original manufacturer's nameplate or stamping, which shall remain on the vessel.

If the alteration requires removal of the original manufacturer's nameplate or stamping, it shall be retained and reattached to the altered vessel. Such relocation shall be noted on the R-1 report subject to the jurisdictional requirements.

Managerial Factors in Handling Code Work

The producer of welded pressure vessels is faced today with many serious problems, not the least of which is managing to stay in business

FORM R-3 REPORT OF PARTS FABRICATED BY WELDING
in accordance with provisions of the National Board Inspection Code

1. Manufactured by ___(1)___ (name of manufacturer) ___(2)___ (Form R No.)
   _____ (address) ___(53)___ (P.O. No. Job No. etc.)
2. Manufactured for ___(29)___ (name of purchaser)
   _____ (address)
3. Design Condition specified by ___(30)___ (name of organization)   Code design by ___(31)___ (name of organization)
4. Design Code ___(32)___ (code type and section) ___(33)___ (code year) ___(34)___ (addenda date) ___(35)___ (formula on which MAWP is based)
5. Identification of Parts

Name of Part	Qty.	Line No.	Manufacturer's Identifying No.	Manufacturer's Drawing No.	MAWP	Shop Hydro psi	Year Built
(36)	(37)	(38)	(39)	(40)	(41)	(13)	(9)

6. Description of Parts

Line No.	(a) Connections other than tubes			Heads or Ends			(b) Tubes		
	Size and Shape	Material Spec No.	Thickness In.	Shape	Thickness In.	Material Spec No.	Diameter In.	Thickness In.	Material Spec No.
(38)	(42)	(43)	(44)	(45)	(46)	(43)	(47)	(48)	(43)

7. Remarks ___(15)___
_____
_____
_____

Figure 1.5 (*Continued*)

Form R-3 (back)

_____(2)_____
(Form R No.)

CERTIFICATE OF SHOP COMPLIANCE

I, ____(16)____, certify that to the best of my knowledge and belief the statements in this report are correct and that all material, fabrication, construction, and workmanship of the described parts conforms to the National Board Inspection Code and standards of construction cited.
National Board "R" Certificate of Authorization No. ____(17)____ expires on ____(18)____, _____
Date ____(19)____, _____ ____(20)____ Signed _____ ____(21)____ _____
                                      (name of R Certificate holder)   (authorized representative)

CERTIFICATE OF SHOP INSPECTION

I, ____(22)____, holding a valid Commission issued by The National Board of Boiler and Pressure Vessel Inspectors and certificate of competency issued by the jurisdiction of ____(23)____ and employed by ____(24)____ of ____(25)____ have inspected the parts described in this report on ____(26)____, _____ and state that to the best of my knowledge and belief the parts comply with the applicable requirements of the National Board Inspection Code.
By signing this certificate, neither the undersigned nor my employer makes any warranty, expressed or implied, concerning the work described in this report. Furthermore, neither the undersigned nor my employer shall be liable in any manner for any personal injury, property damage or loss of any kind arising from or connected with this inspection.
Date ____(19)____, _____ Signed ____(27)____ Commissions ____(28)____
                                               (Inspector)   (National Board (incl endorsements), and jurisdiction, and no.)

Figure 1.5 (*Continued*)

---

and earn a fair profit.* Severe competition exists throughout the plate-fabricating industry, not only because of the growth in the number of fabricating shops but also because of the tremendous technological advances made in recent years. Many of the older, more heavily capitalized plants have been hit hard. Some of these have successfully turned to specialized fields; others produce specialty products that return a good profit when in demand. When demand slackens, these shops often turn to general fabrication, thereby increasing the competitive situation for the general plate fabricator.

For this type of fabricator, who essentially runs a job shop, the profit margin is restricted from the outset because jobs are usually awarded to the lowest bidder. Although there is nothing unusual in this method of competing, it does increase the need to think ahead clearly and to avoid unnecessary costs. For the job shop handling ASME Code vessels as well as non-Code work, the importance of careful planning is greatly increased because of the relatively high costs of Code material and also the variety in techniques and procedures demanded.

A second important factor in the award of a contract is delivery time, and here again planning assumes utmost importance. The buyer of a

---

*The material in this section, Managerial Factors in Handling Code Work, was contributed by C. H. Willer, former editor of *Welding Journal.*

FORM R-4 REPORT SUPPLEMENTARY SHEET
in accordance with provisions of the National Board Inspection Code

1. Work performed by ___(1)___ ___(49)___   ___(2)___ ___(49)___
   (name)   (Form R reference)
   (address)   ___(53)___ ___(49)___
   (P.O. No. Job No. etc.)

2. Owner ___(3 or 29)___ ___(49)___
   (name)

   (address)

3. Location of installation ___(4)___ ___(49)___
   (name)

   (address)

Reference Line No.	Continued from Form R-___ (50)
(51)	(52)

Date ___(19)___, _____ Signed ___(21)___ Name ___(20)___ ___(49)___
(authorized representative) (authorized organization)

Date ___(19)___, _____ Signed ___(27)___ Commissions ___(28)___ ___(49)___
(Inspector) (National Board (incl endorsements), and jurisdiction, and no.)

Figure 1.5 (*Continued*)

Guide for Completing National Board FORM R Reports

1. Name and address of the R Certificate organization which performed the work.

2. Form R Number assigned by the organization which performed the work. (A single, sequential log shall be maintained by the organization for R-1, R-2, and R-3 Forms registered with the National Board).

3. Name and address of the Owner of the pressure retaining item.

4. Name and address of plant or facility where the pressure retaining item is installed.

5. Description of the pressure retaining item, such as boiler or pressure vessel.

6. Name of original manufacturer of the pressure retaining item if a boiler or pressure vessel. If other than a boiler or pressure vessel, complete if known.

7. Serial number of the pressure retaining item as assigned by the original manufacturer.

8. Identification of the pressure retaining item by applicable registration number. If installed in Canada, indicate the Canadian design registration number (CRN), and list the drawing number under "other."

9. Identify the year in which fabrication/construction of the item was completed.

10. Indicate edition and addenda of the NBIC under which this work is being performed.

11. Indicate name, section, and division of original code of construction, including edition and addenda used for work being performed. (For repairs only, indicate "none" if not known.)

12. State exact scope of work, and attach additional data, sketch, Form R-4, etc. as necessary. If additional data is attached, so state.

13. Indicate test pressure applied.

14. To be completed for all welded pressure components added during the work. Indicate part, item number, manufacturer's name, stamped identification, and data report type.

15. Indicate any additional information pertaining to the work involved (eg. routine repairs.) For Form R-3 the part manufacturer is to indicate the extent he has performed any or all of the design function. If only a portion of the design, state which portion.

16. Type or print name of authorized representative of the "R" Certificate Holder.

17. Indicate National Board R Certificate or Authorization number.

18. Indicate month, day, and year that the R certificate expires.

Figure 1.5 (*Continued*)

Origin, Development, and Jurisdiction of ASME Code   25

19. Enter date certified.

20. Name of R Certificate organization which performed the identified work.

21. Signature of authorized representative.

22. Type or print name of Inspector.

23. Indicate Inspector's jurisdiction.

24. Indicate Inspector's employer.

25. Indicate address of Inspector's employer (city and state or province).

26. Indicate month, day, and year of inspection by Inspector.

27. Signature of Inspector.

28. National Board commission number of Inspector, including endorsements, jurisdiction, and certificate of competency numbers.

29. Name and address of organization which purchased the parts for incorporation in the repair or alteration, if known. If built for stock, so state.

30. Name of organization responsible for specifying the code design conditions.

31. Name of organization responsible for performing the code design, if known.

32. Name, section, and division of the design code, if known.

33. Indicate code edition year used for fabrication.

34. Indicate code addenda date used for fabrication.

35. Indicate code paragraph reference for formula used to establish the MAWP, if known.

36. Identify name of part, such as "superheater header."

37. Indicate quantity of named parts.

38. Match line number references for identification of parts and description of parts.

39. Indicate manufacturer's serial number for the named part.

40. Indicate drawing number for the named part.

41. Indicate Maximum Allowable Working Pressure for the part, if known.

42. Use inside diameter for size; indicate shape as square, round, etc.

Figure 1.5 (*Continued*)

43. Indicate the complete material specification number and grade.
44. Indicate nominal thickness of plate and minimum thickness after forming.
45. Indicate shape as flat, dished, ellipsoidal, or hemispherical.
46. Indicate minimum thickness after forming.
47. Indicate outside diameter.
48. Indicate minimum thickness of tubes.
49. Complete information identical to that shown on the Form R to which this sheet is supplementary.
50. Indicate the Form R type. Example: Form R-1, Form R-2, Form R-3
51. Indicate the reference line number from the Form R to which this sheet is supplementary.
52. Complete information for which there was insufficient space on the reference Form R.
53. If applicable, purchase order, job number, etc. assigned by the organization performing the work.

Figure 1.5 (*Continued*)

Code vessel is not necessarily concerned if a producer incurs extra expense through carelessness but is very often acutely aware of delivery time. Good customer relations and even future business depend upon on-time deliveries. Obviously, cost and delivery time are related.

The job shop, then is faced with three fundamental problems in the production of Code work:

1. The ability to fabricate in accordance with the Code
2. The necessity for keeping costs at a minimum
3. The importance of making deliveries on time

The question naturally arises, "What concrete steps can be taken to assure the best fulfillment of these requirements?" A few general suggestions bear mention. First, there is no substitute for reading and understanding the ASME Code. Many protests are heard on this score, even from inspectors and engineers, but a knowledge of those sections of the Code that pertain to the type of work in which one is engaged is a necessity. But even this is not enough. Some vessel requirements are extremely simple; others are not. The vast industrial complex of today has created such a wide variety of pressure vessels that they constitute a perpetual challenge to the ingenuity of the fabricator. For the job shop, close cooperation and understanding are required from start to

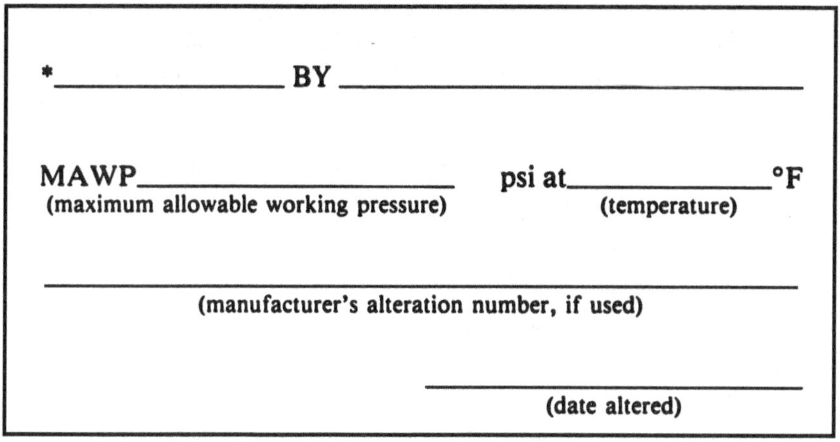

```
*_____ BY _____

MAWP_____   psi at_____°F
(maximum allowable working pressure)   (temperature)

_____
(manufacturer's alteration number, if used)

                        _____
                        (date altered)
```

Figure 1.6 Stamping or nameplate of an altered boiler or pressure vessel. (*Reprinted with permission from the* National Board Inspection Code, *1992 ed., p. 49. Copyright 1992 National Board of Boiler and Pressure Vessel Inspectors.*)

```
 {R}®      _____
            (name of repair firm)

No._____   _____
   (National Board
   Repair symbol
   stamp no.)      _____
                   [date of repair(s)]
```

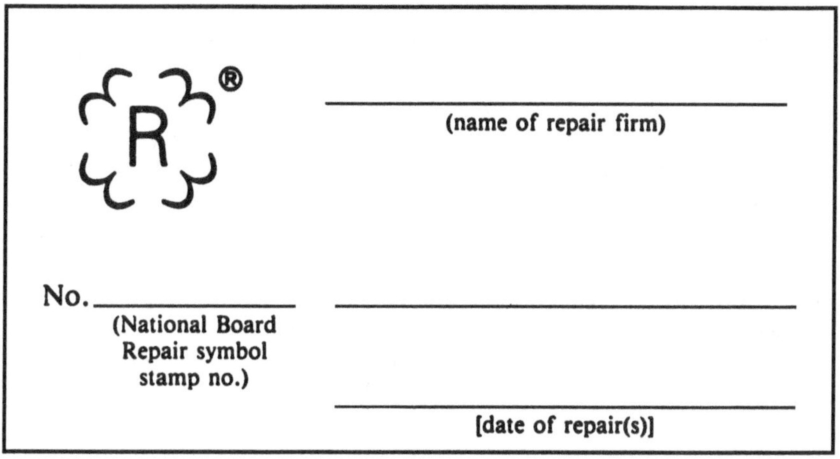

Figure 1.7 Stamping or nameplate of a boiler or pressure vessel repaired by welding. (*Reprinted with permission from the* National Board Inspection Code, *1992 ed., p. 49. Copyright 1992 National Board of Boiler and Pressure Vessel Inspectors.*)

finish. Sales, estimating, design, and purchasing must all be backed up by sound shop experience and a reasonable time schedule.

Particular methods for handling an order vary from one company to another. But within the limits set by customer specification and the rules of the Code, the job shop can seek a competitive advantage and avoid unnecessary costs and time losses by determining the following:

1. An overall production schedule
2. Acceptable alternatives in plate material and components
3. Acceptable alternatives in design geometry
4. Acceptable alternatives in fabricating methods
5. Acceptable sources of supply

Some freedom of choice is usually left to the fabricator, and finding the best means of satisfying the requirements at the least cost is the essence of this phase of the manufacturing process.

Work-Flow Schedules

By the time an order for a Code job is accepted, it has been estimated, bid, and sold. For many reasons, a job-shop estimator does not completely and finally specify all details of material design and fabrication at the time of bidding. After an order is received, a work-flow schedule must be drawn up that allots time for shop fabrication, procurement of materials and components, design drafting, and takeoff. The shop fabrication schedule is based on an estimate of the total quantity of work-hours required, the necessary work-hour sequence (allowing for nonadditive, parallel operations), and finally the stipulated delivery date and the total shop load for the calendar period. This information determines the required delivery date for plate material, purchased components, and shop subassemblies. The lead time for these items sets the date for purchasing and thus for the completion of the design, drafting, and takeoff details initially required. An estimate of the time needed for the latter operations indicates the provisional starting date.

Although this simple computation can be made in many different ways, it is most important that it be made in a specific way, for it forms the basis on which all scheduling and expediting are founded. If the provisional starting date indicates on-time delivery, there is no problem. The sequence of determined dates becomes the production schedule, and expediting is managed accordingly. If the estimated delivery falls considerably ahead of time, the entire schedule can be appropriately delayed or more time can be allotted to any or all of the production steps. In this situation, the cost of maintaining the inventory must be balanced against the future availability of material in a changing market. It must be remembered that although filler jobs are desirable, a reduction in work rate is not. If the estimated delivery falls after the required date, on the other hand, the operations must be reviewed to determine where and how time can be saved. Sometimes a simple solution is quickly found; more often, pressure has to be applied on all concerned, including the ASME Code Inspector; however, this pressure must be applied in a rational way.

It should be realized that the two hard necessities governing job-shop work—cost saving and time saving—are not always compatible. Earlier deliveries of purchased material usually cost more; heavier pressure on a work force usually results in overtime. Code restrictions on design, fabrication, and materials narrow the field of action. Nevertheless, certain routine procedures exist that will result in considerable savings.

Mill Orders

From a cost and delivery angle, the design of a pressure vessel and the purchasing of the plate and heads are often related. Thus the steel mill system of extras, based on the weight of items ordered, varies inversely with the weight, a 10,000-lb item having no extra and lesser quantities taking gradually increased extras according to the published table. The maximum cost difference could amount to many dollars a ton. As an item consists of any number of plates of one width and one thickness of a given quality, it pays to design a vessel with equal-width plates (see Appendix A). Various mills also publish extras based on width, and these tables must be consulted. A buyer might be able to save considerably by arbitrarily increasing a plate width a fraction of an inch.

Tonnage mounts up fast in steel plate. The question is often asked, "Is it less expensive to design a given vessel for a higher test steel with less thickness even though the quality extra incurred by the decreased thickness may be higher?" It sometimes works out this way, by using, say, an SA-515 grade 70 steel instead of an SA-285 grade C steel. Fabrication cost differences must also be considered.

Sometimes the buyer rather than the designer will note that a plate useful for stock inventory can be added to a special item to bring the total weight into a lower class of extras, thus saving money. This is also true of heads. At some mills, discounts for quantity begin with four to six heads, and very sizable discounts are offered for larger quantities. It is also highly desirable to attempt to keep head diameters, radii, and thicknesses to the standards carried in mill stocks.

Mill conditions often affect the length of time required to fill an order. As mills have different rolling practices, schedules, minimum-quantity stipulations, base prices, size limits, and the like, all these details should be discussed with a sales representative before attempting to place an order, especially if the buyer is not experienced. The lead time for most mills in normal periods runs about 45 days, but it is not safe to guess.

Warehouse Orders

If delivery requirements make it impossible to obtain steel from a mill, Code plate can be obtained in various sizes from warehouses. Quick delivery can be obtained of certain size plates, usually the lighter gages up to $\frac{1}{2}$ in thick, 96 in wide, and 10 to 20 ft long. In addition to the higher base price, the buyer must consider the cost of obtaining the plate size desired from the sizes offered. Heads of certain sizes and gages are carried by some warehouses and also are stocked for immediate delivery by some mills. Standard thicknesses of $\frac{3}{16}$, $\frac{1}{4}$, $\frac{5}{16}$, $\frac{3}{8}$, and $\frac{1}{2}$ in are typical of flanged and dished heads with diameters up to 96 in. Special heads are also listed in published catalogs.

Shop Organization

By far the biggest opportunities for saving time and money lie in the handling and organizing of Code work in the shop. Unless shop procedure is properly set up and the work scheduled in advance for inspection, extra cost and delay are likely.

Each visit by a Code Inspector represents a direct cash outlay both for the time used by the Inspector and for the time taken by personnel to show and review the work. Many shops adopt the practice of arranging for periodic visits so that as much work as possible can be grouped for one visit and the availability of the Inspector can be relied upon. Lack of planning often results in a last-minute rush to get ready with possible overtime expense or in an incomplete presentation that necessitates an additional inspection for some overlooked detail. If a job is not carefully watched, it may progress beyond the point at which an Inspector can make a satisfactory investigation of a completed operation.

Similarly, the vessel may be found to contain components for which the Code quality standard cannot be established. Either situation may result in costly rework or even complete loss. Sometimes it seems that in the complexity of a busy job shop it is almost impossible to plan very far ahead. Lack of planning, however, only means that more plans must be made, discarded, revised, and started over again.

Before work is started, the complete specifications should be reviewed, the sequence of work operations determined, and a rough shop schedule including the expected dates of inspection formulated. Without this schedule a final delivery date cannot safely be determined. The job must follow this schedule closely under responsible supervision.

When Code material is received, it should be carefully marked and checked. At least by the time of layout, the reference numbers should be checked against material certification and the required physical and chemical properties proved. At this time also, material should be checked for thickness and surface defects. This simple routine should help to prevent unnecessary and costly rework and loss of time.

Chapter

2

Descriptive Guide to the ASME Code Section VIII, Division 1, Pressure Vessels

An attempt is made in this chapter to assist those wishing to find specific information quickly without having to wade through the large fund of specifications and recommendations encompassed by the Code. Figures and tables outline designs, materials, or performance requirements, indicate conditions for which precautions are necessary, and list Code paragraphs to be read for the proper information. Experience has proven that many of these items are often overlooked. Also included is a handy reference chart (Fig. 2.1) that graphically illustrates the various parts of a pressure vessel and the Code paragraphs that apply to each. Table 2.1 lists the various classes of materials from Section VIII, Division 1, the covering Code section part, applicable references for the stress value tables found in Section II, Part D, and additional helpful hints. It should be remembered, however, that there is no substitute for reading the Code itself.

Requirements for Establishing Design Thickness Based on Degree of Radiography (Code Pars. UW-11 and UW-12)

The requirements based on the degree of radiographic examination of butt welds in pressure vessels constructed to the requirements of Section VIII, Division 1 of the *ASME Boiler and Pressure Vessel Code* have been the subject of numerous inquiries for interpretation from manufacturers, designers, Authorized Inspectors, and users of these vessels. Most of the questions have been concerned with the examination requirements for circumferential butt welds attaching seamless heads to shell sections.

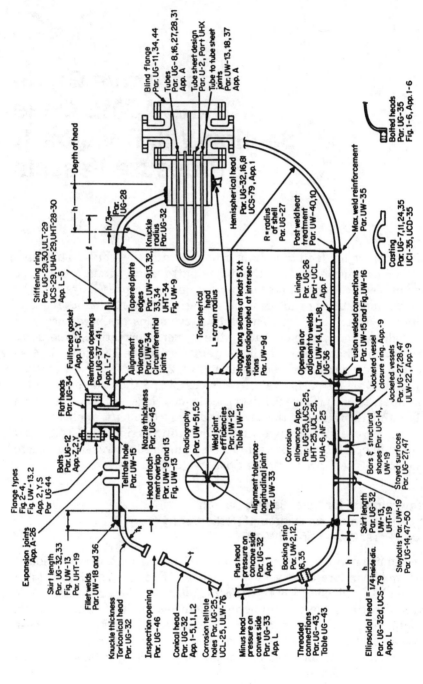

Figure 2.1 Reference chart for ASME Pressure Vessel Code, Section VIII, Division 1.

TABLE 2.1 Classes of Materials

Material	Covering Code part	Applicable Code stress value tables	Remarks
Carbon and low-alloy steels	UCS	Code Section II, Part D, Table 1A	Basis for establishing stress values—Code Appendix P, UG-23
			Low-temperature service requires use of notch-tough materials—Code Pars. UCS-65, UCS-66, UCS-67, UCS-68, UG-84
			Code Figs. UCS-66, UCS-66.1, UCS-66.2
			In high-temperature operation, creep strength is essential
			Design temperature—Code Par. UG-20
			Design pressure—Code Par. UG-21, Fn. 8
			Temperature above 800°F may cause carbide phase of carbon steel to convert to graphite
			Pipe and tubes—Code Pars. UG-8, UG-10, UG-16, UG-31, UCS-9, UCS-27
			Creep and rupture properties—Code Par. UCS-151
Nonferrous metals	UNF	Code Section II, Part D, Table 1B	Basis for establishing values—Code, Appendix P, UG-23
			Metal characteristics—Code Par. UNF, Appendix NF, NF-1 to NF-14
			Low-temperature operation—Code Par. UNF-65
			Nonferrous castings—Code Par. UNF-8
High-alloy steels	UHA	Code Section II, Part D, Table 1A	Selection and treatment of austenitic chromium–nickel steels—Code Par. UHA-11, UHA Appendix HA, UHA-100 to UHA-109

TABLE 2.1 Classes of Materials (*Continued*)

Material	Covering Code part	Applicable Code stress value tables	Remarks
High-alloy steels (*cont.*)		Code Section II, Part D, Table 1A	Inspection and tests—Code Pars. UHA-34, UHA-50, UHA-51, UHA-52
			Liquid penetrant examination required if shell thickness exceeds ¾ in—all 36% nickel steel welds—Code Par. UHA-34
			Low-temperature service—Code Pars. UHA-51, UG-84
			High-alloy castings—Code Pars. UHA-8, UG-7
			Code Par. Ug-7
Castings			Code Pars. UG-11, UG-24, UCS-8—Code Appendix 7
Cast iron	UCI	UCI-23	Vessels not permitted to contain lethal or flammable substances—Code Par. UCI-2
			Selection of materials—Code Pars. UCI-5, UCI-12, UG-11, UCS-10, UCS-11, UCI-3, UCI-1, UG-10
			Inspection and tests—Code Pars. UCI-90, UCI-99, UCI-101, UCI-3
			Repairs in cast-iron materials—Code Par. UCI-78
Dual cast iron	UCI		Code Pars. UCI-1, UCI-23, UCI-29
Integrally clad plate, weld metal overlay, or applied linings	UCL	(See Code Pars. UCL-11, UCL 23.)	Suggest careful study of entire metal UCL section Selection of materials—Code Selection of materials—Code Pars. UCL-1, UCL-3, UCL-10, UCS-5, UF-5, ULW-5, UCL-11, UCL-12, UG-10
			Qualification of welding procedure—Code Pars. UCL-40 to -46
			Postweld heat treatment—Code Pars. UCL-34, UCS-56 (including cautionary footnote)
			Inspection and test—Code Pars. UCL-50, UCL-51, UCL-52
			Spot radiography required if cladding is included in computing required thickness—Code Par. UCL-23(c)
			Use of linings—Code Par. UG-26 and Code Appendix F

TABLE 2.1 Classes of Materials (*Continued*)

Material	Covering Code part	Applicable Code stress value tables	Remarks
Welded and seamless pipe and tubes (carbon and low-alloy steels)	UCS	Code Section II, Part D, Table 1A	Thickness under internal pressure—Code Par. UG-27
			Thickness under external pressure—Code Par. UG-28
			Provide additional thickness when tubes are threaded and when corrosion, erosion, or wear caused by cleaning is expected —Code Par. UG-31
			For calculating thickness required, minimum pipe wall thickness is 87.5 percent of nominal wall thickness
			30-in maximum on welded pipe made by open-hearth, basic oxygen, or electric-furnace process—Code Par. USC-27
Welded and seamless pipe (high-alloy steels)	UHA	Code Table 1A	
Forgings	UF	Code Section II, Part D, Table 1A	Materials—Code Pars. UG-6, UG-7, UG-11, UF-6, UCS-7 and Section II, Part D, Table 1A
			Welding—Code Par. UF-32 (see also Section IX Code Par. QW-250 and Variables, Code Pars. QW-404.12, QW-406.3, QW-407.2, QW-409.1 when welding forgings)
Low-temperature materials	ULT	ULT-23	Operation at very low temperatures, requires use of notch-tough materials
Layered construction	ULW		Vessels having a shell and/or heads made up of two or more separate layers—Code Par. ULW-2
Ferritic steels with tensile properties enhanced by heat treatment	UHT	Code Table 1A	Scope—Code Par. UHT-1 Marking on plate or stamping, use "low-stress" stamps—Code Par. UHT-86

The Code has always been clear regarding requirements for butt welds attaching shell sections which have fully radiographed, longitudinal, butt welds, and when a joint efficiency of 1.00 is used in the calculation of shell thickness. A revision given in the Summer 1969 Addenda provided, for the first time, a requirement for radiographic examination of the butt weld attaching a seamless head to a seamless section in order to be consistent with the rules for butt-welded sections. However, the unwillingness to accept this new requirement generated further inquiries. Consequently, the Committee reopened the subject for possible further revision.

In an effort to achieve uniformity, *The Transactions of the ASME Series J Journal of Pressure Vessel Technology,* in the February 1975 issue, and subsequent follow-up article in February 1977, included examples by G. M. Eisenberg, Senior Pressure Vessel Engineering Administrator, clarifying design factors that were used as a guide. The National Board of Boiler and Pressure Vessel Inspectors also published these examples in their April 1975 *National Board Bulletin.*

The degrees of radiographic examination shown in Code Par. UW-11 include full, spot, and no radiography requirements. The factors for allowable stress and joint efficiency (quality factors) to be used in calculations for shell sections and head thicknesses are set forth in Code Par. UW-12 and Code Table UW-12 (see Table 2.2). The use of spot radiography, in lieu of partial radiography on Category B or C butt welds, provided a relaxation of the previous requirements. [Refer to Code Par. UW-11(a)(5)(b); partial radiography is no longer addressed by Section VIII, Division 1.]

The Code details the joint efficiency requirements and the percentage of the tabulated allowable stress which shall be used in head calculations and shell calculations for both circumferential (hoop) and longitudinal (axial) stresses.

In the 1986 Addenda to ASME Section VIII, Division 1, extensive revisions to the past rules that affect the minimum design thickness requirements used in the design of Code pressure parts were made. These new revisions have continued to evolve slowly to the rules which are in effect today. Additionally, these new requirements adversely affect how the maximum allowable working pressure or design thicknesses of materials used in Code construction is determined. The new rules are used to determine the method for selection of joint efficiencies and/or stress multipliers and how to apply them in the design of Section VIII, Division 1 pressure vessels.

Under these new requirements the designer must consider each ASME Code Section VIII, Division 1, Table UW-12 joint separately for the value of E which will be utilized in the formulas of Code Pars. UG-27, UG-32, UG-34, and UG-37, and Code Appendix 1, Appendix 9-6, and Appendix 13. The symbol E is formally known to mean *joint effi-*

TABLE 2.2 Welded Joint Efficiency Values

Type of joint and radiography	Efficiency allowed, percent	Code reference
Double-welded butt joints (Type 1)		Par. UW-11
Fully radiographed	100	Pars. UW-51, UW-35
Spot-radiographed	85	Pars. UW-12, UW-52
No radiograph	70	Table UW-12
Single-welded butt joints (backing strip left in place) (Type 2)		Par. UW-52 Par. UCS-25
Fully radiographed	90	Par. UW-51
Spot-radiographed	80	Par. UW-52
No radiograph	65	
Single-welded butt joints no backing strip (Type 3) limited to circumferential joints only, not over ⅝ in thick and not over 24-in ouside diameter	60	Table UW-12
Fillet weld lap joints and single-welded butt circumferential joints		Table UW-12
Seamless vessel sections or heads (spot-radiographed)	100	Par. UW-12(d)
Seamless vessel sections or heads (no radiography)	85	Par. UW-12(d)

ciency or *efficiency*. Today these terms may also be known as a *stress multiplier* which can be either in the form of a joint efficiency or stress reduction (quality) factor.

Under the revised rules, the designer would be allowed to employ spot radiography of the circumferential joints in a vessel. This could possibly be the case even if the designer had elected to perform full radiography of the intersecting longitudinal butt weld. For example, one approach may be to fully radiograph the longitudinal butt joints of a vessel utilizing $E=1.00$ in the formula. The designer may then choose to apply a lesser degree of radiographic examination (for example, spot radiography) for the circumferential butt joints using an efficiency $E = 0.85$ (see Table 2.3).

It should be noted with caution that the Code only allows one level of radiographic examination difference between longitudinal butt joints (or seamless equivalents) and the intersecting circumferential Table UW-12 butt joints in vessels, vessel sections, and heads [see Code Par. UW-11(a)(5)(b)].

The Code now recognizes that seamless vessel sections (for example, pipe used for shells of vessels) or seamless heads and their intersecting Category B and C butt joints are considered equivalent to welded parts of the same geometry. This rule applies only when the longitudinal joint is welded from both the inside and outside of the weld joint preparation or when some other process is used which will ensure that the butt joint has complete penetration at the joint.

TABLE 2.3 Radiographic Requirements

Remarks	Code reference
Full Radiography	
Full radiograph all joints over 1½ in (required radiographing on lesser thickness)	Pars. UW-11(a)2, UCS-57, Table UCS-57
Circumferential butt joints of nozzles and communicating chambers with diameter not exceeding 10 in or with thickness not exceeding 1⅛ in do not require radiography	Par. UW-11(a)4
Vessels containing lethal substances must be fully radiographed	Pars. UW-11(a)1, UW-2
Efficiency of fully radiographed joints	Par. UW-12, Table UW-12
Radiographic acceptance standards	Par. UW-51
Surface weld buildup	Pars. UW-42, UCS-56(f)(5)
Carbon and low-alloy steel materials	Pars. UCS-57, UCS-19
High-alloy steel materials	Par. UHA-33
Clad-plate materials	Par. UCL-35
Nonferrous materials	Pars. UNF-57, UNF-91
Ferritic steels whose tensile properties have been enhanced by heat treatment	Par. UHT-57
Layered vessel inner shells and heads	Par. ULW-51
Welded joints in layers of layered vessels	Par. ULW-52(d)
Butt welded joints in layered vessels	Par. ULW-54
Low-temperature vessels	Par. ULT-57

For calculations involving the circumferential stresses of seamless parts, or for thicknesses of seamless shells or heads, the value of E may equal 1.00 when the spot radiographic requirements of Code Par. UW-11(a)(5)(b) are met. Additionally, the value of E may equal 0.85 and may be utilized in the formula if the designer elects not to perform any radiography. The value of $E = 0.85$ for circumferential butt joints and longitudinal butt joints which connect seamless vessels, vessel sections, and heads, may be used provided that these welds are Type 3, 4, 5, or 6 joints as derived from Code Table UW-12.

The stress reduction factors of the pre-1986 Addenda of 0.80 and 0.85 are no longer applicable. Under the new rules there is only one stress reduction factor less than 1.00. This is a factor of 0.85, known now as a *quality factor*. This quality factor only applies to seamless vessels, vessel sections, and heads of Code Par. UW-12(d).

When pressure-welding processes are to be utilized in the production of Code pressure vessels, such as flash induction welding, pressure gas welding, and inertia and continuous drive friction welding, the value

TABLE 2.3 Radiographic Requirements (*Continued*)

Remarks	Code reference
Spot Radiography	
General recommendations	Par. UW-11(b)
Spot-radiographing required for higher weld joint efficiencies	Pars. UW-12, UW-52, Table UW-12
Minimum number of spots	Par. UW-52(b)
Minimum length of spot radiograph, 6 in	Par. UW-52(c)
Surface weld buildup	Par. UW-42
If cladding is included in computing required thickness, spot radiograph is mandatory	Par. UCL-23(c)
Vessels having 50 ft or less of main seams require one spot examination; larger vessels require one spot for each 50 ft of welding	Par. UW-52(b)1
Additional spots may be selected to examine welding of each welder or welding operator	Par. UW-52(b)2
Repair welding defects in material; forgings	Pars. UF-37, UCD-78
General requirements for castings	Par. UG-24
Low-temperature vessels	Par. ULT-57
High-alloy steel vessels	Par. UHA-33
Ferritic steel whose tensile properties have been enhanced by heat treatment	Par. UHT-57
Joints welded with austenitic chromium-nickel steel	Pars. UCL-35, UCL-36
Repair welds in layered construction	Pars. ULW-57, ULW-51

$E = 0.80$ may continue to be used in Code calculations provided the welding process used is permitted by the rules of Section VIII, Division 1 for the materials to be welded and provided that the quality of such welds are tested. Such test specimens shall be representative of the production welding on each vessel. These specimens may be taken from the shell of the vessel or from a separate "runoff" tab attached as a protrusion of the longitudinal joint of the vessel. If there is no longitudinal joint, a separate test plate of the same base material, thickness, and welding materials which will be used on the production piece shall be used. Each test piece shall be tested with one reduced section tension test specimen and two side bend test specimens which shall meet the requirements of ASME Code Section IX Pars. QW-150 and QW-160.

As a result of this new joint efficiency requirement allowing a per-joint quality factor, many vessels now built will be utilizing the RT-4 when marking the ASME Code data plate (see Fig 2.18). This marking applies only when a part of the vessel satisfies the requirements of UW-11(a) or when RT-1 [the complete vessel satisfies the full radiographic requirements of UW-11(a)], RT-2 [the complete vessel satisfies the radiograph requirements of UW-11(a)(5) and the spot radiographic requirements of UW-11(a)(5)(b)], or RT-3 [the complete vessel satisfies the spot radiographic requirements of UW-11(b)] do not apply.

The Remarks section of the Manufacturer's Data Report must also list the various combinations of joint efficiencies, E. In cases where complex vessels are designed with various sections, a joint efficiency map may be generated and documented on the U-4 form of the ASME Manufacturer's Data Report Supplementary Sheet.

Figures 2.2 to 2.5 (courtesy of the American Society of Mechanical Engineers) depict flow charts which indicate the required joint efficiencies for various components as well as clearly outlining where the designer may find the applicable joint efficiencies for various weld categories in the ASME Code Section VIII, Division 1.

Use of Internal-Pressure Cylindrical Shell Thickness Tables (Tables 2.4a to 2.4f)

These tables may be applied to seamless shells (joint efficiency E = 1.00). These tables may also be applied to shells with double-welded butt longitudinal joints that have been fully radiographed (joint efficiency = 1.00), spot-radiographed (joint efficiency = 0.85), or not radiographed (joint efficiency = 0.70). They may also be applied to single-welded butt joints with a backing strip that are fully radiographed (joint efficiency = 0.90), spot-radiographed (joint efficiency = 0.80), or not radiographed (joint efficiency = 0.65). The required shell thickness may be determined by the following steps:

1. Determine the appropriate table corresponding to the desired joint efficiency. Note that the value of joint efficiency for each table is printed in the upper left-hand corner.
2. Locate the desired diameter. The diameter values are printed across the top of the page for each table.
3. Locate the desired pressure. The pressure values are located in the left column.
4. Read the thickness corresponding to the diameter and pressure.

Each table is based on an allowable stress value of 13,800 psi. For other stress values, multiply the thickness read from the table by either the following constant factors or factors read from Fig. 2.6.

Stress, psi	11,300	12,500	13,800	15,000	16,300	17,500	18,800
Constant	1.224	1.105	1.000	0.919	0.846	0.787	0.733

Thickness equations for cylindrical shells may be found in Table 2.5. For corrosion allowance in spot-radiographed or fully radiographed carbon and low-alloy steel vessels containing air, steam, or water less

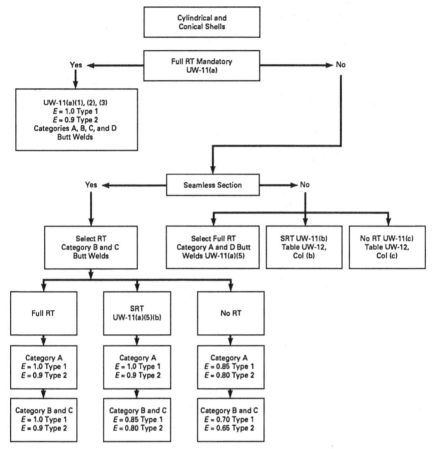

Figure 2.2 Joint efficiency and weld joint type—cylinders and cones (Code Fig. L-1.4-1). (*Courtesy of The American Society of Mechanical Engineers.*)

than ¼ in thick, add ⅙ of the calculated plate thickness (see Code Par. UCS-25). No corrosion allowance is necessary for vessels that have not been radiographed (see Code Table UW-12, column C).

The ASME has increased the allowable stresses for many of the materials in the lower temperature range, after considerable study determined that the safety and reliability of vessels in such service conditions are acceptable. These changes have made it possible to use thinner materials in many cases and thus decrease the weight of certain vessels. The ASME's revisions do not affect the charts in this book because the higher stresses will still be properly correlated with the

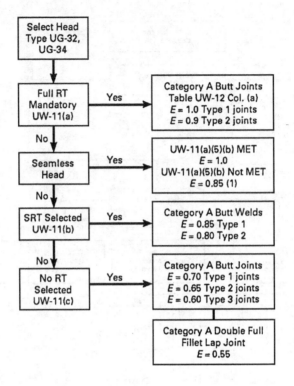

NOTES:
(1) See UW-12(d) when head-to-shell attachment weld is Type No. 3, 4, 5, or 6.

Figure 2.3 Joint efficiencies for formed heads and unstayed flat heads and covers with categories A and D butt welds. (*Courtesy of The American Society of Mechanical Engineers.*)

material thicknesses required. However, should a constant factor for an allowable stress be needed that is not given in Fig. 2.7, the following formula may be used:

$$\text{Constant factor} = \frac{14{,}178}{(\text{allowable stress})^{1.003}}$$

Use of Simplified External-Pressure Cylindrical Shell Thickness Charts (Figs. 2.7 and 2.8)

These charts give at a glance the shell thickness required for carbon-steel and type 304 stainless steel vessels under external pressure, thereby eliminating the many steps required by Code Par. UG-28; Code Section II, Part D, Subpart 3 for carbon or low-alloy steel with a minimum yield strength of 30,000 psi and over and type 405 and 410 stainless steels; and for stainless steel type 304 also using Code Par. UG-28; Code Section II,

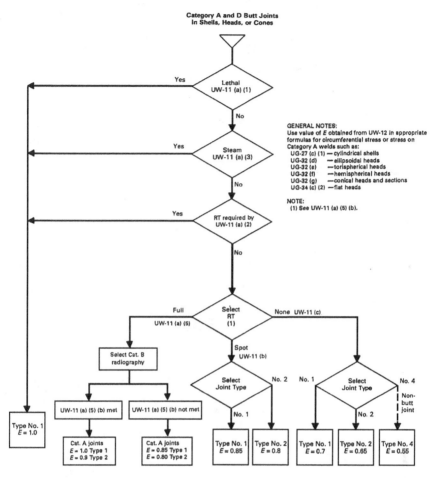

Figure 2.4 Joint efficiencies for categories A and D welded joints in shells, heads, or cones (Code Fig. L-1.4-3). (*Courtesy of The American Society of Mechanical Engineers.*)

Part D, Subpart 3. The temperature limits of Code Par. UG-20(c) shall be taken into consideration. Table 2.6 sets forth the required calculations for cylindrical and spherical shells under external pressure.

To obtain the required thickness from Figs. 2.7 and 2.8, (1) place a rule horizontally at the intersection of the applicable L/D_0 and pressure lines on the left side of the chart, and (2) read the thickness value where the rule crosses the vertical D_0 line on the right-hand side of the chart. Other common problems may also be easily solved. The allowable external pressure of an existing (or proposed) vessel may be determined by (1) entering the chart on the right-hand side at the proper thickness and diameter and (2) moving horizontally to the left and reading the pressure at the vertical L/D_0 line. A similar procedure may

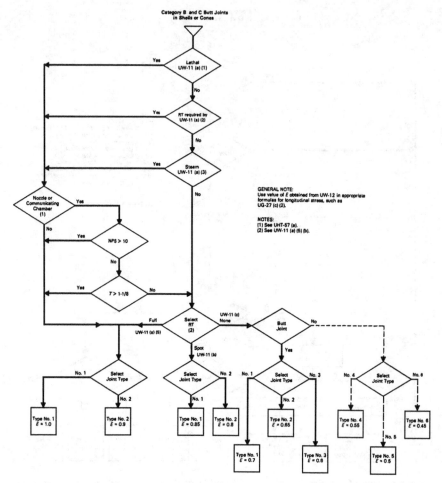

Figure 2.5 Joint efficiencies for categories B and C welded joints in shells or cones (Code Fig. L-1.4-4). (*Courtesy of The American Society of Mechanical Engineers.*)

be used to find the maximum allowable effective length L for the spacing of stiffening rings at a specified pressure (see Figs. 2.7 and 2.8).

Use of Internal-Pressure–Thickness Tables for Ellipsoidal and Torispherical Heads (Tables 2.7*a* to 2.7*l*)

The head thickness tables may be applied to ellipsoidal 2:1 heads or torispherical heads. They may be applied to seamless heads (joint efficiency = 1.0), double-welded butt heads that have been fully radiographed (joint efficiency = 1.0), spot-radiographed (joint efficiency =

TABLE 2.4a Shell Thickness Tables in Inches: Joint Efficiency=1.0

SHELL THICKNESS TABLES IN INCHES

JOINT EFF.=1.00 ALLOWABLE STRESS VALUE= 15800. PSI

INSIDE DIAMETER, IN.

PSI	12.	18.	24.	30.	36.	42.	48.	54.	60.	66.	72.	78.	84.	90.	96.	102.	108.	114.	120.	126.	132.
15.	0.007	0.010	0.013	0.016	0.020	0.023	0.026	0.029	0.033	0.036	0.039	0.042	0.046	0.049	0.052	0.055	0.059	0.062	0.065	0.069	0.072
20.	0.009	0.013	0.017	0.022	0.026	0.030	0.035	0.039	0.044	0.048	0.052	0.057	0.061	0.065	0.070	0.074	0.078	0.083	0.087	0.091	0.096
25.	0.011	0.016	0.022	0.027	0.033	0.038	0.044	0.049	0.054	0.060	0.065	0.071	0.076	0.082	0.087	0.092	0.098	0.103	0.109	0.114	0.120
30.	0.013	0.020	0.026	0.033	0.039	0.046	0.052	0.059	0.065	0.072	0.078	0.085	0.091	0.098	0.104	0.111	0.118	0.124	0.131	0.137	0.144
35.	0.015	0.023	0.030	0.038	0.046	0.053	0.061	0.069	0.076	0.084	0.091	0.099	0.107	0.114	0.122	0.130	0.137	0.145	0.152	0.160	0.168
40.	0.017	0.026	0.035	0.044	0.052	0.061	0.070	0.078	0.087	0.096	0.105	0.113	0.122	0.131	0.139	0.148	0.157	0.166	0.174	0.183	0.192
45.	0.020	0.029	0.039	0.049	0.059	0.069	0.078	0.088	0.098	0.108	0.118	0.127	0.137	0.147	0.157	0.167	0.176	0.186	0.196	0.206	0.216
50.	0.022	0.033	0.044	0.054	0.065	0.076	0.087	0.098	0.109	0.120	0.131	0.142	0.153	0.163	0.174	0.185	0.196	0.207	0.218	0.229	0.240
55.	0.024	0.036	0.048	0.060	0.072	0.084	0.096	0.108	0.120	0.132	0.144	0.156	0.168	0.180	0.192	0.205	0.216	0.228	0.240	0.252	0.264
60.	0.026	0.039	0.052	0.065	0.078	0.092	0.105	0.118	0.131	0.144	0.157	0.170	0.183	0.196	0.209	0.222	0.235	0.248	0.262	0.275	0.288
65.	0.028	0.043	0.057	0.071	0.085	0.099	0.113	0.128	0.142	0.156	0.170	0.184	0.198	0.213	0.227	0.241	0.255	0.269	0.283	0.298	0.312
70.	0.031	0.046	0.061	0.076	0.092	0.107	0.122	0.137	0.153	0.168	0.183	0.198	0.214	0.229	0.244	0.259	0.275	0.290	0.305	0.321	0.336
75.	0.033	0.050	0.065	0.082	0.098	0.115	0.131	0.147	0.164	0.180	0.196	0.213	0.229	0.245	0.262	0.278	0.294	0.311	0.327	0.344	0.360
80.	0.035	0.052	0.070	0.087	0.105	0.122	0.140	0.157	0.175	0.192	0.209	0.227	0.244	0.262	0.279	0.297	0.314	0.332	0.349	0.366	0.384
85.	0.037	0.056	0.074	0.093	0.111	0.130	0.148	0.167	0.185	0.204	0.223	0.241	0.260	0.278	0.297	0.315	0.334	0.353	0.371	0.390	0.408
90.	0.039	0.059	0.079	0.098	0.118	0.138	0.157	0.177	0.197	0.216	0.236	0.256	0.275	0.295	0.315	0.334	0.354	0.373	0.393	0.413	0.432
95.	0.041	0.062	0.083	0.104	0.124	0.145	0.166	0.187	0.208	0.228	0.249	0.270	0.291	0.312	0.332	0.353	0.373	0.394	0.415	0.436	0.456
100.	0.044	0.066	0.087	0.109	0.131	0.153	0.174	0.196	0.218	0.240	0.262	0.284	0.306	0.327	0.349	0.371	0.393	0.415	0.437	0.459	0.480
105.	0.046	0.069	0.092	0.115	0.138	0.161	0.183	0.206	0.229	0.252	0.275	0.298	0.321	0.344	0.367	0.390	0.413	0.436	0.459	0.482	0.504
110.	0.048	0.072	0.096	0.120	0.144	0.168	0.192	0.216	0.240	0.264	0.288	0.313	0.336	0.360	0.384	0.408	0.433	0.457	0.481	0.505	0.529
115.	0.050	0.075	0.101	0.126	0.151	0.176	0.201	0.226	0.251	0.276	0.302	0.327	0.352	0.377	0.402	0.427	0.453	0.478	0.503	0.528	0.553
120.	0.052	0.079	0.105	0.131	0.157	0.184	0.210	0.236	0.262	0.288	0.315	0.341	0.367	0.393	0.420	0.446	0.472	0.498	0.524	0.551	0.577
125.	0.055	0.082	0.109	0.137	0.164	0.191	0.219	0.246	0.273	0.301	0.328	0.355	0.383	0.410	0.437	0.464	0.492	0.519	0.546	0.574	0.601
130.	0.057	0.085	0.114	0.142	0.171	0.199	0.227	0.256	0.284	0.313	0.341	0.369	0.398	0.426	0.455	0.483	0.512	0.540	0.568	0.597	0.625
135.	0.059	0.089	0.118	0.148	0.177	0.207	0.236	0.266	0.295	0.325	0.354	0.384	0.413	0.443	0.472	0.502	0.531	0.561	0.590	0.620	0.649
140.	0.061	0.092	0.122	0.153	0.184	0.214	0.245	0.276	0.306	0.337	0.367	0.398	0.429	0.459	0.490	0.521	0.551	0.582	0.612	0.643	0.674
145.	0.063	0.095	0.127	0.159	0.190	0.222	0.254	0.285	0.317	0.349	0.381	0.412	0.444	0.476	0.508	0.539	0.571	0.603	0.634	0.666	0.698
150.	0.066	0.098	0.131	0.164	0.197	0.230	0.263	0.295	0.328	0.361	0.394	0.427	0.460	0.492	0.525	0.558	0.591	0.624	0.656	0.689	0.722
160.	0.070	0.105	0.140	0.175	0.210	0.245	0.280	0.315	0.350	0.385	0.420	0.455	0.490	0.525	0.560	0.595	0.630	0.665	0.701	0.736	0.771
170.	0.074	0.112	0.149	0.186	0.223	0.261	0.298	0.335	0.372	0.410	0.447	0.484	0.521	0.559	0.596	0.633	0.670	0.707	0.745	0.782	0.819
180.	0.079	0.118	0.158	0.197	0.237	0.276	0.316	0.355	0.395	0.434	0.474	0.513	0.552	0.592	0.631	0.671	0.710	0.750	0.789	0.828	0.868
190.	0.083	0.125	0.167	0.208	0.250	0.292	0.333	0.375	0.416	0.458	0.500	0.541	0.583	0.625	0.666	0.708	0.750	0.791	0.833	0.875	0.916
200.	0.088	0.132	0.175	0.219	0.263	0.307	0.351	0.395	0.439	0.482	0.526	0.570	0.614	0.658	0.702	0.746	0.790	0.833	0.877	0.921	0.965
210.	0.092	0.138	0.184	0.230	0.276	0.323	0.369	0.415	0.461	0.507	0.553	0.599	0.645	0.691	0.737	0.783	0.829	0.875	0.921	0.968	1.014
220.	0.097	0.145	0.193	0.241	0.290	0.338	0.386	0.435	0.483	0.531	0.579	0.628	0.676	0.724	0.773	0.821	0.869	0.917	0.966	1.014	1.062
230.	0.101	0.152	0.202	0.253	0.303	0.354	0.404	0.455	0.505	0.556	0.606	0.657	0.707	0.758	0.808	0.859	0.909	0.960	1.010	1.061	1.111
240.	0.105	0.158	0.211	0.264	0.316	0.369	0.422	0.475	0.527	0.580	0.633	0.686	0.738	0.791	0.844	0.896	0.949	1.002	1.054	1.107	1.160
250.	0.110	0.165	0.220	0.275	0.330	0.385	0.440	0.495	0.549	0.604	0.659	0.714	0.769	0.824	0.879	0.934	0.989	1.044	1.099	1.154	1.209
260.	0.114	0.172	0.229	0.286	0.343	0.400	0.457	0.515	0.572	0.629	0.686	0.743	0.800	0.858	0.915	0.972	1.029	1.086	1.143	1.201	1.258
270.	0.119	0.178	0.238	0.297	0.356	0.416	0.475	0.535	0.594	0.653	0.713	0.772	0.832	0.891	0.950	1.010	1.069	1.128	1.188	1.247	1.307
280.	0.123	0.185	0.246	0.308	0.370	0.431	0.493	0.555	0.616	0.678	0.739	0.801	0.863	0.924	0.986	1.048	1.109	1.171	1.232	1.294	1.356
290.	0.128	0.192	0.255	0.319	0.383	0.447	0.511	0.575	0.638	0.702	0.766	0.830	0.894	0.958	1.022	1.085	1.149	1.213	1.277	1.341	1.405
300.	0.132	0.198	0.264	0.330	0.396	0.463	0.529	0.595	0.661	0.727	0.793	0.859	0.925	0.991	1.057	1.123	1.189	1.256	1.322	1.388	1.454
310.	0.137	0.205	0.273	0.342	0.410	0.478	0.546	0.615	0.683	0.751	0.820	0.888	0.956	1.025	1.093	1.161	1.230	1.298	1.366	1.435	1.503
320.	0.141	0.212	0.282	0.353	0.423	0.494	0.564	0.635	0.705	0.776	0.847	0.917	0.988	1.058	1.129	1.199	1.270	1.340	1.411	1.481	1.552
330.	0.146	0.218	0.291	0.364	0.437	0.509	0.582	0.655	0.728	0.800	0.873	0.946	1.019	1.092	1.165	1.237	1.310	1.383	1.456	1.528	1.601
340.	0.150	0.225	0.300	0.375	0.450	0.525	0.600	0.675	0.750	0.825	0.900	0.975	1.050	1.125	1.200	1.275	1.350	1.425	1.500	1.575	1.650
350.	0.155	0.232	0.309	0.386	0.464	0.541	0.618	0.695	0.773	0.850	0.927	1.004	1.082	1.159	1.236	1.313	1.391	1.468	1.545	1.623	1.700
360.	0.159	0.239	0.318	0.398	0.477	0.557	0.636	0.716	0.795	0.875	0.954	1.034	1.113	1.193	1.272	1.352	1.431	1.511	1.590	1.670	1.749
370.	0.163	0.245	0.327	0.409	0.490	0.572	0.654	0.736	0.817	0.899	0.981	1.063	1.144	1.226	1.308	1.390	1.471	1.553	1.635	1.717	1.798

TABLE 2.4b Shell Thickness Tables in Inches: Joint Efficiency=0.90

SHELL THICKNESS TABLES IN INCHES

JOINT EFF.=0.90 ALLOWABLE STRESS VALUE= 13800. PSI

INSIDE DIAMETER, IN.

PSI	12.	18.	24.	30.	36.	42.	48.	54.	60.	66.	72.	78.	84.	90.	96.	102.	108.	114.	120.	126.	132.
15.	0.007	0.011	0.015	0.018	0.022	0.025	0.029	0.033	0.036	0.040	0.044	0.047	0.051	0.054	0.058	0.062	0.065	0.069	0.073	0.076	0.080
20.	0.010	0.015	0.019	0.024	0.029	0.034	0.039	0.044	0.048	0.053	0.058	0.063	0.068	0.073	0.077	0.082	0.087	0.092	0.097	0.102	0.106
25.	0.012	0.018	0.024	0.030	0.036	0.042	0.048	0.054	0.060	0.067	0.073	0.079	0.085	0.091	0.097	0.103	0.109	0.115	0.121	0.127	0.133
30.	0.015	0.022	0.029	0.036	0.044	0.051	0.058	0.065	0.073	0.080	0.087	0.094	0.102	0.109	0.116	0.123	0.131	0.138	0.145	0.152	0.160
35.	0.017	0.025	0.034	0.042	0.051	0.059	0.068	0.076	0.085	0.093	0.102	0.110	0.119	0.127	0.136	0.144	0.152	0.161	0.169	0.178	0.186
40.	0.019	0.029	0.039	0.048	0.058	0.068	0.077	0.087	0.097	0.106	0.116	0.126	0.135	0.145	0.155	0.165	0.174	0.184	0.194	0.203	0.213
45.	0.022	0.033	0.044	0.054	0.065	0.076	0.087	0.098	0.109	0.120	0.131	0.142	0.153	0.164	0.175	0.185	0.196	0.207	0.218	0.229	0.240
50.	0.024	0.036	0.048	0.061	0.073	0.085	0.097	0.109	0.121	0.133	0.145	0.157	0.169	0.182	0.194	0.206	0.218	0.230	0.242	0.254	0.266
55.	0.027	0.040	0.053	0.067	0.080	0.093	0.107	0.120	0.133	0.147	0.160	0.173	0.186	0.200	0.213	0.226	0.240	0.253	0.266	0.280	0.293
60.	0.029	0.044	0.058	0.073	0.087	0.102	0.116	0.131	0.145	0.160	0.174	0.189	0.203	0.218	0.232	0.247	0.262	0.276	0.291	0.305	0.320
65.	0.031	0.047	0.063	0.079	0.094	0.110	0.126	0.142	0.157	0.173	0.189	0.205	0.220	0.236	0.252	0.268	0.284	0.299	0.315	0.331	0.347
70.	0.034	0.051	0.068	0.085	0.102	0.119	0.136	0.153	0.170	0.187	0.204	0.221	0.238	0.255	0.272	0.289	0.306	0.322	0.339	0.356	0.373
75.	0.036	0.055	0.073	0.091	0.109	0.127	0.145	0.164	0.182	0.200	0.218	0.236	0.255	0.273	0.291	0.309	0.327	0.346	0.364	0.382	0.400
80.	0.039	0.058	0.078	0.097	0.116	0.136	0.155	0.175	0.194	0.213	0.233	0.252	0.272	0.291	0.310	0.330	0.349	0.369	0.388	0.407	0.427
85.	0.041	0.062	0.082	0.103	0.124	0.144	0.165	0.186	0.206	0.227	0.247	0.268	0.289	0.309	0.330	0.350	0.371	0.392	0.412	0.433	0.454
90.	0.044	0.066	0.087	0.109	0.131	0.153	0.175	0.197	0.218	0.240	0.262	0.284	0.306	0.328	0.349	0.371	0.393	0.415	0.437	0.458	0.480
95.	0.046	0.069	0.092	0.115	0.138	0.161	0.184	0.207	0.230	0.254	0.277	0.300	0.323	0.346	0.369	0.392	0.415	0.438	0.461	0.484	0.507
100.	0.049	0.073	0.097	0.121	0.146	0.170	0.194	0.218	0.243	0.267	0.291	0.316	0.340	0.364	0.388	0.413	0.437	0.461	0.485	0.510	0.534
105.	0.051	0.076	0.102	0.127	0.153	0.178	0.204	0.229	0.255	0.280	0.306	0.331	0.357	0.382	0.408	0.433	0.459	0.484	0.510	0.535	0.561
110.	0.053	0.080	0.107	0.134	0.160	0.187	0.214	0.240	0.267	0.294	0.321	0.347	0.374	0.401	0.427	0.454	0.481	0.508	0.534	0.561	0.588
115.	0.056	0.084	0.112	0.140	0.168	0.196	0.223	0.251	0.279	0.307	0.335	0.363	0.391	0.419	0.447	0.475	0.503	0.531	0.559	0.587	0.615
120.	0.058	0.087	0.117	0.146	0.175	0.204	0.233	0.262	0.292	0.321	0.350	0.379	0.408	0.437	0.466	0.496	0.525	0.554	0.583	0.612	0.641
125.	0.061	0.091	0.122	0.152	0.182	0.213	0.243	0.273	0.304	0.334	0.365	0.395	0.425	0.456	0.486	0.516	0.547	0.577	0.608	0.638	0.668
130.	0.063	0.095	0.126	0.158	0.190	0.221	0.253	0.284	0.316	0.348	0.379	0.411	0.442	0.474	0.506	0.537	0.569	0.600	0.632	0.664	0.695
135.	0.066	0.098	0.131	0.164	0.197	0.230	0.263	0.295	0.328	0.361	0.394	0.427	0.460	0.492	0.525	0.558	0.591	0.624	0.656	0.689	0.722
140.	0.068	0.102	0.136	0.170	0.204	0.238	0.273	0.306	0.340	0.375	0.409	0.443	0.477	0.511	0.545	0.579	0.613	0.647	0.681	0.715	0.749
145.	0.071	0.106	0.141	0.176	0.212	0.247	0.282	0.317	0.353	0.388	0.423	0.459	0.494	0.529	0.565	0.600	0.635	0.670	0.705	0.741	0.776
150.	0.073	0.109	0.146	0.182	0.219	0.255	0.292	0.328	0.365	0.401	0.438	0.474	0.511	0.547	0.584	0.620	0.657	0.693	0.730	0.766	0.803
160.	0.078	0.117	0.156	0.195	0.234	0.273	0.312	0.351	0.389	0.428	0.467	0.506	0.545	0.584	0.623	0.662	0.701	0.740	0.779	0.818	0.857
170.	0.083	0.124	0.166	0.207	0.248	0.290	0.331	0.373	0.414	0.455	0.497	0.538	0.580	0.621	0.662	0.704	0.745	0.787	0.828	0.869	0.911
180.	0.088	0.132	0.175	0.219	0.263	0.307	0.351	0.395	0.439	0.482	0.526	0.570	0.614	0.658	0.702	0.746	0.789	0.833	0.877	0.921	0.965
190.	0.093	0.139	0.185	0.232	0.278	0.324	0.371	0.417	0.463	0.510	0.556	0.602	0.648	0.695	0.741	0.787	0.834	0.880	0.926	0.973	1.019
200.	0.098	0.146	0.195	0.244	0.293	0.341	0.390	0.439	0.488	0.537	0.585	0.634	0.683	0.732	0.780	0.829	0.878	0.927	0.976	1.024	1.073
210.	0.102	0.154	0.205	0.256	0.307	0.359	0.410	0.461	0.512	0.564	0.615	0.666	0.717	0.769	0.820	0.871	0.922	0.974	1.025	1.076	1.127
220.	0.107	0.161	0.215	0.269	0.322	0.376	0.430	0.483	0.537	0.591	0.645	0.698	0.752	0.806	0.859	0.913	0.967	1.021	1.074	1.128	1.182
230.	0.112	0.168	0.225	0.281	0.337	0.393	0.449	0.506	0.562	0.618	0.674	0.730	0.787	0.843	0.899	0.955	1.011	1.067	1.124	1.180	1.236
240.	0.117	0.176	0.235	0.293	0.352	0.411	0.469	0.528	0.587	0.645	0.704	0.762	0.821	0.880	0.938	0.997	1.056	1.114	1.173	1.232	1.290
250.	0.122	0.183	0.244	0.306	0.367	0.428	0.489	0.550	0.611	0.672	0.733	0.795	0.856	0.917	0.978	1.039	1.100	1.161	1.222	1.284	1.345
260.	0.127	0.191	0.254	0.318	0.382	0.445	0.509	0.572	0.636	0.700	0.763	0.827	0.890	0.954	1.018	1.081	1.145	1.208	1.272	1.336	1.399
270.	0.132	0.198	0.264	0.330	0.396	0.463	0.529	0.595	0.661	0.727	0.793	0.859	0.925	0.991	1.057	1.123	1.189	1.256	1.322	1.388	1.454
280.	0.137	0.206	0.274	0.343	0.411	0.480	0.548	0.617	0.686	0.754	0.823	0.891	0.960	1.028	1.097	1.166	1.234	1.303	1.371	1.440	1.508
290.	0.142	0.213	0.284	0.355	0.426	0.497	0.568	0.639	0.710	0.781	0.852	0.924	0.995	1.066	1.137	1.208	1.279	1.350	1.421	1.492	1.563
300.	0.147	0.221	0.294	0.368	0.441	0.515	0.588	0.662	0.735	0.809	0.882	0.956	1.029	1.103	1.176	1.250	1.324	1.397	1.471	1.544	1.618
310.	0.152	0.228	0.304	0.380	0.456	0.532	0.608	0.684	0.760	0.836	0.912	0.988	1.064	1.140	1.216	1.292	1.368	1.444	1.520	1.596	1.672
320.	0.157	0.236	0.314	0.393	0.471	0.550	0.628	0.707	0.785	0.864	0.942	1.021	1.099	1.178	1.256	1.335	1.413	1.492	1.570	1.649	1.727
330.	0.162	0.243	0.324	0.405	0.486	0.567	0.648	0.729	0.810	0.891	0.972	1.053	1.134	1.215	1.296	1.377	1.458	1.539	1.620	1.701	1.782
340.	0.167	0.251	0.334	0.417	0.501	0.584	0.668	0.751	0.835	0.918	1.002	1.085	1.169	1.252	1.336	1.419	1.503	1.586	1.670	1.753	1.837
350.	0.172	0.258	0.344	0.430	0.516	0.602	0.688	0.774	0.860	0.946	1.032	1.118	1.204	1.290	1.376	1.462	1.548	1.634	1.720	1.806	1.892
360.	0.177	0.265	0.354	0.442	0.531	0.619	0.708	0.796	0.885	0.973	1.062	1.150	1.239	1.327	1.416	1.504	1.593	1.681	1.770	1.858	1.947
370.	0.182	0.273	0.364	0.455	0.546	0.637	0.728	0.819	0.910	1.001	1.092	1.183	1.274	1.365	1.456	1.547	1.638	1.729	1.820	1.911	2.002

TABLE 2.4c Shell Thickness Tables in Inches: Joint Efficiency=0.85

SHELL THICKNESS TABLES IN INCHES

JOINT EFF.=0.85) ALLOWABLE STRESS VALUE= 13800. PSI

PSI	\multicolumn{21}{c}{INSIDE DIAMETER, IN.}
	12.	18.	24.	30.	36.	42.	48.	54.	60.	66.	72.	78.	84.	90.	96.	102.	108.	114.	120.	126.	132.
15.	0.008	0.012	0.015	0.019	0.023	0.027	0.031	0.035	0.038	0.042	0.046	0.050	0.054	0.058	0.061	0.065	0.069	0.073	0.077	0.081	0.084
20.	0.010	0.015	0.020	0.026	0.031	0.036	0.041	0.046	0.051	0.056	0.061	0.067	0.072	0.077	0.082	0.087	0.092	0.097	0.102	0.108	0.113
25.	0.013	0.019	0.026	0.032	0.038	0.045	0.051	0.058	0.064	0.070	0.077	0.083	0.090	0.096	0.102	0.109	0.115	0.122	0.128	0.134	0.141
30.	0.015	0.023	0.031	0.038	0.046	0.054	0.061	0.069	0.077	0.085	0.092	0.100	0.108	0.115	0.123	0.131	0.138	0.146	0.154	0.161	0.169
35.	0.018	0.027	0.036	0.045	0.054	0.063	0.072	0.081	0.090	0.099	0.108	0.117	0.126	0.135	0.143	0.152	0.161	0.170	0.179	0.188	0.197
40.	0.021	0.031	0.041	0.051	0.062	0.072	0.082	0.092	0.103	0.113	0.123	0.133	0.144	0.154	0.164	0.174	0.185	0.195	0.205	0.215	0.226
45.	0.023	0.035	0.046	0.058	0.069	0.081	0.092	0.104	0.115	0.127	0.138	0.150	0.161	0.173	0.185	0.196	0.208	0.219	0.231	0.242	0.254
50.	0.026	0.038	0.051	0.064	0.077	0.090	0.103	0.115	0.128	0.141	0.154	0.167	0.179	0.192	0.205	0.218	0.231	0.244	0.256	0.269	0.282
55.	0.028	0.042	0.056	0.071	0.085	0.099	0.113	0.127	0.141	0.155	0.169	0.183	0.197	0.212	0.226	0.240	0.254	0.268	0.282	0.296	0.310
60.	0.031	0.046	0.062	0.077	0.092	0.108	0.123	0.139	0.154	0.169	0.185	0.200	0.215	0.231	0.246	0.262	0.277	0.292	0.308	0.323	0.339
65.	0.033	0.050	0.067	0.083	0.100	0.117	0.133	0.150	0.167	0.183	0.200	0.217	0.234	0.250	0.267	0.284	0.300	0.317	0.334	0.351	0.367
70.	0.036	0.054	0.072	0.090	0.108	0.126	0.144	0.162	0.180	0.198	0.216	0.234	0.252	0.270	0.289	0.307	0.325	0.343	0.361	0.379	0.397
75.	0.039	0.058	0.077	0.096	0.116	0.135	0.154	0.173	0.193	0.212	0.231	0.250	0.270	0.289	0.308	0.328	0.347	0.366	0.385	0.405	0.424
80.	0.041	0.062	0.082	0.103	0.123	0.144	0.164	0.185	0.206	0.226	0.247	0.267	0.288	0.308	0.329	0.349	0.370	0.390	0.411	0.431	0.452
85.	0.044	0.066	0.087	0.109	0.131	0.153	0.175	0.197	0.218	0.240	0.262	0.284	0.306	0.328	0.349	0.371	0.393	0.415	0.437	0.458	0.480
90.	0.046	0.069	0.092	0.116	0.139	0.162	0.185	0.208	0.231	0.254	0.277	0.301	0.324	0.347	0.370	0.393	0.416	0.439	0.462	0.486	0.509
95.	0.049	0.073	0.098	0.122	0.146	0.171	0.195	0.220	0.244	0.269	0.293	0.317	0.342	0.366	0.391	0.415	0.439	0.464	0.488	0.513	0.537
100.	0.051	0.077	0.103	0.129	0.154	0.180	0.206	0.231	0.257	0.283	0.308	0.334	0.360	0.385	0.411	0.437	0.463	0.488	0.514	0.540	0.566
105.	0.054	0.081	0.108	0.135	0.162	0.189	0.216	0.243	0.270	0.297	0.324	0.351	0.378	0.405	0.432	0.459	0.486	0.513	0.540	0.567	0.594
110.	0.057	0.085	0.113	0.141	0.170	0.198	0.226	0.255	0.283	0.311	0.340	0.368	0.396	0.425	0.453	0.481	0.509	0.538	0.566	0.594	0.622
115.	0.059	0.089	0.118	0.148	0.178	0.207	0.237	0.266	0.296	0.325	0.355	0.385	0.414	0.444	0.473	0.503	0.532	0.562	0.592	0.621	0.651
120.	0.062	0.093	0.124	0.154	0.185	0.216	0.247	0.278	0.309	0.340	0.371	0.401	0.432	0.463	0.494	0.525	0.556	0.587	0.618	0.648	0.679
125.	0.064	0.096	0.129	0.161	0.193	0.225	0.257	0.290	0.322	0.354	0.386	0.418	0.450	0.483	0.515	0.547	0.579	0.611	0.644	0.676	0.708
130.	0.067	0.100	0.134	0.167	0.201	0.234	0.268	0.301	0.335	0.368	0.402	0.435	0.469	0.502	0.536	0.569	0.602	0.636	0.669	0.703	0.736
135.	0.070	0.104	0.139	0.174	0.209	0.243	0.278	0.313	0.348	0.382	0.417	0.452	0.487	0.522	0.556	0.591	0.626	0.661	0.695	0.730	0.765
140.	0.072	0.108	0.144	0.180	0.216	0.252	0.289	0.325	0.361	0.397	0.433	0.469	0.505	0.541	0.577	0.613	0.649	0.685	0.721	0.757	0.793
145.	0.075	0.112	0.149	0.187	0.224	0.261	0.299	0.336	0.374	0.411	0.448	0.486	0.523	0.560	0.598	0.635	0.673	0.710	0.747	0.785	0.822
150.	0.077	0.116	0.155	0.193	0.232	0.271	0.309	0.348	0.387	0.425	0.464	0.503	0.541	0.580	0.619	0.657	0.696	0.735	0.773	0.812	0.851
160.	0.083	0.124	0.165	0.206	0.248	0.289	0.330	0.371	0.413	0.454	0.495	0.536	0.578	0.619	0.660	0.701	0.743	0.784	0.825	0.866	0.908
170.	0.088	0.132	0.175	0.219	0.263	0.307	0.351	0.395	0.439	0.482	0.526	0.570	0.614	0.658	0.702	0.746	0.790	0.833	0.877	0.921	0.965
180.	0.093	0.139	0.186	0.232	0.279	0.325	0.372	0.418	0.465	0.512	0.558	0.605	0.651	0.698	0.744	0.791	0.837	0.884	0.930	0.977	1.024
190.	0.098	0.147	0.196	0.245	0.294	0.343	0.393	0.442	0.491	0.540	0.589	0.638	0.687	0.737	0.786	0.835	0.884	0.933	0.982	1.032	1.081
200.	0.103	0.155	0.207	0.258	0.310	0.362	0.414	0.465	0.517	0.569	0.621	0.672	0.724	0.776	0.828	0.879	0.931	0.983	1.035	1.086	1.138
210.	0.109	0.163	0.217	0.271	0.326	0.380	0.434	0.489	0.543	0.597	0.651	0.706	0.760	0.814	0.869	0.923	0.977	1.032	1.086	1.140	1.194
220.	0.114	0.171	0.228	0.285	0.341	0.398	0.455	0.512	0.569	0.626	0.683	0.740	0.797	0.854	0.911	0.967	1.024	1.081	1.138	1.195	1.252
230.	0.119	0.179	0.238	0.298	0.357	0.417	0.476	0.536	0.595	0.655	0.714	0.774	0.833	0.893	0.952	1.012	1.071	1.131	1.190	1.250	1.310
240.	0.124	0.186	0.249	0.311	0.373	0.435	0.497	0.559	0.621	0.684	0.746	0.808	0.870	0.932	0.994	1.056	1.119	1.181	1.243	1.305	1.367
250.	0.130	0.194	0.259	0.324	0.389	0.453	0.518	0.583	0.648	0.712	0.777	0.842	0.907	0.972	1.036	1.101	1.166	1.231	1.295	1.340	1.425
260.	0.135	0.202	0.270	0.337	0.404	0.472	0.539	0.607	0.674	0.741	0.809	0.876	0.943	1.011	1.078	1.146	1.213	1.280	1.348	1.415	1.483
270.	0.140	0.210	0.280	0.350	0.420	0.490	0.560	0.630	0.700	0.770	0.840	0.910	0.980	1.050	1.120	1.190	1.260	1.330	1.400	1.470	1.540
280.	0.145	0.218	0.291	0.363	0.436	0.509	0.581	0.654	0.727	0.799	0.872	0.944	1.017	1.090	1.162	1.235	1.308	1.380	1.453	1.526	1.598
290.	0.151	0.226	0.301	0.376	0.452	0.527	0.602	0.678	0.753	0.828	0.903	0.979	1.054	1.129	1.205	1.280	1.355	1.430	1.506	1.581	1.656
300.	0.156	0.234	0.312	0.390	0.468	0.545	0.623	0.701	0.779	0.857	0.935	1.013	1.091	1.169	1.247	1.325	1.403	1.481	1.558	1.636	1.714
310.	0.161	0.242	0.322	0.403	0.483	0.564	0.644	0.725	0.806	0.886	0.967	1.047	1.128	1.208	1.289	1.370	1.450	1.531	1.611	1.692	1.772
320.	0.166	0.250	0.333	0.416	0.499	0.582	0.666	0.749	0.832	0.915	0.998	1.082	1.165	1.248	1.331	1.414	1.498	1.581	1.664	1.747	1.830
330.	0.172	0.258	0.343	0.429	0.515	0.601	0.687	0.773	0.858	0.944	1.030	1.116	1.202	1.288	1.374	1.459	1.545	1.631	1.717	1.803	1.889
340.	0.177	0.265	0.354	0.442	0.531	0.619	0.708	0.796	0.885	0.973	1.062	1.150	1.239	1.327	1.416	1.504	1.593	1.681	1.770	1.858	1.947
350.	0.182	0.273	0.365	0.456	0.547	0.638	0.729	0.820	0.911	1.003	1.094	1.185	1.276	1.367	1.458	1.549	1.641	1.732	1.823	1.914	2.005
360.	0.188	0.281	0.375	0.469	0.563	0.657	0.750	0.844	0.938	1.032	1.126	1.219	1.313	1.407	1.501	1.595	1.688	1.782	1.876	1.970	2.064
370.	0.193	0.289	0.386	0.482	0.579	0.675	0.772	0.868	0.965	1.061	1.157	1.254	1.350	1.447	1.543	1.640	1.736	1.833	1.929	2.026	2.122

TABLE 2.4d Shell Thickness Tables in Inches: Joint Efficiency=0.80

SWELL THICKNESS TABLES IN INCHES

JOINT EFF.=0.80   ALLOWABLE STRESS VALUE= 13800. PSI

INSIDE DIAMETER, IN.

PSI	12.	18.	24.	30.	36.	42.	48.	54.	60.	66.	72.	78.	84.	90.	96.	102.	108.	114.	120.	126.	132.
15.	0.008	0.012	0.016	0.020	0.024	0.029	0.033	0.037	0.041	0.045	0.049	0.053	0.057	0.061	0.065	0.069	0.073	0.078	0.082	0.086	0.090
20.	0.011	0.016	0.022	0.027	0.033	0.038	0.044	0.049	0.054	0.060	0.065	0.071	0.076	0.082	0.087	0.092	0.098	0.103	0.109	0.114	0.120
25.	0.014	0.020	0.027	0.034	0.041	0.048	0.054	0.061	0.068	0.075	0.082	0.088	0.095	0.102	0.109	0.116	0.122	0.129	0.136	0.143	0.150
30.	0.016	0.024	0.033	0.041	0.049	0.057	0.065	0.073	0.082	0.090	0.098	0.106	0.114	0.122	0.131	0.139	0.147	0.155	0.163	0.171	0.180
35.	0.019	0.028	0.038	0.048	0.057	0.067	0.076	0.086	0.095	0.105	0.114	0.124	0.133	0.143	0.152	0.162	0.172	0.181	0.191	0.200	0.210
40.	0.022	0.033	0.044	0.054	0.065	0.076	0.087	0.098	0.109	0.120	0.131	0.142	0.153	0.163	0.174	0.185	0.196	0.207	0.218	0.229	0.240
45.	0.025	0.037	0.049	0.061	0.074	0.086	0.098	0.110	0.123	0.135	0.147	0.159	0.172	0.184	0.196	0.208	0.221	0.233	0.245	0.257	0.270
50.	0.027	0.041	0.054	0.068	0.082	0.095	0.109	0.123	0.136	0.150	0.163	0.177	0.191	0.204	0.218	0.232	0.245	0.259	0.272	0.286	0.300
55.	0.030	0.045	0.060	0.075	0.090	0.105	0.120	0.135	0.150	0.165	0.180	0.195	0.210	0.225	0.240	0.255	0.270	0.285	0.300	0.315	0.330
60.	0.033	0.049	0.065	0.082	0.098	0.115	0.131	0.147	0.164	0.180	0.196	0.213	0.229	0.246	0.262	0.278	0.294	0.311	0.327	0.344	0.360
65.	0.035	0.053	0.071	0.089	0.106	0.124	0.142	0.160	0.177	0.195	0.213	0.230	0.248	0.266	0.284	0.301	0.319	0.337	0.355	0.372	0.390
70.	0.038	0.057	0.076	0.095	0.115	0.134	0.153	0.172	0.191	0.210	0.229	0.248	0.267	0.286	0.306	0.325	0.344	0.363	0.382	0.401	0.420
75.	0.041	0.061	0.082	0.102	0.123	0.143	0.164	0.184	0.205	0.225	0.246	0.266	0.287	0.307	0.328	0.348	0.369	0.389	0.409	0.430	0.450
80.	0.044	0.066	0.087	0.109	0.131	0.153	0.175	0.197	0.219	0.240	0.262	0.284	0.306	0.328	0.349	0.372	0.393	0.415	0.437	0.459	0.481
85.	0.046	0.070	0.093	0.116	0.139	0.162	0.186	0.209	0.232	0.255	0.278	0.302	0.325	0.348	0.371	0.394	0.418	0.441	0.464	0.487	0.511
90.	0.049	0.074	0.098	0.123	0.147	0.172	0.197	0.221	0.246	0.270	0.295	0.319	0.344	0.369	0.393	0.418	0.442	0.467	0.492	0.516	0.541
95.	0.052	0.078	0.104	0.130	0.156	0.182	0.208	0.234	0.260	0.286	0.311	0.337	0.363	0.389	0.415	0.441	0.467	0.493	0.519	0.545	0.571
100.	0.055	0.082	0.109	0.137	0.164	0.191	0.219	0.246	0.273	0.301	0.328	0.355	0.383	0.410	0.437	0.464	0.492	0.519	0.546	0.574	0.601
105.	0.057	0.086	0.115	0.143	0.172	0.201	0.230	0.258	0.287	0.316	0.344	0.373	0.402	0.430	0.459	0.488	0.517	0.545	0.574	0.603	0.631
110.	0.060	0.090	0.120	0.150	0.180	0.210	0.241	0.271	0.301	0.331	0.361	0.391	0.421	0.451	0.481	0.511	0.541	0.571	0.601	0.632	0.662
115.	0.063	0.094	0.126	0.157	0.189	0.220	0.252	0.283	0.314	0.346	0.377	0.409	0.440	0.472	0.503	0.535	0.566	0.597	0.629	0.660	0.692
120.	0.066	0.098	0.131	0.164	0.197	0.230	0.263	0.295	0.328	0.361	0.394	0.427	0.460	0.492	0.525	0.558	0.591	0.624	0.656	0.689	0.722
125.	0.068	0.103	0.137	0.171	0.205	0.239	0.274	0.308	0.342	0.376	0.410	0.445	0.479	0.513	0.547	0.581	0.616	0.650	0.684	0.718	0.752
130.	0.071	0.107	0.142	0.178	0.213	0.249	0.285	0.320	0.356	0.391	0.427	0.463	0.498	0.534	0.569	0.605	0.640	0.676	0.712	0.747	0.783
135.	0.074	0.111	0.148	0.185	0.222	0.259	0.296	0.333	0.370	0.407	0.443	0.480	0.517	0.554	0.591	0.628	0.665	0.702	0.739	0.776	0.813
140.	0.077	0.115	0.153	0.192	0.230	0.268	0.307	0.345	0.383	0.422	0.460	0.498	0.537	0.575	0.613	0.652	0.690	0.728	0.767	0.805	0.843
145.	0.079	0.119	0.159	0.199	0.238	0.278	0.318	0.357	0.397	0.437	0.477	0.516	0.556	0.596	0.635	0.675	0.715	0.755	0.794	0.834	0.874
150.	0.082	0.123	0.164	0.205	0.247	0.288	0.329	0.370	0.411	0.452	0.493	0.534	0.575	0.616	0.657	0.699	0.740	0.781	0.822	0.863	0.904
160.	0.088	0.132	0.175	0.219	0.263	0.307	0.351	0.395	0.439	0.482	0.526	0.570	0.614	0.658	0.702	0.746	0.789	0.833	0.877	0.921	0.965
170.	0.093	0.140	0.187	0.233	0.280	0.326	0.373	0.420	0.466	0.513	0.560	0.606	0.653	0.699	0.746	0.793	0.839	0.886	0.933	0.979	1.026
180.	0.099	0.148	0.198	0.247	0.296	0.346	0.395	0.445	0.494	0.543	0.593	0.642	0.692	0.741	0.790	0.840	0.889	0.939	0.988	1.037	1.087
190.	0.104	0.157	0.209	0.261	0.313	0.365	0.417	0.470	0.522	0.574	0.626	0.678	0.730	0.783	0.835	0.887	0.939	0.991	1.044	1.096	1.148
200.	0.110	0.165	0.220	0.275	0.330	0.385	0.440	0.495	0.549	0.604	0.659	0.714	0.769	0.824	0.879	0.934	0.989	1.044	1.099	1.154	1.209
210.	0.115	0.173	0.231	0.289	0.346	0.404	0.462	0.520	0.577	0.635	0.693	0.750	0.808	0.866	0.924	0.981	1.039	1.097	1.154	1.212	1.270
220.	0.121	0.182	0.242	0.303	0.363	0.424	0.484	0.545	0.605	0.666	0.726	0.787	0.847	0.908	0.968	1.029	1.089	1.150	1.210	1.271	1.331
230.	0.127	0.190	0.253	0.316	0.380	0.443	0.506	0.570	0.633	0.696	0.759	0.823	0.886	0.949	1.013	1.076	1.139	1.203	1.266	1.329	1.392
240.	0.132	0.198	0.264	0.330	0.396	0.463	0.529	0.595	0.661	0.727	0.793	0.859	0.925	0.991	1.057	1.123	1.189	1.256	1.322	1.388	1.454
250.	0.138	0.207	0.275	0.344	0.413	0.482	0.551	0.620	0.689	0.758	0.826	0.895	0.964	1.033	1.102	1.171	1.240	1.309	1.377	1.446	1.515
260.	0.143	0.215	0.287	0.358	0.430	0.502	0.573	0.645	0.717	0.788	0.860	0.932	1.003	1.075	1.147	1.218	1.290	1.362	1.433	1.505	1.577
270.	0.149	0.223	0.298	0.372	0.447	0.521	0.596	0.670	0.745	0.819	0.894	0.968	1.042	1.117	1.191	1.266	1.340	1.415	1.489	1.564	1.638
280.	0.155	0.232	0.309	0.386	0.464	0.541	0.618	0.695	0.773	0.850	0.927	1.004	1.082	1.159	1.236	1.313	1.391	1.468	1.545	1.623	1.700
290.	0.160	0.240	0.320	0.400	0.480	0.560	0.641	0.721	0.801	0.881	0.961	1.041	1.121	1.201	1.281	1.361	1.441	1.521	1.601	1.681	1.761
300.	0.166	0.249	0.331	0.414	0.497	0.580	0.663	0.746	0.829	0.912	0.994	1.077	1.160	1.243	1.326	1.409	1.492	1.575	1.657	1.740	1.823
310.	0.171	0.257	0.343	0.428	0.514	0.600	0.685	0.771	0.857	0.943	1.028	1.114	1.200	1.285	1.371	1.457	1.542	1.628	1.714	1.799	1.885
320.	0.177	0.265	0.354	0.442	0.531	0.619	0.708	0.796	0.885	0.973	1.062	1.150	1.239	1.327	1.416	1.504	1.593	1.681	1.770	1.858	1.947
330.	0.183	0.274	0.365	0.457	0.548	0.639	0.730	0.822	0.913	1.004	1.096	1.187	1.278	1.370	1.461	1.552	1.644	1.735	1.826	1.918	2.009
340.	0.188	0.282	0.377	0.471	0.565	0.659	0.753	0.847	0.941	1.035	1.130	1.224	1.318	1.412	1.506	1.600	1.694	1.788	1.883	1.977	2.071
350.	0.194	0.291	0.388	0.485	0.582	0.679	0.776	0.873	0.970	1.066	1.163	1.260	1.357	1.454	1.551	1.648	1.745	1.842	1.939	2.036	2.133
360.	0.200	0.299	0.399	0.499	0.598	0.698	0.798	0.898	0.998	1.098	1.197	1.297	1.397	1.497	1.596	1.696	1.796	1.896	1.996	2.095	2.195
370.	0.205	0.308	0.410	0.513	0.616	0.718	0.821	0.923	1.026	1.129	1.231	1.334	1.436	1.539	1.642	1.744	1.847	1.950	2.052	2.155	2.257

TABLE 2.7e Shell Thickness Tables in Inches: Joint Efficiency = 0.70

SHELL THICKNESS TABLES IN INCHES

JOINT EFF.=0.70 ALLOWABLE STRESS VALUE= 13800. PSI

INSIDE DIAMETER, IN.

PSI	12.	18.	24.	30.	36.	42.	48.	54.	60.	66.	72.	78.	84.	90.	96.	102.	108.	114.	120.	126.	132.
15.	0.009	0.014	0.019	0.023	0.028	0.033	0.037	0.042	0.047	0.051	0.056	0.061	0.065	0.070	0.075	0.079	0.084	0.089	0.093	0.098	0.103
20.	0.012	0.019	0.025	0.031	0.037	0.044	0.050	0.056	0.062	0.068	0.075	0.081	0.087	0.093	0.100	0.106	0.112	0.118	0.124	0.131	0.137
25.	0.016	0.023	0.031	0.039	0.047	0.054	0.062	0.070	0.078	0.086	0.093	0.101	0.109	0.117	0.124	0.132	0.140	0.148	0.156	0.163	0.171
30.	0.019	0.028	0.037	0.047	0.056	0.065	0.075	0.084	0.093	0.103	0.112	0.121	0.131	0.140	0.149	0.159	0.168	0.177	0.187	0.196	0.205
35.	0.022	0.033	0.044	0.054	0.065	0.076	0.087	0.098	0.109	0.120	0.131	0.142	0.153	0.163	0.174	0.185	0.196	0.207	0.218	0.229	0.240
40.	0.025	0.037	0.050	0.062	0.075	0.087	0.100	0.112	0.125	0.137	0.149	0.162	0.174	0.187	0.199	0.212	0.224	0.237	0.249	0.262	0.274
45.	0.028	0.042	0.056	0.070	0.084	0.098	0.112	0.126	0.140	0.154	0.168	0.182	0.196	0.210	0.224	0.238	0.252	0.266	0.280	0.294	0.308
50.	0.031	0.047	0.062	0.078	0.093	0.109	0.125	0.140	0.156	0.171	0.187	0.202	0.218	0.234	0.249	0.265	0.280	0.296	0.312	0.327	0.343
55.	0.034	0.051	0.069	0.086	0.103	0.120	0.137	0.154	0.171	0.189	0.206	0.223	0.240	0.257	0.274	0.291	0.309	0.326	0.343	0.360	0.377
60.	0.037	0.056	0.075	0.094	0.112	0.131	0.150	0.168	0.187	0.206	0.224	0.243	0.262	0.281	0.299	0.318	0.337	0.355	0.374	0.393	0.411
65.	0.041	0.061	0.081	0.101	0.122	0.142	0.162	0.182	0.203	0.223	0.243	0.263	0.284	0.304	0.324	0.345	0.365	0.385	0.405	0.426	0.446
70.	0.044	0.066	0.087	0.109	0.131	0.153	0.175	0.197	0.218	0.240	0.262	0.284	0.306	0.328	0.349	0.371	0.393	0.415	0.437	0.459	0.480
75.	0.047	0.070	0.094	0.117	0.140	0.164	0.187	0.211	0.234	0.257	0.281	0.304	0.328	0.351	0.374	0.398	0.421	0.445	0.468	0.491	0.515
80.	0.050	0.075	0.100	0.125	0.150	0.175	0.200	0.225	0.250	0.275	0.300	0.325	0.350	0.375	0.400	0.424	0.449	0.474	0.499	0.524	0.549
85.	0.053	0.080	0.106	0.133	0.159	0.186	0.212	0.239	0.265	0.292	0.318	0.345	0.372	0.398	0.425	0.451	0.478	0.504	0.531	0.557	0.584
90.	0.056	0.084	0.112	0.141	0.169	0.197	0.225	0.253	0.281	0.309	0.337	0.365	0.394	0.422	0.450	0.478	0.506	0.534	0.562	0.590	0.618
95.	0.059	0.089	0.119	0.148	0.178	0.208	0.237	0.267	0.297	0.326	0.356	0.386	0.415	0.445	0.475	0.505	0.534	0.564	0.594	0.623	0.653
100.	0.063	0.094	0.125	0.156	0.188	0.219	0.250	0.281	0.313	0.344	0.375	0.406	0.438	0.469	0.500	0.531	0.563	0.594	0.625	0.656	0.688
105.	0.066	0.098	0.131	0.164	0.197	0.230	0.263	0.295	0.328	0.361	0.394	0.427	0.460	0.492	0.525	0.558	0.591	0.624	0.656	0.689	0.722
110.	0.069	0.103	0.138	0.172	0.206	0.241	0.275	0.310	0.344	0.378	0.413	0.447	0.482	0.516	0.550	0.585	0.619	0.654	0.688	0.722	0.757
115.	0.072	0.108	0.144	0.180	0.216	0.252	0.288	0.324	0.360	0.396	0.432	0.468	0.504	0.540	0.576	0.612	0.647	0.683	0.719	0.755	0.791
120.	0.075	0.113	0.150	0.188	0.225	0.263	0.300	0.338	0.375	0.413	0.451	0.488	0.526	0.563	0.601	0.638	0.676	0.713	0.751	0.788	0.826
125.	0.078	0.117	0.156	0.196	0.235	0.274	0.313	0.352	0.391	0.430	0.469	0.509	0.548	0.587	0.626	0.665	0.704	0.743	0.782	0.822	0.861
130.	0.081	0.122	0.163	0.204	0.244	0.285	0.326	0.366	0.407	0.448	0.488	0.529	0.570	0.611	0.651	0.692	0.733	0.773	0.814	0.855	0.895
135.	0.085	0.127	0.169	0.212	0.254	0.296	0.338	0.381	0.423	0.465	0.507	0.550	0.592	0.634	0.676	0.719	0.761	0.803	0.846	0.888	0.930
140.	0.088	0.132	0.175	0.219	0.263	0.307	0.351	0.395	0.439	0.482	0.526	0.570	0.614	0.658	0.702	0.746	0.789	0.833	0.877	0.921	0.965
145.	0.091	0.136	0.182	0.227	0.273	0.318	0.364	0.409	0.455	0.500	0.545	0.591	0.636	0.682	0.727	0.772	0.818	0.863	0.909	0.954	1.000
150.	0.094	0.141	0.188	0.235	0.282	0.329	0.376	0.423	0.470	0.517	0.564	0.611	0.658	0.705	0.752	0.799	0.846	0.893	0.940	0.987	1.034
160.	0.100	0.151	0.201	0.251	0.301	0.351	0.402	0.452	0.502	0.552	0.602	0.652	0.703	0.753	0.803	0.853	0.903	0.954	1.004	1.054	1.104
170.	0.107	0.160	0.213	0.267	0.320	0.374	0.427	0.480	0.534	0.587	0.640	0.694	0.747	0.800	0.854	0.907	0.961	1.014	1.067	1.121	1.174
180.	0.113	0.170	0.226	0.283	0.339	0.396	0.452	0.509	0.566	0.622	0.678	0.735	0.791	0.848	0.905	0.961	1.018	1.074	1.131	1.187	1.244
190.	0.119	0.179	0.239	0.299	0.358	0.418	0.478	0.537	0.597	0.657	0.717	0.776	0.836	0.896	0.955	1.015	1.075	1.135	1.194	1.254	1.314
200.	0.126	0.189	0.252	0.315	0.377	0.440	0.503	0.566	0.629	0.692	0.755	0.818	0.881	0.943	1.006	1.069	1.132	1.195	1.258	1.321	1.384
210.	0.132	0.198	0.264	0.330	0.396	0.463	0.529	0.595	0.661	0.727	0.793	0.859	0.925	0.991	1.057	1.123	1.189	1.256	1.322	1.388	1.454
220.	0.139	0.208	0.277	0.346	0.416	0.485	0.554	0.623	0.693	0.762	0.831	0.901	0.970	1.039	1.108	1.178	1.247	1.316	1.385	1.455	1.524
230.	0.145	0.217	0.290	0.362	0.435	0.507	0.580	0.652	0.725	0.797	0.870	0.942	1.014	1.087	1.159	1.232	1.304	1.377	1.449	1.522	1.594
240.	0.151	0.227	0.303	0.378	0.454	0.530	0.605	0.681	0.757	0.832	0.908	0.984	1.059	1.135	1.211	1.286	1.362	1.438	1.513	1.589	1.665
250.	0.158	0.237	0.315	0.394	0.473	0.552	0.631	0.710	0.789	0.868	0.946	1.025	1.104	1.183	1.262	1.341	1.420	1.498	1.577	1.656	1.735
260.	0.164	0.246	0.328	0.410	0.492	0.574	0.657	0.739	0.821	0.903	0.985	1.067	1.149	1.231	1.313	1.395	1.477	1.559	1.641	1.723	1.806
270.	0.171	0.256	0.341	0.426	0.512	0.597	0.682	0.768	0.853	0.938	1.023	1.109	1.194	1.279	1.364	1.450	1.535	1.620	1.706	1.791	1.876
280.	0.177	0.265	0.354	0.442	0.531	0.619	0.708	0.796	0.885	0.973	1.062	1.150	1.239	1.327	1.416	1.504	1.593	1.681	1.770	1.858	1.947
290.	0.183	0.275	0.367	0.459	0.550	0.642	0.734	0.825	0.917	1.009	1.101	1.192	1.284	1.376	1.467	1.559	1.651	1.743	1.834	1.926	2.018
300.	0.190	0.285	0.380	0.475	0.570	0.665	0.759	0.854	0.949	1.044	1.139	1.234	1.329	1.424	1.519	1.613	1.708	1.803	1.898	1.993	2.088
310.	0.196	0.294	0.393	0.491	0.589	0.687	0.785	0.883	0.982	1.080	1.178	1.276	1.374	1.472	1.571	1.669	1.767	1.865	1.963	2.061	2.160
320.	0.203	0.304	0.406	0.507	0.608	0.710	0.811	0.913	1.014	1.115	1.217	1.318	1.420	1.521	1.622	1.724	1.825	1.926	2.028	2.129	2.231
330.	0.209	0.314	0.419	0.523	0.628	0.732	0.837	0.942	1.046	1.151	1.256	1.360	1.465	1.570	1.674	1.779	1.884	1.988	2.093	2.197	2.302
340.	0.216	0.324	0.431	0.539	0.647	0.755	0.863	0.971	1.079	1.187	1.294	1.402	1.510	1.618	1.726	1.834	1.942	2.049	2.157	2.265	2.373
350.	0.222	0.333	0.444	0.556	0.667	0.778	0.889	1.000	1.111	1.222	1.333	1.444	1.555	1.667	1.778	1.889	2.000	2.111	2.222	2.333	2.444
360.	0.229	0.343	0.457	0.572	0.686	0.801	0.915	1.029	1.144	1.258	1.372	1.487	1.601	1.715	1.830	1.944	2.058	2.173	2.287	2.402	2.516
370.	0.235	0.353	0.470	0.588	0.706	0.823	0.941	1.058	1.176	1.294	1.411	1.529	1.647	1.764	1.882	1.999	2.117	2.235	2.352	2.470	2.587

TABLE 2.4f Shell Thickness Tables in Inches: Joint Efficiency=0.65

SHELL THICKNESS TABLES IN INCHES

JOINT EFF.=0.65   ALLOWABLE STRESS VALUE= 13800. PSI

PSI	12.	18.	24.	30.	36.	42.	48.	54.	60.	66.	72.	78.	84.	90.	96.	102.	108.	114.	120.	126.	132.
15.	0.010	0.015	0.020	0.025	0.030	0.035	0.040	0.045	0.050	0.055	0.060	0.065	0.070	0.075	0.080	0.085	0.090	0.095	0.100	0.105	0.110
20.	0.013	0.020	0.027	0.033	0.040	0.047	0.054	0.060	0.067	0.074	0.080	0.087	0.094	0.100	0.107	0.114	0.121	0.127	0.134	0.141	0.147
25.	0.017	0.025	0.034	0.042	0.050	0.059	0.067	0.075	0.084	0.092	0.101	0.109	0.117	0.126	0.134	0.142	0.151	0.159	0.168	0.176	0.184
30.	0.020	0.030	0.040	0.050	0.060	0.070	0.080	0.090	0.101	0.111	0.121	0.131	0.141	0.151	0.161	0.171	0.181	0.191	0.201	0.211	0.221
35.	0.023	0.035	0.047	0.059	0.070	0.082	0.094	0.106	0.117	0.129	0.141	0.153	0.164	0.176	0.188	0.199	0.211	0.223	0.235	0.246	0.258
40.	0.027	0.040	0.054	0.067	0.080	0.094	0.107	0.121	0.134	0.148	0.161	0.174	0.188	0.201	0.215	0.228	0.241	0.255	0.268	0.282	0.295
45.	0.030	0.045	0.060	0.075	0.090	0.106	0.121	0.136	0.151	0.166	0.181	0.196	0.211	0.226	0.242	0.257	0.272	0.287	0.302	0.317	0.332
50.	0.034	0.050	0.067	0.084	0.101	0.117	0.134	0.151	0.168	0.185	0.201	0.218	0.235	0.252	0.269	0.285	0.302	0.319	0.336	0.353	0.369
55.	0.037	0.055	0.074	0.092	0.111	0.129	0.148	0.166	0.185	0.203	0.222	0.240	0.258	0.277	0.295	0.314	0.332	0.351	0.369	0.388	0.406
60.	0.040	0.060	0.081	0.101	0.121	0.141	0.161	0.181	0.201	0.222	0.242	0.262	0.282	0.302	0.322	0.343	0.363	0.383	0.403	0.423	0.443
65.	0.044	0.066	0.087	0.109	0.131	0.153	0.175	0.197	0.218	0.240	0.262	0.284	0.306	0.328	0.349	0.371	0.393	0.415	0.437	0.459	0.480
70.	0.047	0.071	0.094	0.118	0.141	0.165	0.188	0.212	0.235	0.259	0.282	0.306	0.329	0.353	0.376	0.400	0.423	0.447	0.470	0.494	0.517
75.	0.050	0.076	0.101	0.126	0.151	0.176	0.202	0.227	0.252	0.277	0.303	0.328	0.353	0.378	0.403	0.429	0.454	0.479	0.504	0.529	0.555
80.	0.054	0.081	0.108	0.134	0.161	0.188	0.215	0.242	0.269	0.296	0.323	0.350	0.377	0.403	0.430	0.457	0.484	0.511	0.538	0.565	0.592
85.	0.057	0.086	0.114	0.143	0.172	0.200	0.229	0.257	0.286	0.314	0.343	0.372	0.400	0.429	0.457	0.486	0.515	0.543	0.572	0.600	0.629
90.	0.061	0.091	0.121	0.151	0.182	0.212	0.242	0.273	0.303	0.333	0.363	0.394	0.424	0.454	0.485	0.515	0.545	0.575	0.606	0.636	0.666
95.	0.064	0.096	0.128	0.160	0.192	0.224	0.256	0.288	0.320	0.352	0.384	0.416	0.448	0.480	0.512	0.544	0.576	0.608	0.640	0.672	0.704
100.	0.067	0.101	0.135	0.168	0.202	0.236	0.270	0.303	0.337	0.371	0.405	0.438	0.472	0.506	0.540	0.574	0.607	0.641	0.675	0.709	0.743
105.	0.071	0.106	0.142	0.177	0.212	0.248	0.283	0.318	0.354	0.389	0.424	0.460	0.495	0.530	0.566	0.601	0.637	0.672	0.707	0.743	0.778
110.	0.074	0.111	0.148	0.185	0.222	0.259	0.296	0.334	0.371	0.408	0.445	0.482	0.519	0.556	0.593	0.630	0.667	0.704	0.741	0.779	0.816
115.	0.078	0.116	0.155	0.194	0.233	0.271	0.310	0.349	0.388	0.426	0.465	0.504	0.543	0.581	0.620	0.659	0.698	0.736	0.775	0.814	0.853
120.	0.081	0.121	0.162	0.202	0.243	0.283	0.324	0.364	0.405	0.445	0.486	0.526	0.567	0.607	0.647	0.688	0.728	0.769	0.809	0.850	0.890
125.	0.084	0.126	0.169	0.211	0.253	0.295	0.337	0.379	0.422	0.464	0.506	0.548	0.590	0.632	0.675	0.717	0.759	0.801	0.843	0.885	0.927
130.	0.088	0.132	0.175	0.219	0.263	0.307	0.351	0.395	0.439	0.482	0.526	0.570	0.614	0.658	0.702	0.746	0.789	0.833	0.877	0.921	0.965
135.	0.091	0.137	0.182	0.228	0.273	0.319	0.364	0.410	0.456	0.501	0.547	0.592	0.638	0.683	0.729	0.775	0.820	0.866	0.911	0.957	1.002
140.	0.095	0.142	0.189	0.236	0.284	0.331	0.378	0.425	0.473	0.520	0.567	0.614	0.662	0.709	0.756	0.804	0.851	0.898	0.945	0.993	1.040
145.	0.098	0.147	0.196	0.245	0.294	0.343	0.392	0.441	0.490	0.539	0.588	0.637	0.686	0.735	0.784	0.832	0.881	0.930	0.979	1.028	1.077
150.	0.101	0.152	0.203	0.253	0.304	0.355	0.405	0.456	0.507	0.557	0.608	0.659	0.709	0.760	0.811	0.861	0.912	0.963	1.014	1.064	1.115
160.	0.108	0.162	0.216	0.270	0.325	0.379	0.433	0.487	0.541	0.595	0.649	0.703	0.757	0.811	0.865	0.920	0.974	1.028	1.082	1.136	1.190
170.	0.115	0.173	0.230	0.288	0.345	0.403	0.460	0.518	0.575	0.633	0.690	0.748	0.805	0.863	0.920	0.978	1.035	1.093	1.150	1.208	1.265
180.	0.122	0.183	0.244	0.305	0.366	0.427	0.487	0.548	0.609	0.670	0.731	0.792	0.853	0.914	0.975	1.036	1.097	1.158	1.219	1.280	1.341
190.	0.129	0.193	0.257	0.322	0.386	0.451	0.515	0.579	0.644	0.708	0.772	0.837	0.901	0.965	1.030	1.094	1.159	1.223	1.287	1.352	1.416
200.	0.136	0.203	0.271	0.339	0.407	0.475	0.542	0.610	0.678	0.746	0.814	0.881	0.949	1.017	1.085	1.153	1.220	1.288	1.356	1.424	1.492
210.	0.142	0.213	0.285	0.356	0.427	0.499	0.570	0.641	0.712	0.784	0.855	0.926	0.997	1.069	1.140	1.211	1.282	1.353	1.425	1.496	1.567
220.	0.149	0.224	0.299	0.373	0.448	0.523	0.597	0.672	0.747	0.821	0.896	0.971	1.045	1.120	1.195	1.270	1.344	1.419	1.494	1.568	1.643
230.	0.156	0.234	0.313	0.391	0.469	0.547	0.625	0.703	0.781	0.859	0.938	1.016	1.094	1.172	1.250	1.328	1.406	1.484	1.563	1.641	1.719
240.	0.163	0.245	0.326	0.408	0.489	0.571	0.653	0.734	0.816	0.897	0.979	1.061	1.142	1.224	1.305	1.387	1.468	1.550	1.632	1.713	1.795
250.	0.170	0.255	0.340	0.425	0.510	0.595	0.680	0.765	0.850	0.935	1.020	1.105	1.190	1.276	1.361	1.446	1.531	1.616	1.701	1.786	1.871
260.	0.177	0.265	0.354	0.442	0.531	0.619	0.708	0.796	0.885	0.973	1.062	1.150	1.239	1.327	1.416	1.504	1.593	1.681	1.770	1.858	1.947
270.	0.184	0.276	0.368	0.460	0.552	0.644	0.736	0.828	0.920	1.012	1.104	1.196	1.287	1.379	1.471	1.563	1.655	1.747	1.839	1.931	2.023
280.	0.191	0.286	0.382	0.477	0.573	0.668	0.763	0.859	0.954	1.050	1.145	1.241	1.336	1.431	1.527	1.622	1.718	1.813	1.909	2.004	2.100
290.	0.198	0.297	0.396	0.495	0.593	0.692	0.791	0.890	0.989	1.088	1.187	1.286	1.385	1.484	1.583	1.681	1.780	1.879	1.978	2.077	2.176
300.	0.205	0.307	0.410	0.512	0.614	0.717	0.819	0.922	1.024	1.126	1.229	1.331	1.433	1.536	1.638	1.741	1.843	1.945	2.048	2.150	2.253
310.	0.212	0.318	0.423	0.529	0.635	0.741	0.847	0.953	1.059	1.165	1.270	1.376	1.482	1.588	1.694	1.800	1.906	2.012	2.117	2.223	2.329
320.	0.219	0.328	0.437	0.547	0.656	0.766	0.875	0.984	1.094	1.203	1.312	1.422	1.531	1.640	1.750	1.859	1.969	2.078	2.187	2.297	2.406
330.	0.226	0.339	0.451	0.564	0.677	0.790	0.903	1.016	1.129	1.242	1.354	1.467	1.580	1.693	1.806	1.919	2.031	2.144	2.257	2.370	2.483
340.	0.233	0.349	0.465	0.582	0.698	0.815	0.931	1.047	1.164	1.280	1.396	1.513	1.629	1.745	1.862	1.978	2.094	2.211	2.327	2.444	2.560
350.	0.240	0.360	0.479	0.599	0.719	0.839	0.959	1.079	1.199	1.318	1.438	1.558	1.678	1.798	1.918	2.038	2.158	2.277	2.397	2.517	2.637
360.	0.247	0.370	0.493	0.617	0.740	0.864	0.987	1.110	1.234	1.357	1.480	1.604	1.727	1.851	1.974	2.097	2.221	2.344	2.467	2.591	2.714
370.	0.254	0.381	0.508	0.634	0.761	0.888	1.015	1.142	1.269	1.396	1.523	1.650	1.776	1.903	2.030	2.157	2.284	2.411	2.538	2.665	2.791

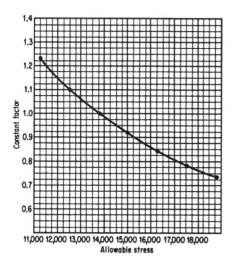

Figure 2.6 Constant factor versus stress.

0.85), or not radiographed (joint efficiency = 0.70). They may also be applied to single-welded butt joints with a backing strip that are fully radiographed (joint efficiency = 0.90), spot-radiographed (joint efficiency = 0.80), or not radiographed (joint efficiency = 0.65). The required head thickness may be determined by the following steps.

1. Determine the appropriate table corresponding to the desired joint efficiency. Note that the value of joint efficiency for each table is printed in the upper left-hand corner.
2. Locate the desired diameter for ellipsoidal heads or inside crown radius for torispherical heads. The diameter or inside crown radius values are printed across the top of the page for each table.
3. Locate the desired pressure. The pressure values are located in the left column.
4. Read the thickness corresponding to the diameter for ellipsoidal heads or inside crown radius for torispherical heads and pressure. Each table is based on an allowable stress value of 13,800 psi. For other stress values, multiply the thickness read from the table by either the following constant factors or factors read from Figs. 2.7 and 2.8.

Stress, psi	11,300	12,500	13,800	15,000	16,300	17,500	18,800
Constant	1.224	1.105	1.000	0.919	0.846	0.787	0.733

TABLE 2.5 Calculations for Cylindrical and Spherical Shells under Internal Pressure

Equations	Remarks	Code reference
Cylindrical Shells under Internal Pressure		
$t = \dfrac{PR}{SE - 0.6P}$	Equations used when t is less than $\tfrac{1}{2}R$ or P is less than $0.385SE$	Par. UG-27
or $P = \dfrac{SEt}{R + 0.6t}$ in which: t = minimum thickness, in P = allowable pressure, psi S = allowable stress, psi E = joint efficiency, percent R = inside radius, in	Circumferential joint formula in Code Par. UG-27 applies only if the joint efficiency is less than one-half the longitudinal joint efficiency or if the loadings described in Code Par. UG-22 cause bending or tension [see examples in Code Appendix L-2(a) and (b)]	Pars. UG-27, UG-22 Appendix L-2(a) and (b)
	Thick cylindrical shells	Appendix 1-2
	Formulas in terms of the outside radius may be used in place of those in Par. UG-27	Appendix 1-1 to 1-3
	Out-of-roundness measure	Par. UG-80
	Welded joint efficiencies	Par. UW-12 Table UW-12
	Stress values of materials (use appropriate table)	Code Section II, Part D, Tables UCI-23, UCD-23, ULT-23
	Corrosion allowance for carbon-steel vessels	Par. UCS-25
Spherical Shells under Internal Pressure		
$t = \dfrac{PR}{2SE - 0.2P}$	Equations used when t is less than $0.356R$ or P is less than $0.665SE$ (unknowns same as above)	Par. UG-27
or $P = \dfrac{2SEt}{R + 0.2t}$	Thick spherical shells Stress values of materials	Appendix 1-3 Code Section II, Part D, Tables UCI-23, UCD-23, ULT-23

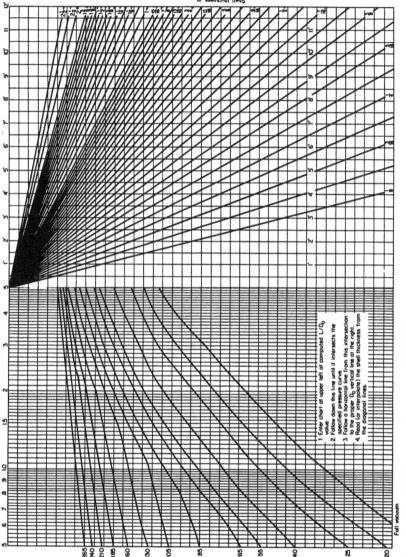

Figure 2.7 Simplified external-pressure shell thickness chart: Carbon-steel yield strength is 30,000 to 38,000 psi; temperature, 300°F.

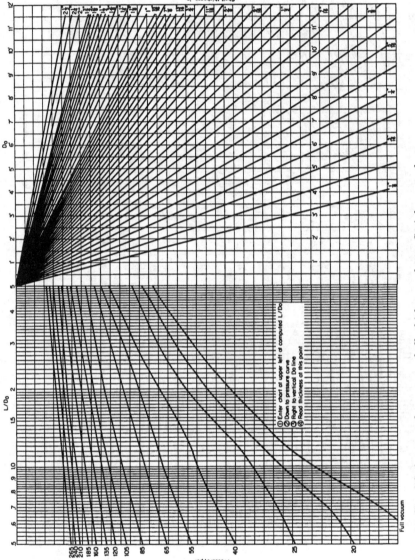

Figure 2.8 Simplified external-pressure shell thickness chart: Stainless steel (18 Cr–8 Ni) Type 304 temperature, 400°F.

TABLE 2.6 Calculations for Cylindrical and Spherical Shells under External Pressure

Equations	Remarks	Code reference
Cylindrical and Spherical Shells under External Pressure		
Vessels intended for service under external working pressure of 15 psi or less		Par. UG-28(f)
Use equations in Code Pars. UG-28(c) and (d) and thickness charts for the applicable material (figure numbers of charts listed under Code references)	Carbon-steel shells	Pars. UG-28, UCS-28 Figs. UG-28, UG-29.1, UG-29.2 Code Section II, Part D, Subpart 3
For stiffening rings, use equations in Code Pars. UG-29, UG-30	Nonferrous metal shells	Pars. UG-28, UNF-28 Code Section II, Part D, Subpart 3
	High-alloy steel shells	Pars. UG-28, UHA-28 Code Section II, Part D, Subpart 3
	Cast-iron shells	Pars. UG-28, UCI-28 Fig. UCI-28
	Clad-steel shells	Pars. UG-28, UCL-26
	Corrugated shells	Par. UCS-28(c)*
	See Code Appendix L for explanation of thickness charts for various materials (introductory note) and for applications of Code formulas and rules	Appendix L
	Basis for establishing external pressure charts	Code Section II, Part D, Subpart 3
Stiffening Rings for Cylindrical Shells under External Pressure		
		Pars. UG-28, UG-29 Figs. UG-29.1, UG-29.2
	Avoid stress	Appendix G
	Stiffening rings may be placed outside or inside a vessel	Par. UG-30 Figs. UG-29.1, UG-30, UG-29.2
	External material charts and tabular values	Code Section II, Part D, Subpart 3

* Rules in Par. PFT-19 of Power Boiler Code may also be used.

TABLE 2.6 Calculations for Cylindrical and Spherical Shells under External Pressure (*Continued*)

Equations	Remarks	Code reference
Stiffening Rings for Cylindrical Shells under External Pressure (*Cont.*)		
	Stiffening rings may be attached to shell by continuous or intermittent welding	Par. UG-30

	Total length of intermittent welding on each side of stiffening ring shall be:
	(a) Rings on outside
	   Not less than one-half outside circumference of the vessel
	(b) Rings on inside
	   Not less than one-third of the circumference of the vessel
	Maximum spacing of intermittent weld 8× thickness of shell for external rings 12× thickness of shell for internal rings
		Fig. UG-30

	Drainage—Opening at bottom for drainage	Fig. UG-29.1
	Vent—Opening at top for vent	Fig. UG-29.1
	Maximum arc of shell left unsupported	Figs. UG-29.1, UG-29.2

Thinning due to the forming process and corrosion allowances are not included in these tables. The thickness equations for ellipsoidal and torispherical (flanged and dished) heads may be found in Table 2.8 (see also Code Par. UW-12). Table 2.9 sets forth the requirements and gives the location of the desired design calculations for various materials which may be planned to be used for Code construction.

Elements of Joint Design for Heads

As stated in the beginning of this chapter regarding the new design thickness stress multipliers, a number of important factors must be considered in determining the thickness required for heads under internal pressure, such as stress values, quality factors, and joint efficiencies, according to the degree of radiographic examination. Many designers have difficulty in determining these factors.

(*Text continues on page 72.*)

TABLE 2.7a Torispherical Head Thickness Tables in Inches: Joint Efficiency=1.0

TORISPHERICAL HEAD THICKNESS TABLES IN INCHES

JOINT EFF.=1.00 ALLOWABLE STRESS VALUE =13800. PSI

INSIDE CROWN RADIUS, INCHES

PSI	12.	18.	24.	30.	36.	42.	48.	54.	60.	66.	72.	78.	84.	90.	96.	102.	108.	114.	120.	126.	132.
15.	0.012	0.017	0.023	0.029	0.035	0.040	0.046	0.052	0.058	0.063	0.069	0.075	0.081	0.087	0.092	0.098	0.104	0.110	0.115	0.121	0.127
20.	0.015	0.023	0.031	0.038	0.046	0.054	0.062	0.069	0.077	0.085	0.092	0.100	0.108	0.116	0.123	0.131	0.139	0.146	0.154	0.162	0.169
25.	0.019	0.029	0.038	0.048	0.058	0.067	0.077	0.087	0.096	0.106	0.115	0.125	0.135	0.144	0.154	0.164	0.173	0.183	0.192	0.202	0.212
30.	0.023	0.035	0.046	0.058	0.069	0.081	0.092	0.104	0.115	0.127	0.139	0.150	0.162	0.173	0.185	0.196	0.208	0.219	0.231	0.242	0.254
35.	0.027	0.040	0.054	0.067	0.081	0.094	0.108	0.121	0.135	0.148	0.162	0.175	0.189	0.202	0.216	0.229	0.242	0.256	0.269	0.283	0.296
40.	0.031	0.046	0.062	0.077	0.092	0.108	0.123	0.139	0.154	0.169	0.185	0.200	0.216	0.231	0.246	0.262	0.277	0.293	0.308	0.323	0.339
45.	0.035	0.052	0.069	0.087	0.104	0.121	0.139	0.156	0.173	0.191	0.208	0.225	0.243	0.260	0.277	0.294	0.312	0.329	0.346	0.364	0.381
50.	0.038	0.058	0.077	0.096	0.115	0.135	0.154	0.173	0.192	0.212	0.231	0.250	0.269	0.289	0.308	0.327	0.346	0.366	0.385	0.404	0.423
55.	0.042	0.064	0.085	0.106	0.127	0.148	0.169	0.191	0.212	0.233	0.254	0.275	0.296	0.318	0.339	0.360	0.381	0.402	0.423	0.445	0.466
60.	0.046	0.069	0.092	0.115	0.139	0.162	0.185	0.208	0.231	0.254	0.277	0.300	0.323	0.346	0.370	0.393	0.416	0.439	0.462	0.485	0.508
65.	0.050	0.075	0.100	0.125	0.150	0.175	0.200	0.225	0.250	0.275	0.300	0.325	0.350	0.375	0.400	0.425	0.450	0.475	0.500	0.525	0.550
70.	0.054	0.081	0.108	0.135	0.162	0.189	0.216	0.243	0.269	0.296	0.323	0.350	0.377	0.404	0.431	0.458	0.485	0.512	0.539	0.566	0.593
75.	0.058	0.087	0.115	0.144	0.173	0.202	0.231	0.260	0.289	0.318	0.346	0.375	0.404	0.433	0.462	0.491	0.520	0.549	0.577	0.606	0.635
80.	0.062	0.092	0.123	0.154	0.185	0.216	0.246	0.277	0.308	0.339	0.370	0.400	0.431	0.462	0.493	0.524	0.554	0.585	0.616	0.647	0.678
85.	0.065	0.098	0.131	0.163	0.196	0.229	0.262	0.295	0.327	0.360	0.393	0.425	0.458	0.491	0.524	0.556	0.589	0.622	0.655	0.687	0.720
90.	0.069	0.104	0.139	0.173	0.208	0.243	0.277	0.312	0.347	0.381	0.416	0.450	0.485	0.520	0.554	0.589	0.624	0.658	0.693	0.728	0.762
95.	0.073	0.110	0.146	0.183	0.219	0.256	0.293	0.329	0.366	0.402	0.439	0.476	0.512	0.549	0.585	0.622	0.658	0.695	0.732	0.768	0.805
100.	0.077	0.116	0.154	0.193	0.231	0.270	0.308	0.347	0.385	0.424	0.462	0.501	0.539	0.578	0.616	0.655	0.693	0.732	0.770	0.809	0.847
105.	0.081	0.121	0.162	0.202	0.243	0.283	0.323	0.364	0.404	0.445	0.485	0.526	0.566	0.606	0.647	0.687	0.728	0.768	0.809	0.849	0.890
110.	0.085	0.127	0.169	0.212	0.254	0.297	0.339	0.381	0.424	0.466	0.508	0.551	0.593	0.635	0.678	0.720	0.762	0.805	0.847	0.890	0.932
115.	0.089	0.133	0.177	0.221	0.266	0.310	0.354	0.399	0.443	0.487	0.531	0.576	0.620	0.664	0.709	0.753	0.797	0.841	0.886	0.930	0.974
120.	0.092	0.139	0.185	0.231	0.277	0.324	0.370	0.416	0.462	0.508	0.555	0.601	0.647	0.693	0.740	0.786	0.832	0.878	0.924	0.970	1.017
125.	0.096	0.144	0.193	0.241	0.289	0.337	0.385	0.433	0.481	0.530	0.578	0.626	0.674	0.722	0.770	0.818	0.867	0.915	0.963	1.011	1.059
130.	0.100	0.150	0.200	0.250	0.300	0.350	0.401	0.451	0.501	0.551	0.601	0.651	0.701	0.751	0.801	0.851	0.901	0.951	1.001	1.051	1.102
135.	0.104	0.156	0.208	0.260	0.312	0.364	0.416	0.468	0.520	0.572	0.624	0.676	0.728	0.780	0.832	0.884	0.936	0.988	1.040	1.092	1.144
140.	0.108	0.162	0.216	0.270	0.324	0.377	0.431	0.485	0.539	0.593	0.647	0.701	0.755	0.809	0.863	0.917	0.971	1.025	1.078	1.132	1.186
145.	0.112	0.168	0.223	0.279	0.335	0.391	0.447	0.503	0.559	0.614	0.670	0.726	0.782	0.838	0.894	0.949	1.005	1.061	1.117	1.173	1.229
150.	0.115	0.173	0.231	0.289	0.347	0.404	0.462	0.520	0.578	0.636	0.693	0.751	0.809	0.867	0.925	0.982	1.040	1.098	1.156	1.213	1.271
155.	0.119	0.179	0.239	0.298	0.358	0.418	0.477	0.537	0.597	0.657	0.716	0.776	0.836	0.896	0.955	1.015	1.075	1.134	1.194	1.254	1.314
160.	0.123	0.185	0.247	0.308	0.370	0.431	0.493	0.555	0.616	0.678	0.740	0.801	0.863	0.925	0.986	1.048	1.109	1.171	1.233	1.294	1.356
165.	0.127	0.191	0.254	0.318	0.381	0.445	0.508	0.572	0.636	0.699	0.763	0.826	0.890	0.954	1.017	1.081	1.144	1.208	1.271	1.335	1.399
170.	0.131	0.196	0.262	0.327	0.393	0.458	0.524	0.589	0.655	0.720	0.786	0.851	0.917	0.982	1.048	1.113	1.179	1.244	1.310	1.375	1.441
175.	0.135	0.202	0.270	0.337	0.404	0.472	0.539	0.607	0.674	0.741	0.809	0.876	0.944	1.011	1.079	1.146	1.213	1.281	1.348	1.416	1.483
180.	0.139	0.208	0.277	0.347	0.416	0.485	0.555	0.624	0.694	0.763	0.832	0.902	0.971	1.040	1.110	1.179	1.248	1.318	1.387	1.456	1.526
185.	0.143	0.214	0.285	0.357	0.428	0.499	0.570	0.641	0.713	0.784	0.855	0.927	0.998	1.069	1.140	1.212	1.283	1.354	1.425	1.497	1.568
190.	0.146	0.220	0.293	0.366	0.439	0.512	0.586	0.659	0.732	0.805	0.879	0.952	1.025	1.098	1.171	1.245	1.318	1.391	1.464	1.537	1.611
200.	0.154	0.231	0.308	0.385	0.462	0.539	0.617	0.694	0.771	0.848	0.925	1.002	1.079	1.156	1.233	1.310	1.387	1.464	1.541	1.618	1.696
210.	0.162	0.243	0.324	0.405	0.486	0.566	0.647	0.728	0.809	0.890	0.971	1.052	1.133	1.214	1.295	1.376	1.457	1.538	1.618	1.699	1.780
220.	0.170	0.254	0.339	0.424	0.509	0.594	0.678	0.763	0.848	0.933	1.017	1.102	1.187	1.272	1.357	1.441	1.526	1.611	1.696	1.781	1.865
230.	0.177	0.266	0.355	0.443	0.532	0.621	0.709	0.798	0.886	0.975	1.064	1.152	1.241	1.330	1.418	1.507	1.596	1.684	1.773	1.862	1.950
240.	0.185	0.278	0.370	0.463	0.555	0.648	0.740	0.833	0.925	1.018	1.110	1.203	1.295	1.388	1.480	1.573	1.665	1.758	1.850	1.943	2.035
250.	0.193	0.289	0.385	0.482	0.578	0.675	0.771	0.867	0.964	1.060	1.156	1.253	1.349	1.446	1.542	1.638	1.735	1.831	1.927	2.024	2.120
260.	0.200	0.301	0.401	0.501	0.601	0.702	0.802	0.902	1.002	1.103	1.203	1.303	1.403	1.503	1.604	1.704	1.804	1.904	2.005	2.105	2.205
270.	0.208	0.312	0.416	0.520	0.625	0.729	0.833	0.937	1.041	1.145	1.249	1.353	1.457	1.561	1.666	1.770	1.874	1.978	2.082	2.186	2.290
280.	0.216	0.324	0.432	0.540	0.648	0.756	0.864	0.972	1.080	1.188	1.295	1.403	1.511	1.619	1.727	1.835	1.943	2.051	2.159	2.267	2.375
290.	0.224	0.335	0.447	0.559	0.671	0.783	0.895	1.006	1.118	1.230	1.342	1.454	1.566	1.677	1.789	1.901	2.013	2.125	2.236	2.348	2.460
300.	0.231	0.347	0.463	0.578	0.694	0.810	0.925	1.041	1.157	1.273	1.388	1.504	1.620	1.735	1.851	1.967	2.083	2.198	2.314	2.430	2.545
310.	0.239	0.359	0.478	0.598	0.717	0.837	0.956	1.076	1.196	1.315	1.435	1.554	1.674	1.793	1.913	2.032	2.152	2.271	2.391	2.511	2.630
320.	0.247	0.370	0.494	0.617	0.740	0.864	0.987	1.111	1.234	1.358	1.481	1.604	1.728	1.851	1.975	2.098	2.221	2.345	2.468	2.592	2.715
330.	0.255	0.382	0.509	0.636	0.764	0.891	1.018	1.146	1.273	1.400	1.527	1.655	1.782	1.909	2.037	2.164	2.291	2.418	2.546	2.673	2.800
340.	0.262	0.393	0.525	0.656	0.787	0.918	1.049	1.180	1.311	1.443	1.574	1.705	1.836	1.967	2.098	2.230	2.361	2.492	2.623	2.754	2.885
350.	0.270	0.405	0.540	0.675	0.810	0.945	1.080	1.215	1.350	1.485	1.620	1.755	1.890	2.025	2.160	2.295	2.430	2.565	2.700	2.835	2.970
360.	0.278	0.417	0.556	0.694	0.833	0.972	1.111	1.250	1.389	1.528	1.667	1.805	1.944	2.083	2.222	2.361	2.500	2.639	2.778	2.916	3.055
370.	0.286	0.428	0.571	0.714	0.857	0.999	1.142	1.285	1.428	1.570	1.713	1.856	1.999	2.141	2.284	2.427	2.570	2.712	2.855	2.998	3.141

TABLE 2.7b Torispherical Head Thickness Tables in Inches: Joint Efficiency=0.90

TORISPHERICAL HEAD THICKNESS TABLES IN INCHES

JOINT EFF.=0.90 ALLOWABLE STRESS VALUE =13800. PSI

INSIDE CROWN RADIUS, INCHES

PSI	12.	18.	24.	30.	36.	42.	48.	54.	60.	66.	72.	78.	84.	90.	96.	102.	108.	114.	120.	126.	132.
15.	0.013	0.019	0.026	0.032	0.038	0.045	0.051	0.058	0.064	0.071	0.077	0.083	0.090	0.096	0.103	0.109	0.115	0.122	0.128	0.135	0.141
20.	0.017	0.026	0.034	0.043	0.051	0.060	0.068	0.077	0.086	0.094	0.103	0.111	0.120	0.128	0.137	0.145	0.154	0.162	0.171	0.180	0.188
25.	0.021	0.032	0.043	0.053	0.064	0.075	0.086	0.096	0.107	0.118	0.128	0.139	0.150	0.160	0.171	0.182	0.192	0.203	0.214	0.225	0.235
30.	0.026	0.038	0.051	0.064	0.077	0.090	0.103	0.115	0.128	0.141	0.154	0.167	0.180	0.192	0.205	0.218	0.231	0.244	0.257	0.269	0.282
35.	0.030	0.045	0.060	0.075	0.090	0.105	0.120	0.135	0.150	0.165	0.180	0.195	0.210	0.225	0.239	0.254	0.269	0.284	0.299	0.314	0.329
40.	0.034	0.051	0.068	0.086	0.103	0.120	0.137	0.154	0.171	0.188	0.205	0.222	0.239	0.257	0.274	0.291	0.308	0.325	0.342	0.359	0.376
45.	0.038	0.058	0.077	0.096	0.115	0.135	0.154	0.173	0.192	0.212	0.231	0.250	0.269	0.289	0.308	0.327	0.346	0.366	0.385	0.404	0.423
50.	0.043	0.064	0.086	0.107	0.128	0.150	0.171	0.192	0.214	0.235	0.257	0.278	0.299	0.321	0.342	0.364	0.385	0.406	0.428	0.449	0.470
55.	0.047	0.071	0.094	0.118	0.141	0.165	0.188	0.212	0.235	0.259	0.282	0.306	0.329	0.353	0.376	0.400	0.423	0.447	0.470	0.494	0.518
60.	0.051	0.077	0.103	0.128	0.154	0.180	0.205	0.231	0.257	0.282	0.308	0.334	0.359	0.385	0.411	0.436	0.462	0.488	0.513	0.539	0.565
65.	0.056	0.083	0.111	0.139	0.167	0.195	0.222	0.250	0.278	0.306	0.334	0.361	0.389	0.417	0.445	0.473	0.500	0.528	0.556	0.584	0.612
70.	0.060	0.090	0.120	0.150	0.180	0.210	0.240	0.269	0.299	0.329	0.359	0.389	0.419	0.449	0.479	0.509	0.539	0.569	0.599	0.629	0.659
75.	0.064	0.096	0.128	0.160	0.193	0.225	0.257	0.289	0.321	0.353	0.385	0.417	0.449	0.481	0.513	0.545	0.578	0.610	0.642	0.674	0.706
80.	0.068	0.103	0.137	0.171	0.205	0.240	0.274	0.308	0.342	0.376	0.411	0.445	0.479	0.513	0.548	0.582	0.616	0.650	0.684	0.719	0.753
85.	0.073	0.109	0.145	0.182	0.218	0.255	0.291	0.327	0.364	0.400	0.436	0.473	0.509	0.545	0.582	0.618	0.655	0.691	0.727	0.764	0.800
90.	0.077	0.116	0.154	0.193	0.231	0.270	0.308	0.347	0.385	0.424	0.462	0.501	0.539	0.578	0.616	0.655	0.693	0.732	0.770	0.809	0.847
95.	0.081	0.122	0.163	0.203	0.244	0.285	0.325	0.366	0.406	0.447	0.488	0.528	0.569	0.610	0.650	0.691	0.732	0.772	0.813	0.854	0.894
100.	0.086	0.128	0.171	0.214	0.257	0.300	0.342	0.385	0.428	0.471	0.513	0.556	0.599	0.642	0.685	0.727	0.770	0.813	0.856	0.899	0.941
105.	0.090	0.135	0.180	0.225	0.270	0.315	0.359	0.404	0.449	0.494	0.539	0.584	0.629	0.674	0.719	0.764	0.809	0.854	0.899	0.944	0.988
110.	0.094	0.141	0.188	0.235	0.282	0.329	0.377	0.424	0.471	0.518	0.565	0.612	0.659	0.706	0.753	0.800	0.847	0.894	0.941	0.988	1.036
115.	0.098	0.148	0.197	0.246	0.295	0.344	0.394	0.443	0.492	0.541	0.591	0.640	0.689	0.738	0.787	0.837	0.886	0.935	0.984	1.033	1.083
120.	0.103	0.154	0.205	0.257	0.308	0.359	0.411	0.462	0.514	0.565	0.616	0.668	0.719	0.770	0.822	0.873	0.924	0.976	1.027	1.078	1.130
125.	0.107	0.160	0.214	0.267	0.321	0.374	0.428	0.481	0.535	0.588	0.642	0.695	0.749	0.802	0.856	0.909	0.963	1.016	1.070	1.123	1.177
130.	0.111	0.167	0.223	0.278	0.334	0.389	0.445	0.501	0.556	0.612	0.668	0.723	0.779	0.835	0.890	0.946	1.001	1.057	1.113	1.168	1.224
135.	0.116	0.173	0.231	0.289	0.347	0.404	0.462	0.520	0.578	0.636	0.693	0.751	0.809	0.867	0.924	0.982	1.040	1.098	1.156	1.213	1.271
140.	0.120	0.180	0.240	0.300	0.360	0.420	0.479	0.539	0.599	0.659	0.719	0.779	0.839	0.899	0.959	1.019	1.079	1.139	1.198	1.258	1.318
145.	0.124	0.186	0.248	0.310	0.373	0.435	0.497	0.559	0.621	0.683	0.745	0.807	0.869	0.931	0.993	1.055	1.117	1.179	1.241	1.304	1.366
150.	0.128	0.193	0.257	0.321	0.385	0.449	0.514	0.578	0.642	0.706	0.770	0.835	0.899	0.963	1.027	1.092	1.156	1.220	1.284	1.348	1.413
160.	0.137	0.205	0.274	0.342	0.411	0.479	0.548	0.616	0.685	0.754	0.822	0.891	0.959	1.028	1.096	1.165	1.233	1.302	1.370	1.439	1.507
170.	0.146	0.218	0.291	0.364	0.437	0.509	0.582	0.655	0.728	0.801	0.874	0.946	1.019	1.092	1.165	1.238	1.311	1.384	1.457	1.530	1.603
180.	0.154	0.231	0.308	0.385	0.462	0.539	0.617	0.694	0.771	0.848	0.925	1.002	1.079	1.156	1.233	1.310	1.387	1.464	1.541	1.618	1.696
190.	0.163	0.244	0.325	0.407	0.488	0.569	0.651	0.732	0.813	0.895	0.976	1.057	1.139	1.220	1.302	1.383	1.464	1.544	1.627	1.708	1.790
200.	0.171	0.257	0.343	0.428	0.514	0.600	0.685	0.771	0.857	0.942	1.028	1.114	1.199	1.285	1.371	1.457	1.542	1.628	1.714	1.799	1.885
210.	0.180	0.270	0.360	0.450	0.540	0.630	0.719	0.809	0.899	0.989	1.079	1.169	1.259	1.349	1.439	1.529	1.619	1.709	1.799	1.889	1.979
220.	0.188	0.283	0.377	0.471	0.565	0.660	0.754	0.848	0.942	1.036	1.131	1.225	1.319	1.413	1.508	1.602	1.696	1.790	1.884	1.979	2.073
230.	0.197	0.296	0.394	0.493	0.591	0.690	0.788	0.887	0.985	1.084	1.182	1.281	1.379	1.478	1.576	1.675	1.773	1.872	1.970	2.069	2.167
240.	0.206	0.308	0.411	0.514	0.617	0.720	0.822	0.925	1.028	1.131	1.234	1.336	1.439	1.542	1.645	1.748	1.851	1.953	2.056	2.159	2.262
250.	0.214	0.321	0.428	0.535	0.643	0.750	0.857	0.964	1.071	1.178	1.285	1.392	1.499	1.606	1.714	1.821	1.928	2.035	2.142	2.249	2.356
260.	0.223	0.334	0.446	0.557	0.668	0.780	0.891	1.003	1.114	1.225	1.337	1.448	1.559	1.671	1.782	1.894	2.005	2.116	2.228	2.339	2.451
270.	0.231	0.347	0.463	0.578	0.694	0.810	0.925	1.041	1.157	1.273	1.388	1.504	1.620	1.735	1.851	1.967	2.082	2.198	2.314	2.429	2.545
280.	0.240	0.360	0.480	0.600	0.720	0.840	0.960	1.080	1.200	1.320	1.440	1.560	1.680	1.800	1.920	2.040	2.160	2.280	2.400	2.520	2.640
290.	0.249	0.373	0.497	0.621	0.746	0.870	0.994	1.118	1.243	1.367	1.491	1.615	1.740	1.864	1.988	2.113	2.237	2.361	2.486	2.610	2.734
300.	0.257	0.386	0.514	0.643	0.771	0.900	1.029	1.157	1.286	1.414	1.543	1.671	1.800	1.929	2.057	2.186	2.314	2.443	2.571	2.700	2.829
310.	0.266	0.399	0.531	0.664	0.797	0.930	1.063	1.195	1.328	1.461	1.594	1.727	1.860	1.993	2.126	2.259	2.392	2.524	2.657	2.790	2.923
320.	0.274	0.411	0.549	0.686	0.823	0.960	1.097	1.234	1.372	1.509	1.646	1.783	1.920	2.057	2.195	2.332	2.469	2.606	2.743	2.880	3.018
330.	0.283	0.424	0.566	0.707	0.849	0.990	1.131	1.273	1.415	1.556	1.698	1.839	1.980	2.122	2.263	2.405	2.546	2.688	2.829	2.971	3.112
340.	0.292	0.437	0.583	0.729	0.875	1.020	1.166	1.312	1.458	1.603	1.749	1.895	2.041	2.186	2.332	2.478	2.624	2.769	2.915	3.061	3.207
350.	0.300	0.450	0.600	0.750	0.900	1.050	1.200	1.351	1.501	1.651	1.801	1.951	2.101	2.251	2.401	2.551	2.701	2.851	3.001	3.151	3.301
360.	0.309	0.463	0.617	0.772	0.926	1.081	1.235	1.389	1.544	1.698	1.852	2.007	2.161	2.315	2.470	2.624	2.778	2.933	3.087	3.242	3.396
370.	0.317	0.476	0.635	0.793	0.952	1.111	1.269	1.428	1.587	1.745	1.904	2.063	2.221	2.380	2.539	2.697	2.856	3.015	3.173	3.332	3.491

TABLE 2.7c Torispherical Head Thickness Tables in Inches: Joint Efficiency=0.85

TORISPHERICAL HEAD THICKNESS TABLES IN INCHES

JOINT EFF.=0.85   ALLOWABLE STRESS VALUE =13800. PSI

PSI	12.	18.	24.	30.	36.	42.	48.	54.	60.	66.	72.	78.	84.	90.	96.	102.	108.	114.	120.	126.	132.
15.	0.014	0.020	0.027	0.034	0.041	0.048	0.054	0.061	0.068	0.075	0.081	0.088	0.095	0.102	0.109	0.115	0.122	0.129	0.136	0.143	0.149
20.	0.018	0.027	0.036	0.045	0.054	0.063	0.072	0.081	0.091	0.100	0.109	0.118	0.127	0.136	0.145	0.154	0.163	0.172	0.181	0.190	0.199
25.	0.023	0.034	0.045	0.057	0.068	0.079	0.091	0.102	0.113	0.125	0.136	0.147	0.158	0.170	0.181	0.192	0.204	0.215	0.226	0.238	0.249
30.	0.027	0.041	0.054	0.068	0.082	0.095	0.109	0.122	0.136	0.149	0.163	0.177	0.190	0.204	0.217	0.231	0.245	0.258	0.272	0.285	0.299
35.	0.032	0.048	0.063	0.079	0.095	0.111	0.127	0.143	0.158	0.174	0.190	0.206	0.222	0.238	0.254	0.269	0.285	0.301	0.317	0.333	0.349
40.	0.036	0.054	0.072	0.091	0.109	0.127	0.145	0.163	0.181	0.199	0.217	0.235	0.254	0.272	0.290	0.308	0.326	0.344	0.362	0.380	0.398
45.	0.041	0.061	0.082	0.102	0.122	0.143	0.163	0.183	0.204	0.224	0.245	0.265	0.285	0.306	0.326	0.346	0.367	0.387	0.408	0.428	0.448
50.	0.045	0.068	0.091	0.113	0.136	0.159	0.181	0.204	0.226	0.249	0.272	0.294	0.317	0.340	0.362	0.385	0.408	0.430	0.453	0.476	0.498
55.	0.050	0.075	0.100	0.125	0.149	0.174	0.199	0.224	0.249	0.274	0.299	0.324	0.349	0.374	0.399	0.423	0.448	0.473	0.498	0.523	0.548
60.	0.054	0.082	0.109	0.136	0.163	0.190	0.217	0.245	0.272	0.299	0.326	0.353	0.380	0.408	0.435	0.462	0.489	0.516	0.544	0.571	0.598
65.	0.059	0.088	0.118	0.147	0.177	0.206	0.236	0.265	0.294	0.324	0.353	0.383	0.412	0.442	0.471	0.500	0.530	0.559	0.589	0.618	0.648
70.	0.063	0.095	0.127	0.159	0.190	0.222	0.254	0.285	0.317	0.349	0.380	0.412	0.444	0.476	0.507	0.539	0.571	0.602	0.634	0.666	0.698
75.	0.068	0.102	0.136	0.170	0.204	0.238	0.272	0.306	0.340	0.374	0.408	0.442	0.476	0.510	0.544	0.578	0.612	0.645	0.679	0.713	0.747
80.	0.072	0.109	0.145	0.181	0.217	0.254	0.290	0.326	0.362	0.399	0.435	0.471	0.507	0.544	0.580	0.616	0.652	0.689	0.725	0.761	0.797
85.	0.077	0.116	0.154	0.193	0.231	0.270	0.308	0.347	0.385	0.424	0.462	0.501	0.539	0.578	0.616	0.655	0.693	0.732	0.770	0.809	0.847
90.	0.082	0.122	0.163	0.204	0.245	0.285	0.326	0.367	0.408	0.449	0.489	0.530	0.571	0.612	0.652	0.693	0.734	0.775	0.815	0.856	0.897
95.	0.086	0.129	0.172	0.215	0.258	0.301	0.344	0.387	0.430	0.473	0.516	0.560	0.603	0.646	0.689	0.732	0.775	0.818	0.861	0.904	0.947
100.	0.091	0.136	0.181	0.227	0.272	0.317	0.362	0.408	0.453	0.498	0.544	0.589	0.634	0.680	0.725	0.770	0.816	0.861	0.906	0.951	0.997
105.	0.095	0.143	0.190	0.238	0.285	0.333	0.381	0.428	0.476	0.523	0.571	0.618	0.666	0.714	0.761	0.809	0.856	0.904	0.951	0.999	1.047
110.	0.100	0.150	0.199	0.249	0.299	0.349	0.399	0.449	0.498	0.548	0.598	0.648	0.698	0.748	0.797	0.847	0.897	0.947	0.997	1.047	1.097
115.	0.104	0.156	0.208	0.261	0.313	0.365	0.417	0.469	0.521	0.573	0.625	0.677	0.730	0.782	0.834	0.886	0.938	0.990	1.042	1.094	1.146
120.	0.109	0.163	0.218	0.272	0.326	0.381	0.435	0.489	0.544	0.598	0.653	0.707	0.761	0.816	0.870	0.924	0.979	1.033	1.088	1.142	1.196
125.	0.113	0.170	0.227	0.283	0.340	0.397	0.453	0.510	0.566	0.623	0.680	0.736	0.793	0.850	0.906	0.963	1.020	1.076	1.133	1.190	1.246
130.	0.118	0.177	0.236	0.295	0.353	0.412	0.471	0.530	0.589	0.648	0.707	0.766	0.825	0.884	0.943	1.002	1.060	1.119	1.178	1.237	1.296
135.	0.122	0.184	0.245	0.306	0.367	0.428	0.490	0.551	0.612	0.673	0.734	0.795	0.857	0.918	0.979	1.040	1.101	1.162	1.224	1.285	1.346
140.	0.127	0.190	0.254	0.317	0.381	0.444	0.508	0.571	0.635	0.698	0.761	0.825	0.888	0.952	1.015	1.079	1.142	1.206	1.269	1.332	1.396
145.	0.131	0.197	0.263	0.329	0.394	0.460	0.526	0.591	0.657	0.723	0.789	0.854	0.920	0.986	1.052	1.117	1.183	1.249	1.314	1.380	1.446
150.	0.136	0.204	0.272	0.340	0.408	0.476	0.544	0.612	0.680	0.748	0.816	0.884	0.952	1.020	1.088	1.156	1.224	1.292	1.360	1.428	1.496
160.	0.145	0.218	0.290	0.363	0.435	0.508	0.580	0.653	0.725	0.798	0.870	0.943	1.015	1.088	1.160	1.233	1.306	1.378	1.451	1.523	1.596
170.	0.154	0.231	0.308	0.385	0.462	0.540	0.617	0.694	0.771	0.848	0.925	1.002	1.079	1.156	1.233	1.310	1.387	1.465	1.542	1.619	1.696
180.	0.163	0.245	0.326	0.408	0.490	0.571	0.653	0.734	0.816	0.898	0.979	1.061	1.143	1.224	1.306	1.387	1.469	1.551	1.632	1.714	1.796
190.	0.172	0.258	0.345	0.431	0.517	0.603	0.689	0.775	0.861	0.947	1.034	1.120	1.206	1.292	1.378	1.465	1.551	1.637	1.723	1.809	1.895
200.	0.181	0.272	0.363	0.453	0.544	0.635	0.726	0.816	0.907	0.998	1.088	1.179	1.270	1.360	1.451	1.542	1.632	1.723	1.814	1.905	1.995
210.	0.190	0.286	0.381	0.476	0.571	0.667	0.762	0.857	0.952	1.048	1.143	1.238	1.333	1.429	1.524	1.619	1.714	1.809	1.905	2.000	2.095
220.	0.200	0.299	0.399	0.499	0.598	0.698	0.798	0.898	0.998	1.098	1.197	1.297	1.397	1.497	1.596	1.696	1.796	1.896	1.996	2.096	2.195
230.	0.209	0.313	0.417	0.522	0.626	0.730	0.835	0.939	1.043	1.148	1.252	1.356	1.461	1.565	1.669	1.773	1.878	1.982	2.086	2.191	2.295
240.	0.218	0.327	0.435	0.544	0.653	0.762	0.871	0.980	1.089	1.198	1.306	1.415	1.524	1.633	1.742	1.851	1.960	2.068	2.177	2.286	2.395
250.	0.227	0.340	0.454	0.567	0.680	0.794	0.907	1.021	1.134	1.248	1.361	1.474	1.588	1.701	1.815	1.928	2.041	2.155	2.268	2.382	2.495
260.	0.236	0.354	0.472	0.590	0.708	0.826	0.944	1.062	1.180	1.298	1.415	1.533	1.651	1.769	1.887	2.005	2.123	2.241	2.359	2.477	2.595
270.	0.245	0.368	0.490	0.613	0.735	0.858	0.980	1.103	1.225	1.348	1.470	1.593	1.715	1.838	1.960	2.083	2.205	2.327	2.450	2.572	2.695
280.	0.255	0.382	0.509	0.636	0.763	0.890	1.016	1.143	1.270	1.397	1.524	1.651	1.779	1.906	2.033	2.160	2.287	2.414	2.541	2.668	2.795
290.	0.263	0.395	0.527	0.658	0.790	0.921	1.053	1.184	1.316	1.448	1.579	1.711	1.842	1.974	2.106	2.237	2.369	2.500	2.632	2.764	2.895
300.	0.273	0.409	0.545	0.681	0.817	0.953	1.089	1.225	1.362	1.498	1.634	1.770	1.906	2.042	2.178	2.315	2.451	2.587	2.723	2.859	2.995
310.	0.281	0.422	0.563	0.704	0.844	0.985	1.126	1.267	1.407	1.548	1.689	1.829	1.970	2.111	2.251	2.392	2.533	2.673	2.814	2.955	3.095
320.	0.291	0.436	0.581	0.726	0.872	1.017	1.162	1.307	1.453	1.598	1.743	1.888	2.034	2.179	2.324	2.469	2.615	2.760	2.905	3.050	3.196
330.	0.300	0.449	0.599	0.749	0.899	1.049	1.198	1.348	1.498	1.648	1.798	1.947	2.097	2.247	2.397	2.547	2.697	2.846	2.996	3.146	3.296
340.	0.309	0.463	0.617	0.772	0.926	1.081	1.235	1.389	1.544	1.698	1.852	2.007	2.161	2.315	2.470	2.624	2.778	2.933	3.087	3.242	3.396
350.	0.318	0.477	0.635	0.795	0.953	1.112	1.271	1.430	1.589	1.748	1.907	2.066	2.225	2.384	2.543	2.702	2.860	3.019	3.178	3.337	3.496
360.	0.327	0.490	0.654	0.817	0.981	1.144	1.308	1.471	1.635	1.798	1.962	2.125	2.289	2.452	2.615	2.779	2.942	3.106	3.269	3.433	3.596
370.	0.336	0.504	0.672	0.840	1.008	1.176	1.344	1.512	1.680	1.848	2.016	2.184	2.352	2.520	2.688	2.856	3.024	3.192	3.360	3.528	3.697

TABLE 2.7d Torispherical Head Thickness Tables in Inches: Joint Efficiency=0.80

TORISPHERICAL HEAD THICKNESS TABLES IN INCHES

JOINT EFF.=0.80   ALLOWABLE STRESS VALUE =13800. PSI

PSI	12.	18.	24.	30.	36.	42.	48.	54.	60.	66.	72.	78.	84.	90.	96.	102.	108.	114.	120.	126.	132.
15.	0.014	0.022	0.029	0.036	0.043	0.051	0.058	0.065	0.072	0.079	0.087	0.094	0.101	0.108	0.115	0.123	0.130	0.137	0.144	0.152	0.159
20.	0.019	0.029	0.038	0.048	0.058	0.067	0.077	0.087	0.096	0.106	0.115	0.125	0.135	0.144	0.154	0.164	0.173	0.183	0.192	0.202	0.212
25.	0.024	0.036	0.048	0.060	0.072	0.084	0.096	0.108	0.120	0.132	0.144	0.156	0.168	0.180	0.192	0.204	0.216	0.229	0.241	0.253	0.265
30.	0.029	0.043	0.058	0.072	0.087	0.101	0.115	0.130	0.144	0.159	0.173	0.188	0.202	0.216	0.231	0.245	0.260	0.274	0.289	0.303	0.318
35.	0.034	0.051	0.067	0.084	0.101	0.118	0.135	0.152	0.168	0.185	0.202	0.219	0.236	0.253	0.269	0.286	0.303	0.320	0.337	0.354	0.370
40.	0.038	0.058	0.077	0.096	0.115	0.135	0.154	0.173	0.192	0.212	0.231	0.250	0.269	0.289	0.308	0.327	0.346	0.366	0.385	0.404	0.423
45.	0.043	0.065	0.087	0.108	0.130	0.152	0.173	0.195	0.217	0.238	0.260	0.281	0.303	0.325	0.346	0.368	0.390	0.411	0.433	0.455	0.476
50.	0.048	0.072	0.096	0.120	0.144	0.168	0.192	0.217	0.241	0.265	0.289	0.313	0.337	0.361	0.385	0.409	0.433	0.457	0.481	0.505	0.529
55.	0.053	0.079	0.106	0.132	0.159	0.185	0.212	0.238	0.265	0.291	0.318	0.344	0.371	0.397	0.423	0.450	0.476	0.503	0.529	0.556	0.582
60.	0.058	0.087	0.115	0.144	0.173	0.202	0.231	0.260	0.289	0.318	0.346	0.375	0.404	0.433	0.462	0.491	0.520	0.549	0.577	0.606	0.635
65.	0.063	0.094	0.125	0.156	0.188	0.219	0.250	0.282	0.313	0.344	0.375	0.407	0.438	0.469	0.501	0.532	0.563	0.594	0.626	0.657	0.688
70.	0.067	0.101	0.135	0.168	0.202	0.236	0.270	0.303	0.337	0.371	0.404	0.438	0.472	0.505	0.539	0.573	0.606	0.640	0.674	0.707	0.741
75.	0.072	0.108	0.144	0.180	0.217	0.253	0.289	0.325	0.361	0.397	0.433	0.469	0.505	0.541	0.578	0.614	0.650	0.686	0.722	0.758	0.794
80.	0.077	0.116	0.154	0.193	0.231	0.270	0.308	0.347	0.385	0.424	0.462	0.501	0.539	0.578	0.616	0.655	0.693	0.732	0.770	0.809	0.847
85.	0.082	0.123	0.164	0.205	0.245	0.286	0.327	0.368	0.409	0.450	0.491	0.532	0.573	0.614	0.655	0.696	0.736	0.777	0.818	0.859	0.900
90.	0.087	0.130	0.173	0.217	0.260	0.303	0.347	0.390	0.433	0.477	0.520	0.563	0.607	0.650	0.693	0.736	0.780	0.823	0.866	0.910	0.953
95.	0.091	0.137	0.183	0.229	0.274	0.320	0.366	0.412	0.457	0.503	0.549	0.595	0.640	0.686	0.732	0.777	0.823	0.869	0.915	0.960	1.006
100.	0.096	0.144	0.193	0.241	0.289	0.337	0.385	0.433	0.481	0.530	0.578	0.626	0.674	0.722	0.770	0.818	0.867	0.915	0.963	1.011	1.059
105.	0.101	0.152	0.202	0.253	0.303	0.354	0.404	0.455	0.506	0.556	0.607	0.657	0.708	0.758	0.809	0.859	0.910	0.960	1.011	1.062	1.112
110.	0.106	0.159	0.212	0.265	0.318	0.371	0.424	0.477	0.530	0.583	0.636	0.688	0.741	0.794	0.847	0.900	0.953	1.006	1.059	1.112	1.165
115.	0.111	0.166	0.221	0.277	0.332	0.388	0.443	0.498	0.554	0.609	0.664	0.720	0.775	0.831	0.886	0.941	0.997	1.052	1.107	1.163	1.218
120.	0.116	0.173	0.231	0.289	0.347	0.404	0.462	0.520	0.578	0.636	0.693	0.751	0.809	0.867	0.924	0.982	1.040	1.098	1.156	1.213	1.271
125.	0.120	0.181	0.241	0.301	0.361	0.421	0.482	0.542	0.602	0.662	0.722	0.782	0.843	0.903	0.963	1.023	1.083	1.144	1.204	1.264	1.324
130.	0.125	0.188	0.250	0.313	0.376	0.438	0.501	0.563	0.626	0.689	0.751	0.814	0.876	0.939	1.002	1.064	1.127	1.189	1.252	1.315	1.377
135.	0.130	0.195	0.260	0.325	0.390	0.455	0.520	0.585	0.650	0.715	0.780	0.845	0.910	0.975	1.040	1.105	1.170	1.235	1.300	1.365	1.430
140.	0.135	0.202	0.270	0.337	0.405	0.472	0.539	0.607	0.674	0.742	0.809	0.876	0.944	1.011	1.079	1.146	1.214	1.281	1.348	1.416	1.483
145.	0.140	0.210	0.279	0.349	0.419	0.489	0.559	0.629	0.698	0.768	0.838	0.908	0.978	1.048	1.117	1.187	1.257	1.327	1.397	1.467	1.536
150.	0.144	0.217	0.289	0.361	0.433	0.506	0.578	0.650	0.722	0.795	0.867	0.939	1.011	1.084	1.156	1.228	1.300	1.373	1.445	1.517	1.589
155.	0.149	0.224	0.299	0.373	0.448	0.522	0.597	0.672	0.747	0.821	0.896	0.971	1.045	1.120	1.195	1.269	1.344	1.419	1.493	1.568	1.643
160.	0.154	0.231	0.308	0.385	0.462	0.540	0.617	0.694	0.771	0.848	0.925	1.002	1.079	1.156	1.233	1.310	1.387	1.464	1.542	1.619	1.696
165.	0.159	0.239	0.318	0.398	0.477	0.557	0.636	0.716	0.795	0.875	0.954	1.034	1.113	1.193	1.272	1.351	1.431	1.510	1.590	1.670	1.749
170.	0.164	0.246	0.328	0.409	0.491	0.573	0.655	0.737	0.819	0.901	0.983	1.065	1.146	1.228	1.310	1.392	1.474	1.556	1.638	1.720	1.802
175.	0.169	0.253	0.337	0.422	0.506	0.590	0.674	0.759	0.843	0.928	1.012	1.097	1.181	1.265	1.349	1.434	1.518	1.602	1.686	1.771	1.855
180.	0.173	0.260	0.347	0.434	0.520	0.607	0.694	0.780	0.867	0.954	1.041	1.127	1.214	1.301	1.387	1.474	1.561	1.648	1.734	1.821	1.908
185.	0.183	0.275	0.366	0.458	0.549	0.641	0.732	0.824	0.915	1.007	1.098	1.190	1.282	1.373	1.465	1.556	1.648	1.739	1.831	1.922	2.014
200.	0.193	0.289	0.385	0.482	0.578	0.675	0.771	0.867	0.964	1.060	1.156	1.253	1.349	1.446	1.542	1.638	1.735	1.831	1.927	2.024	2.120
210.	0.202	0.304	0.405	0.506	0.607	0.708	0.810	0.911	1.012	1.113	1.214	1.316	1.417	1.518	1.619	1.720	1.822	1.923	2.024	2.125	2.226
220.	0.212	0.318	0.424	0.530	0.636	0.742	0.848	0.954	1.060	1.166	1.272	1.378	1.484	1.590	1.697	1.802	1.908	2.015	2.121	2.227	2.333
230.	0.222	0.333	0.443	0.554	0.665	0.776	0.887	0.998	1.108	1.219	1.330	1.441	1.552	1.663	1.774	1.884	1.995	2.106	2.217	2.328	2.439
240.	0.231	0.347	0.463	0.578	0.694	0.810	0.925	1.041	1.157	1.273	1.388	1.504	1.620	1.735	1.851	1.967	2.082	2.198	2.314	2.429	2.545
250.	0.241	0.362	0.482	0.603	0.723	0.844	0.964	1.085	1.205	1.326	1.446	1.567	1.687	1.808	1.928	2.049	2.169	2.290	2.410	2.531	2.651
260.	0.251	0.376	0.502	0.627	0.752	0.878	1.003	1.128	1.254	1.379	1.504	1.630	1.755	1.880	2.006	2.131	2.256	2.382	2.507	2.632	2.758
270.	0.260	0.391	0.521	0.651	0.781	0.911	1.041	1.172	1.302	1.432	1.562	1.692	1.823	1.953	2.083	2.213	2.343	2.473	2.604	2.734	2.864
280.	0.270	0.405	0.540	0.675	0.810	0.945	1.080	1.215	1.350	1.485	1.620	1.755	1.890	2.025	2.160	2.295	2.430	2.565	2.700	2.835	2.970
290.	0.280	0.420	0.559	0.699	0.839	0.979	1.119	1.259	1.399	1.538	1.678	1.818	1.958	2.098	2.238	2.377	2.517	2.657	2.797	2.937	3.077
300.	0.289	0.434	0.579	0.723	0.868	1.013	1.157	1.302	1.447	1.592	1.736	1.881	2.026	2.170	2.315	2.460	2.604	2.749	2.894	3.038	3.183
310.	0.299	0.449	0.598	0.748	0.897	1.047	1.196	1.346	1.495	1.645	1.794	1.944	2.093	2.243	2.392	2.542	2.691	2.841	2.990	3.140	3.290
320.	0.309	0.463	0.617	0.772	0.926	1.081	1.235	1.389	1.544	1.698	1.852	2.007	2.161	2.315	2.470	2.624	2.778	2.933	3.087	3.242	3.396
330.	0.318	0.478	0.637	0.796	0.955	1.114	1.274	1.433	1.592	1.751	1.910	2.070	2.229	2.388	2.547	2.706	2.866	3.025	3.184	3.343	3.502
340.	0.328	0.492	0.656	0.820	0.984	1.148	1.313	1.476	1.640	1.804	1.968	2.132	2.297	2.461	2.625	2.789	2.953	3.117	3.281	3.445	3.609
350.	0.338	0.507	0.676	0.844	1.013	1.182	1.351	1.520	1.689	1.858	2.027	2.195	2.364	2.533	2.702	2.871	3.040	3.209	3.378	3.547	3.715
360.	0.347	0.521	0.695	0.869	1.042	1.216	1.390	1.563	1.737	1.911	2.085	2.258	2.432	2.606	2.779	2.953	3.127	3.301	3.474	3.648	3.822
370.	0.357	0.536	0.714	0.893	1.071	1.250	1.428	1.607	1.786	1.964	2.143	2.321	2.500	2.678	2.857	3.036	3.214	3.393	3.571	3.750	3.928

TABLE 2.7e Torispherical Head Thickness Tables in Inches: Joint Efficiency=0.70

TORISPHERICAL HEAD THICKNESS TABLES IN INCHES

JOINT EFF.=0.70   ALLOWABLE STRESS VALUE =13800. PSI

PSI	12.	18.	24.	30.	36.	42.	48.	54.	60.	66.	72.	78.	84.	90.	96.	102.	108.	114.	120.	126.	132.
15.	0.016	0.025	0.033	0.041	0.049	0.058	0.066	0.074	0.082	0.091	0.099	0.107	0.115	0.124	0.132	0.140	0.148	0.157	0.165	0.173	0.181
20.	0.022	0.033	0.044	0.055	0.066	0.077	0.088	0.099	0.110	0.121	0.132	0.143	0.154	0.165	0.176	0.187	0.198	0.209	0.220	0.231	0.242
25.	0.027	0.041	0.055	0.069	0.082	0.096	0.110	0.124	0.137	0.151	0.165	0.179	0.192	0.206	0.220	0.234	0.247	0.261	0.275	0.289	0.302
30.	0.033	0.049	0.066	0.082	0.099	0.115	0.132	0.148	0.165	0.181	0.198	0.214	0.231	0.247	0.264	0.280	0.297	0.313	0.330	0.346	0.363
35.	0.038	0.058	0.077	0.096	0.115	0.135	0.154	0.173	0.192	0.212	0.231	0.250	0.269	0.289	0.308	0.327	0.346	0.366	0.385	0.404	0.423
40.	0.044	0.066	0.088	0.110	0.132	0.154	0.176	0.198	0.220	0.242	0.264	0.286	0.308	0.330	0.352	0.374	0.396	0.418	0.440	0.462	0.484
45.	0.049	0.074	0.099	0.124	0.148	0.173	0.198	0.223	0.247	0.272	0.297	0.322	0.346	0.371	0.396	0.421	0.445	0.470	0.495	0.520	0.544
50.	0.055	0.082	0.110	0.137	0.165	0.192	0.220	0.247	0.275	0.302	0.330	0.357	0.385	0.412	0.440	0.467	0.495	0.522	0.550	0.577	0.605
55.	0.061	0.091	0.121	0.151	0.182	0.212	0.242	0.272	0.303	0.333	0.363	0.393	0.424	0.454	0.484	0.514	0.545	0.575	0.605	0.635	0.666
60.	0.066	0.099	0.132	0.165	0.198	0.231	0.264	0.297	0.330	0.363	0.396	0.429	0.462	0.495	0.528	0.561	0.594	0.627	0.660	0.693	0.726
65.	0.072	0.107	0.143	0.179	0.215	0.250	0.286	0.322	0.358	0.393	0.429	0.465	0.501	0.536	0.572	0.608	0.644	0.679	0.715	0.751	0.787
70.	0.077	0.116	0.154	0.193	0.231	0.270	0.308	0.347	0.385	0.424	0.462	0.501	0.539	0.578	0.616	0.655	0.693	0.732	0.770	0.809	0.847
75.	0.083	0.124	0.165	0.206	0.248	0.289	0.330	0.371	0.413	0.454	0.495	0.536	0.578	0.619	0.660	0.701	0.743	0.784	0.825	0.866	0.908
80.	0.088	0.132	0.176	0.220	0.264	0.308	0.352	0.396	0.440	0.484	0.528	0.572	0.616	0.660	0.704	0.748	0.792	0.836	0.880	0.924	0.968
85.	0.094	0.140	0.187	0.234	0.281	0.327	0.374	0.421	0.468	0.514	0.561	0.608	0.655	0.701	0.748	0.795	0.842	0.889	0.935	0.982	1.029
90.	0.099	0.149	0.198	0.248	0.297	0.347	0.396	0.446	0.495	0.545	0.594	0.644	0.693	0.743	0.792	0.842	0.891	0.941	0.990	1.040	1.089
95.	0.105	0.157	0.209	0.261	0.314	0.366	0.418	0.470	0.523	0.575	0.627	0.680	0.732	0.784	0.836	0.889	0.941	0.993	1.045	1.098	1.150
100.	0.110	0.165	0.220	0.275	0.330	0.385	0.440	0.495	0.550	0.605	0.660	0.715	0.770	0.825	0.880	0.935	0.990	1.045	1.101	1.156	1.211
105.	0.116	0.173	0.231	0.289	0.347	0.404	0.462	0.520	0.578	0.636	0.693	0.751	0.809	0.867	0.924	0.982	1.040	1.098	1.156	1.213	1.271
110.	0.121	0.182	0.242	0.303	0.363	0.424	0.484	0.545	0.605	0.666	0.726	0.787	0.847	0.908	0.969	1.029	1.090	1.150	1.211	1.271	1.332
115.	0.127	0.190	0.253	0.316	0.380	0.443	0.506	0.570	0.633	0.696	0.759	0.823	0.886	0.949	1.013	1.076	1.139	1.203	1.266	1.329	1.392
120.	0.132	0.198	0.264	0.330	0.396	0.462	0.528	0.594	0.660	0.726	0.793	0.859	0.925	0.991	1.057	1.123	1.189	1.255	1.321	1.387	1.453
125.	0.138	0.206	0.275	0.344	0.413	0.482	0.550	0.619	0.688	0.757	0.826	0.894	0.963	1.032	1.101	1.170	1.238	1.307	1.376	1.445	1.514
130.	0.143	0.215	0.286	0.358	0.429	0.501	0.572	0.644	0.716	0.787	0.859	0.930	1.002	1.073	1.145	1.216	1.288	1.360	1.431	1.503	1.574
135.	0.149	0.223	0.297	0.372	0.446	0.520	0.594	0.669	0.743	0.817	0.892	0.966	1.040	1.115	1.189	1.263	1.338	1.412	1.486	1.561	1.635
140.	0.154	0.231	0.308	0.385	0.462	0.539	0.617	0.694	0.771	0.848	0.925	1.002	1.079	1.156	1.233	1.310	1.387	1.464	1.541	1.618	1.696
145.	0.160	0.240	0.319	0.399	0.479	0.559	0.639	0.718	0.798	0.878	0.958	1.038	1.118	1.197	1.277	1.357	1.437	1.517	1.596	1.676	1.756
150.	0.165	0.248	0.330	0.413	0.495	0.578	0.661	0.743	0.826	0.908	0.991	1.074	1.156	1.239	1.321	1.404	1.486	1.569	1.652	1.734	1.817
160.	0.176	0.264	0.352	0.440	0.529	0.617	0.705	0.793	0.881	0.969	1.057	1.145	1.233	1.321	1.410	1.498	1.586	1.674	1.762	1.850	1.938
170.	0.187	0.281	0.374	0.468	0.562	0.655	0.749	0.843	0.936	1.030	1.123	1.217	1.311	1.404	1.498	1.591	1.685	1.779	1.872	1.966	2.059
180.	0.198	0.297	0.397	0.496	0.595	0.694	0.793	0.892	0.991	1.090	1.190	1.289	1.388	1.487	1.586	1.685	1.784	1.883	1.983	2.082	2.181
190.	0.209	0.314	0.419	0.523	0.628	0.733	0.837	0.942	1.047	1.151	1.256	1.361	1.465	1.570	1.675	1.779	1.884	1.988	2.093	2.198	2.302
200.	0.220	0.330	0.441	0.551	0.661	0.771	0.881	0.991	1.102	1.212	1.322	1.432	1.542	1.652	1.763	1.873	1.983	2.093	2.203	2.313	2.424
210.	0.231	0.347	0.463	0.578	0.694	0.810	0.925	1.041	1.157	1.273	1.388	1.504	1.620	1.735	1.851	1.967	2.082	2.198	2.314	2.429	2.545
220.	0.242	0.364	0.485	0.606	0.727	0.848	0.970	1.091	1.212	1.333	1.454	1.576	1.697	1.818	1.939	2.061	2.182	2.303	2.424	2.545	2.667
230.	0.253	0.380	0.507	0.634	0.760	0.887	1.014	1.140	1.267	1.394	1.521	1.647	1.774	1.901	2.028	2.154	2.281	2.408	2.535	2.661	2.788
240.	0.264	0.397	0.529	0.661	0.793	0.926	1.058	1.190	1.322	1.455	1.587	1.719	1.852	1.984	2.116	2.248	2.381	2.513	2.645	2.777	2.910
250.	0.275	0.413	0.551	0.689	0.826	0.964	1.102	1.240	1.378	1.516	1.653	1.791	1.929	2.067	2.205	2.342	2.480	2.618	2.756	2.893	3.031
260.	0.287	0.430	0.573	0.717	0.860	1.003	1.146	1.290	1.433	1.576	1.720	1.863	2.006	2.150	2.293	2.436	2.579	2.723	2.866	3.009	3.153
270.	0.298	0.446	0.595	0.744	0.893	1.042	1.191	1.339	1.488	1.637	1.786	1.935	2.084	2.233	2.381	2.530	2.679	2.828	2.977	3.125	3.274
280.	0.309	0.463	0.617	0.772	0.926	1.081	1.235	1.389	1.544	1.698	1.852	2.007	2.161	2.316	2.470	2.624	2.778	2.933	3.087	3.242	3.396
290.	0.320	0.480	0.640	0.799	0.959	1.119	1.279	1.439	1.599	1.759	1.919	2.079	2.238	2.398	2.558	2.718	2.878	3.038	3.198	3.358	3.518
300.	0.331	0.496	0.662	0.827	0.993	1.158	1.323	1.489	1.654	1.820	1.985	2.150	2.316	2.481	2.647	2.812	2.978	3.143	3.308	3.474	3.639
310.	0.342	0.513	0.684	0.855	1.026	1.197	1.368	1.539	1.710	1.880	2.051	2.222	2.393	2.564	2.735	2.906	3.077	3.248	3.419	3.590	3.761
320.	0.353	0.529	0.706	0.882	1.059	1.235	1.412	1.588	1.765	1.941	2.118	2.294	2.471	2.647	2.824	3.000	3.177	3.353	3.530	3.706	3.883
330.	0.364	0.546	0.728	0.910	1.092	1.274	1.456	1.638	1.820	2.002	2.184	2.366	2.548	2.730	2.912	3.094	3.276	3.458	3.640	3.822	4.004
340.	0.375	0.563	0.750	0.938	1.125	1.313	1.500	1.688	1.876	2.063	2.251	2.438	2.626	2.813	3.001	3.188	3.376	3.564	3.751	3.939	4.126
350.	0.386	0.579	0.772	0.965	1.159	1.352	1.545	1.738	1.931	2.124	2.317	2.510	2.703	2.896	3.089	3.283	3.476	3.669	3.862	4.055	4.248
360.	0.397	0.596	0.795	0.993	1.192	1.390	1.589	1.788	1.986	2.185	2.384	2.582	2.781	2.979	3.178	3.377	3.575	3.774	3.973	4.171	4.370
370.	0.408	0.613	0.817	1.021	1.225	1.429	1.633	1.838	2.042	2.246	2.450	2.654	2.858	3.063	3.267	3.471	3.675	3.879	4.083	4.288	4.492

TABLE 2.7f Torispherical Head Thickness Tables in Inches: Joint Efficiency=0.65

TORISPHERICAL HEAD THICKNESS TABLES IN INCHES

JOINT EFF.=0.65   ALLOWABLE STRESS VALUE =13800. PSI

INSIDE CROWN RADIUS, INCHES

PSI	12.	18.	24.	30.	36.	42.	48.	54.	60.	66.	72.	78.	84.	90.	96.	102.	108.	114.	120.	126.	132.
15.	0.018	0.027	0.036	0.044	0.053	0.062	0.071	0.080	0.089	0.098	0.107	0.115	0.124	0.133	0.142	0.151	0.160	0.169	0.178	0.187	0.195
20.	0.024	0.036	0.047	0.059	0.071	0.083	0.095	0.107	0.118	0.130	0.142	0.154	0.166	0.178	0.189	0.201	0.213	0.225	0.237	0.249	0.261
25.	0.030	0.044	0.059	0.074	0.089	0.104	0.118	0.133	0.148	0.163	0.178	0.192	0.207	0.222	0.237	0.252	0.266	0.281	0.296	0.311	0.326
30.	0.036	0.053	0.071	0.089	0.107	0.124	0.142	0.160	0.178	0.195	0.213	0.231	0.249	0.266	0.284	0.302	0.320	0.338	0.355	0.373	0.391
35.	0.041	0.062	0.083	0.104	0.124	0.145	0.166	0.187	0.207	0.228	0.249	0.269	0.290	0.311	0.332	0.352	0.373	0.394	0.415	0.435	0.456
40.	0.047	0.071	0.095	0.118	0.142	0.166	0.190	0.213	0.237	0.261	0.284	0.308	0.332	0.355	0.379	0.403	0.426	0.450	0.474	0.497	0.521
45.	0.053	0.080	0.107	0.133	0.160	0.187	0.213	0.240	0.267	0.293	0.320	0.346	0.373	0.400	0.426	0.453	0.480	0.506	0.533	0.560	0.586
50.	0.059	0.089	0.118	0.148	0.178	0.207	0.237	0.267	0.296	0.326	0.355	0.385	0.415	0.444	0.474	0.503	0.533	0.563	0.592	0.622	0.652
55.	0.065	0.098	0.130	0.163	0.195	0.228	0.261	0.293	0.326	0.358	0.391	0.424	0.456	0.489	0.521	0.554	0.586	0.619	0.652	0.684	0.717
60.	0.071	0.107	0.142	0.178	0.213	0.249	0.284	0.320	0.355	0.391	0.427	0.462	0.498	0.533	0.569	0.604	0.640	0.675	0.711	0.746	0.782
65.	0.077	0.116	0.154	0.193	0.231	0.270	0.308	0.347	0.385	0.424	0.462	0.501	0.539	0.578	0.616	0.655	0.693	0.732	0.770	0.809	0.847
70.	0.083	0.124	0.166	0.207	0.249	0.290	0.332	0.373	0.415	0.456	0.498	0.539	0.581	0.622	0.664	0.705	0.746	0.788	0.829	0.870	0.912
75.	0.089	0.133	0.178	0.222	0.267	0.311	0.355	0.400	0.444	0.489	0.533	0.578	0.622	0.667	0.711	0.755	0.800	0.844	0.888	0.933	0.978
80.	0.095	0.142	0.190	0.237	0.284	0.332	0.379	0.427	0.474	0.521	0.569	0.616	0.664	0.711	0.758	0.806	0.853	0.901	0.948	0.995	1.043
85.	0.101	0.151	0.201	0.252	0.302	0.353	0.403	0.453	0.504	0.554	0.604	0.655	0.705	0.755	0.806	0.856	0.907	0.957	1.007	1.058	1.108
90.	0.107	0.160	0.213	0.267	0.320	0.373	0.427	0.480	0.533	0.587	0.640	0.693	0.747	0.800	0.853	0.907	0.960	1.013	1.067	1.120	1.173
95.	0.113	0.169	0.225	0.281	0.338	0.394	0.450	0.507	0.563	0.619	0.675	0.732	0.788	0.844	0.901	0.957	1.013	1.070	1.126	1.182	1.239
100.	0.119	0.178	0.237	0.296	0.356	0.415	0.474	0.533	0.593	0.652	0.711	0.770	0.830	0.889	0.948	1.007	1.067	1.126	1.185	1.245	1.304
105.	0.124	0.187	0.249	0.311	0.373	0.436	0.498	0.560	0.622	0.685	0.747	0.809	0.871	0.934	0.996	1.058	1.120	1.183	1.245	1.307	1.369
110.	0.130	0.196	0.261	0.326	0.391	0.456	0.522	0.587	0.652	0.717	0.782	0.848	0.913	0.978	1.043	1.108	1.174	1.239	1.304	1.369	1.434
115.	0.136	0.204	0.273	0.341	0.409	0.477	0.545	0.613	0.682	0.750	0.818	0.886	0.954	1.022	1.091	1.159	1.227	1.295	1.363	1.431	1.500
120.	0.142	0.213	0.285	0.356	0.427	0.498	0.569	0.640	0.711	0.782	0.854	0.925	0.996	1.067	1.138	1.209	1.280	1.352	1.423	1.494	1.565
125.	0.148	0.222	0.296	0.370	0.445	0.519	0.593	0.667	0.741	0.815	0.889	0.963	1.037	1.111	1.186	1.260	1.334	1.408	1.482	1.556	1.630
130.	0.154	0.231	0.308	0.385	0.462	0.539	0.617	0.694	0.771	0.848	0.925	1.002	1.079	1.156	1.233	1.310	1.387	1.464	1.541	1.618	1.696
135.	0.160	0.240	0.320	0.400	0.480	0.560	0.640	0.720	0.800	0.880	0.960	1.040	1.121	1.201	1.281	1.361	1.441	1.521	1.601	1.681	1.761
140.	0.166	0.249	0.332	0.415	0.498	0.581	0.664	0.747	0.830	0.913	0.996	1.079	1.162	1.245	1.328	1.411	1.494	1.577	1.660	1.743	1.826
145.	0.172	0.258	0.344	0.430	0.516	0.602	0.688	0.774	0.860	0.946	1.032	1.118	1.204	1.290	1.376	1.462	1.548	1.634	1.720	1.805	1.891
150.	0.178	0.267	0.356	0.445	0.534	0.623	0.712	0.801	0.889	0.978	1.067	1.156	1.245	1.334	1.423	1.512	1.601	1.690	1.779	1.868	1.957
160.	0.190	0.285	0.380	0.474	0.569	0.664	0.759	0.854	0.949	1.044	1.139	1.234	1.328	1.423	1.518	1.613	1.708	1.803	1.898	1.993	2.087
170.	0.202	0.302	0.403	0.504	0.605	0.706	0.807	0.907	1.008	1.109	1.210	1.311	1.412	1.512	1.613	1.714	1.815	1.916	2.017	2.117	2.218
180.	0.214	0.320	0.427	0.534	0.641	0.747	0.854	0.961	1.068	1.174	1.281	1.388	1.495	1.602	1.708	1.815	1.922	2.029	2.135	2.242	2.349
190.	0.225	0.338	0.451	0.564	0.676	0.789	0.902	1.014	1.127	1.240	1.353	1.465	1.578	1.691	1.803	1.916	2.029	2.142	2.254	2.367	2.480
200.	0.237	0.356	0.475	0.593	0.712	0.831	0.949	1.068	1.187	1.305	1.424	1.543	1.661	1.780	1.899	2.017	2.136	2.255	2.373	2.492	2.611
210.	0.249	0.374	0.498	0.623	0.748	0.872	0.997	1.121	1.246	1.371	1.495	1.620	1.744	1.869	1.994	2.118	2.243	2.368	2.492	2.617	2.741
220.	0.261	0.392	0.522	0.653	0.783	0.914	1.044	1.175	1.306	1.436	1.567	1.697	1.828	1.958	2.089	2.219	2.350	2.481	2.611	2.742	2.872
230.	0.273	0.410	0.546	0.683	0.819	0.956	1.092	1.229	1.365	1.502	1.638	1.775	1.911	2.048	2.184	2.321	2.457	2.594	2.730	2.867	3.003
240.	0.285	0.427	0.570	0.712	0.855	0.997	1.140	1.282	1.425	1.567	1.709	1.852	1.994	2.137	2.279	2.422	2.564	2.707	2.849	2.992	3.134
250.	0.297	0.445	0.594	0.742	0.890	1.039	1.187	1.336	1.484	1.632	1.781	1.929	2.078	2.226	2.375	2.523	2.671	2.820	2.968	3.117	3.265
260.	0.309	0.463	0.617	0.772	0.926	1.081	1.235	1.389	1.544	1.698	1.852	2.007	2.161	2.315	2.470	2.624	2.778	2.933	3.087	3.242	3.396
270.	0.321	0.481	0.641	0.802	0.962	1.122	1.283	1.443	1.603	1.763	1.924	2.084	2.244	2.405	2.565	2.725	2.886	3.046	3.206	3.367	3.527
280.	0.333	0.499	0.665	0.831	0.998	1.164	1.330	1.496	1.663	1.829	1.995	2.162	2.328	2.494	2.660	2.827	2.993	3.159	3.325	3.492	3.658
290.	0.344	0.517	0.689	0.861	1.033	1.206	1.378	1.550	1.722	1.895	2.067	2.239	2.411	2.583	2.756	2.928	3.100	3.272	3.445	3.617	3.789
300.	0.356	0.535	0.713	0.891	1.069	1.247	1.426	1.604	1.782	1.960	2.138	2.316	2.495	2.673	2.851	3.029	3.207	3.386	3.564	3.742	3.920
310.	0.368	0.552	0.737	0.921	1.105	1.289	1.473	1.657	1.841	2.026	2.210	2.394	2.578	2.762	2.946	3.131	3.315	3.499	3.683	3.867	4.051
320.	0.380	0.570	0.760	0.951	1.141	1.331	1.521	1.711	1.901	2.091	2.281	2.471	2.662	2.852	3.042	3.232	3.422	3.612	3.802	3.992	4.182
330.	0.392	0.588	0.784	0.980	1.176	1.373	1.569	1.765	1.961	2.157	2.353	2.549	2.745	2.941	3.137	3.333	3.529	3.725	3.921	4.118	4.314
340.	0.404	0.606	0.808	1.010	1.212	1.414	1.616	1.818	2.020	2.222	2.424	2.626	2.829	3.031	3.233	3.435	3.637	3.839	4.041	4.243	4.445
350.	0.416	0.624	0.832	1.040	1.248	1.456	1.664	1.872	2.080	2.288	2.496	2.704	2.912	3.120	3.328	3.536	3.744	3.952	4.160	4.368	4.576
360.	0.428	0.642	0.856	1.070	1.284	1.498	1.712	1.926	2.140	2.354	2.568	2.782	2.996	3.210	3.424	3.637	3.851	4.065	4.279	4.493	4.707
370.	0.440	0.660	0.880	1.100	1.320	1.540	1.759	1.979	2.199	2.419	2.639	2.859	3.079	3.299	3.519	3.739	3.959	4.179	4.399	4.619	4.839

TABLE 2.7g Ellipsoidal Head Thickness Tables in Inches: Joint Efficiency=1.0

ELLIPSOIDAL HEAD THICKNESS TABLES IN INCHES   MAJOR TO MINOR AXIS RATIO=2:1

JOINT EFF.=1.00   ALLOWABLE STRESS VALUE =13800. PSI

PSI	\multicolumn{21}{c}{INSIDE DIAMETER, IN.}
	12.	18.	24.	30.	36.	42.	48.	54.	60.	66.	72.	78.	84.	90.	96.	102.	108.	114.	120.	126.	132.
15.	0.007	0.010	0.013	0.016	0.020	0.023	0.026	0.029	0.033	0.036	0.039	0.042	0.046	0.049	0.052	0.055	0.059	0.062	0.065	0.068	0.072
20.	0.009	0.013	0.017	0.022	0.026	0.030	0.035	0.039	0.043	0.048	0.052	0.057	0.061	0.065	0.070	0.074	0.078	0.083	0.087	0.091	0.096
25.	0.011	0.016	0.022	0.027	0.033	0.038	0.043	0.049	0.054	0.060	0.065	0.071	0.076	0.082	0.087	0.092	0.098	0.103	0.109	0.114	0.120
30.	0.013	0.020	0.026	0.033	0.039	0.046	0.052	0.059	0.065	0.072	0.078	0.085	0.091	0.098	0.104	0.111	0.117	0.124	0.130	0.137	0.144
35.	0.015	0.023	0.030	0.038	0.046	0.053	0.061	0.068	0.076	0.084	0.091	0.099	0.107	0.114	0.122	0.129	0.137	0.145	0.152	0.160	0.167
40.	0.017	0.026	0.035	0.043	0.052	0.061	0.070	0.078	0.087	0.096	0.104	0.113	0.122	0.130	0.139	0.148	0.157	0.165	0.174	0.183	0.191
45.	0.020	0.029	0.039	0.049	0.059	0.069	0.078	0.088	0.098	0.108	0.117	0.127	0.137	0.147	0.157	0.166	0.176	0.186	0.196	0.206	0.215
50.	0.022	0.033	0.043	0.054	0.065	0.076	0.087	0.098	0.109	0.120	0.130	0.141	0.152	0.163	0.174	0.185	0.196	0.207	0.217	0.228	0.239
55.	0.024	0.036	0.048	0.060	0.072	0.084	0.096	0.108	0.120	0.132	0.144	0.156	0.167	0.179	0.191	0.203	0.215	0.227	0.239	0.251	0.263
60.	0.026	0.039	0.052	0.065	0.078	0.091	0.104	0.117	0.130	0.144	0.157	0.170	0.183	0.196	0.209	0.222	0.235	0.248	0.261	0.274	0.287
65.	0.028	0.042	0.057	0.071	0.085	0.099	0.113	0.127	0.141	0.156	0.170	0.184	0.198	0.212	0.226	0.240	0.254	0.269	0.283	0.297	0.311
70.	0.030	0.046	0.061	0.076	0.091	0.107	0.122	0.137	0.152	0.167	0.183	0.198	0.213	0.228	0.244	0.259	0.274	0.289	0.305	0.320	0.335
75.	0.033	0.049	0.065	0.082	0.098	0.114	0.131	0.147	0.163	0.179	0.196	0.212	0.228	0.245	0.261	0.277	0.294	0.310	0.326	0.343	0.359
80.	0.035	0.052	0.070	0.087	0.104	0.122	0.139	0.157	0.174	0.191	0.209	0.226	0.244	0.261	0.278	0.296	0.313	0.331	0.348	0.365	0.383
85.	0.037	0.055	0.074	0.092	0.111	0.129	0.148	0.166	0.185	0.203	0.222	0.240	0.259	0.277	0.296	0.314	0.333	0.351	0.370	0.388	0.407
90.	0.039	0.059	0.078	0.098	0.117	0.137	0.157	0.176	0.196	0.215	0.235	0.254	0.274	0.294	0.313	0.333	0.352	0.372	0.392	0.411	0.431
95.	0.041	0.062	0.083	0.103	0.124	0.145	0.165	0.186	0.207	0.227	0.248	0.269	0.289	0.310	0.331	0.351	0.372	0.393	0.413	0.434	0.455
100.	0.044	0.065	0.087	0.109	0.130	0.152	0.174	0.196	0.218	0.239	0.261	0.283	0.305	0.326	0.348	0.370	0.392	0.413	0.435	0.457	0.479
105.	0.046	0.069	0.091	0.114	0.137	0.160	0.183	0.206	0.228	0.251	0.274	0.297	0.320	0.343	0.366	0.388	0.411	0.434	0.457	0.480	0.503
110.	0.048	0.072	0.096	0.120	0.144	0.168	0.191	0.215	0.239	0.263	0.287	0.311	0.335	0.359	0.383	0.407	0.431	0.455	0.479	0.503	0.527
115.	0.050	0.075	0.100	0.125	0.150	0.175	0.200	0.225	0.250	0.275	0.300	0.325	0.350	0.375	0.400	0.425	0.450	0.475	0.500	0.525	0.550
120.	0.052	0.078	0.104	0.131	0.157	0.183	0.209	0.235	0.261	0.287	0.313	0.339	0.365	0.392	0.418	0.444	0.470	0.496	0.522	0.548	0.574
125.	0.054	0.082	0.109	0.136	0.163	0.190	0.218	0.245	0.272	0.299	0.326	0.354	0.381	0.408	0.435	0.462	0.490	0.517	0.544	0.571	0.598
130.	0.057	0.085	0.113	0.141	0.170	0.198	0.226	0.255	0.283	0.311	0.339	0.368	0.396	0.424	0.453	0.481	0.509	0.537	0.566	0.594	0.622
135.	0.059	0.088	0.118	0.147	0.176	0.206	0.235	0.264	0.294	0.323	0.353	0.382	0.411	0.441	0.470	0.499	0.529	0.558	0.588	0.617	0.646
140.	0.061	0.091	0.122	0.152	0.183	0.213	0.244	0.274	0.305	0.335	0.366	0.396	0.427	0.457	0.488	0.518	0.549	0.579	0.609	0.640	0.670
145.	0.063	0.095	0.126	0.158	0.189	0.221	0.252	0.284	0.316	0.347	0.379	0.410	0.442	0.473	0.505	0.536	0.568	0.600	0.631	0.663	0.694
150.	0.065	0.098	0.131	0.163	0.196	0.229	0.261	0.294	0.326	0.359	0.392	0.424	0.457	0.490	0.522	0.555	0.588	0.620	0.653	0.686	0.718
160.	0.070	0.104	0.139	0.174	0.209	0.244	0.279	0.313	0.348	0.383	0.418	0.453	0.488	0.522	0.557	0.592	0.627	0.662	0.697	0.731	0.766
170.	0.074	0.111	0.148	0.185	0.222	0.259	0.296	0.333	0.370	0.407	0.444	0.481	0.518	0.555	0.592	0.629	0.666	0.703	0.740	0.777	0.814
180.	0.078	0.118	0.157	0.196	0.235	0.274	0.313	0.353	0.392	0.431	0.470	0.509	0.549	0.588	0.627	0.666	0.705	0.744	0.784	0.823	0.862
190.	0.083	0.124	0.165	0.207	0.248	0.290	0.331	0.372	0.414	0.455	0.496	0.538	0.579	0.620	0.662	0.703	0.745	0.786	0.827	0.869	0.910
200.	0.087	0.131	0.174	0.218	0.261	0.305	0.348	0.392	0.435	0.479	0.522	0.566	0.610	0.653	0.697	0.740	0.784	0.827	0.871	0.914	0.958
210.	0.091	0.137	0.183	0.229	0.274	0.320	0.366	0.411	0.457	0.503	0.549	0.594	0.640	0.686	0.732	0.777	0.823	0.869	0.914	0.960	1.006
220.	0.096	0.144	0.192	0.240	0.287	0.335	0.383	0.431	0.479	0.527	0.575	0.623	0.671	0.719	0.766	0.814	0.862	0.910	0.958	1.006	1.054
230.	0.100	0.150	0.200	0.250	0.301	0.351	0.401	0.451	0.501	0.551	0.601	0.651	0.701	0.751	0.801	0.851	0.902	0.952	1.002	1.052	1.102
240.	0.105	0.157	0.209	0.261	0.314	0.366	0.418	0.470	0.523	0.575	0.627	0.679	0.732	0.784	0.836	0.889	0.941	0.993	1.045	1.098	1.150
250.	0.109	0.163	0.218	0.272	0.327	0.381	0.436	0.490	0.544	0.599	0.653	0.708	0.762	0.817	0.871	0.926	0.980	1.034	1.089	1.143	1.198
260.	0.113	0.170	0.227	0.283	0.340	0.396	0.453	0.510	0.566	0.623	0.680	0.736	0.793	0.849	0.906	0.963	1.019	1.076	1.133	1.189	1.246
270.	0.118	0.176	0.235	0.294	0.353	0.412	0.470	0.529	0.588	0.647	0.706	0.765	0.823	0.882	0.941	1.000	1.059	1.117	1.176	1.235	1.294
280.	0.122	0.183	0.244	0.305	0.366	0.427	0.488	0.549	0.610	0.671	0.732	0.793	0.854	0.915	0.976	1.037	1.098	1.159	1.220	1.281	1.342
290.	0.126	0.190	0.253	0.316	0.379	0.442	0.505	0.569	0.632	0.695	0.758	0.821	0.884	0.948	1.011	1.074	1.137	1.200	1.264	1.327	1.390
300.	0.131	0.196	0.261	0.327	0.392	0.458	0.523	0.588	0.654	0.719	0.784	0.850	0.915	0.980	1.046	1.111	1.176	1.242	1.307	1.373	1.438
310.	0.135	0.203	0.270	0.338	0.405	0.473	0.540	0.608	0.675	0.743	0.811	0.878	0.946	1.013	1.081	1.148	1.216	1.283	1.351	1.418	1.486
320.	0.140	0.210	0.279	0.349	0.418	0.488	0.558	0.627	0.697	0.767	0.837	0.906	0.976	1.046	1.116	1.185	1.255	1.325	1.395	1.464	1.534
330.	0.144	0.216	0.288	0.360	0.431	0.503	0.575	0.647	0.719	0.791	0.863	0.935	1.007	1.079	1.151	1.222	1.294	1.366	1.438	1.510	1.582
340.	0.148	0.222	0.296	0.370	0.445	0.519	0.593	0.667	0.741	0.815	0.889	0.963	1.037	1.111	1.186	1.260	1.334	1.408	1.482	1.556	1.630
350.	0.153	0.229	0.305	0.381	0.458	0.534	0.610	0.687	0.763	0.839	0.915	0.992	1.068	1.144	1.220	1.297	1.373	1.449	1.526	1.602	1.678
360.	0.157	0.235	0.314	0.392	0.471	0.549	0.628	0.706	0.785	0.863	0.942	1.020	1.099	1.177	1.255	1.334	1.412	1.491	1.569	1.648	1.726
370.	0.161	0.242	0.323	0.403	0.484	0.565	0.645	0.726	0.807	0.887	0.968	1.048	1.129	1.210	1.290	1.371	1.452	1.532	1.613	1.694	1.774

TABLE 2.7h Ellipsoidal Head Thickness Tables in Inches: Joint Efficiency=0.90

ELLIPSOIDAL HEAD THICKNESS TABLES IN INCHES   MAJOR TO MINOR AXIS RATIO=2:1

JOINT EFF.=0.90   ALLOWABLE STRESS VALUE =13800. PSI

INSIDE DIAMETER, IN.

PSI	12.	18.	24.	30.	36.	42.	48.	54.	60.	66.	72.	78.	84.	90.	96.	102.	108.	114.	120.	126.	132.
15.	0.007	0.011	0.014	0.018	0.022	0.025	0.029	0.033	0.036	0.040	0.043	0.047	0.051	0.054	0.058	0.062	0.065	0.069	0.072	0.076	0.080
20.	0.010	0.014	0.019	0.024	0.029	0.034	0.039	0.043	0.048	0.053	0.058	0.063	0.068	0.072	0.077	0.082	0.087	0.092	0.097	0.101	0.106
25.	0.012	0.018	0.024	0.030	0.036	0.042	0.048	0.054	0.060	0.066	0.072	0.079	0.085	0.091	0.097	0.103	0.109	0.115	0.121	0.127	0.133
30.	0.014	0.022	0.029	0.036	0.043	0.051	0.058	0.065	0.072	0.080	0.087	0.094	0.101	0.109	0.116	0.123	0.130	0.138	0.145	0.152	0.159
35.	0.017	0.025	0.034	0.042	0.051	0.059	0.068	0.076	0.085	0.093	0.101	0.110	0.118	0.127	0.135	0.144	0.152	0.161	0.169	0.178	0.186
40.	0.019	0.029	0.039	0.048	0.058	0.068	0.077	0.087	0.097	0.106	0.116	0.126	0.135	0.145	0.155	0.164	0.174	0.184	0.193	0.203	0.213
45.	0.022	0.033	0.043	0.054	0.065	0.076	0.087	0.098	0.109	0.120	0.130	0.141	0.152	0.163	0.174	0.185	0.196	0.207	0.217	0.228	0.239
50.	0.024	0.036	0.048	0.060	0.072	0.085	0.097	0.109	0.121	0.133	0.145	0.157	0.169	0.181	0.193	0.205	0.217	0.230	0.242	0.254	0.266
55.	0.027	0.040	0.053	0.066	0.080	0.093	0.106	0.120	0.133	0.146	0.159	0.173	0.186	0.199	0.213	0.226	0.239	0.253	0.266	0.279	0.292
60.	0.029	0.043	0.058	0.072	0.087	0.101	0.116	0.130	0.145	0.159	0.174	0.188	0.203	0.217	0.232	0.246	0.261	0.275	0.290	0.304	0.319
65.	0.031	0.047	0.063	0.079	0.094	0.110	0.126	0.141	0.157	0.173	0.189	0.204	0.220	0.236	0.251	0.267	0.283	0.298	0.314	0.330	0.346
70.	0.034	0.051	0.068	0.085	0.102	0.118	0.135	0.152	0.169	0.186	0.203	0.220	0.237	0.254	0.271	0.288	0.305	0.321	0.338	0.355	0.372
75.	0.036	0.054	0.073	0.091	0.109	0.127	0.145	0.163	0.181	0.199	0.218	0.236	0.254	0.272	0.290	0.308	0.326	0.344	0.363	0.381	0.399
80.	0.039	0.058	0.077	0.097	0.116	0.135	0.155	0.174	0.193	0.213	0.232	0.251	0.271	0.290	0.309	0.329	0.348	0.367	0.387	0.406	0.425
85.	0.041	0.062	0.082	0.103	0.123	0.144	0.164	0.185	0.205	0.226	0.247	0.267	0.288	0.308	0.329	0.349	0.370	0.390	0.411	0.431	0.452
90.	0.044	0.065	0.087	0.109	0.131	0.152	0.174	0.196	0.218	0.239	0.261	0.283	0.305	0.326	0.348	0.370	0.392	0.413	0.435	0.457	0.479
95.	0.046	0.069	0.092	0.115	0.138	0.161	0.184	0.207	0.230	0.253	0.276	0.299	0.322	0.344	0.367	0.390	0.413	0.436	0.459	0.482	0.505
100.	0.048	0.073	0.097	0.121	0.145	0.169	0.193	0.218	0.242	0.266	0.290	0.314	0.338	0.363	0.387	0.411	0.435	0.459	0.483	0.508	0.532
105.	0.051	0.076	0.102	0.127	0.152	0.178	0.203	0.228	0.254	0.279	0.305	0.330	0.355	0.381	0.406	0.432	0.457	0.482	0.508	0.533	0.558
110.	0.053	0.080	0.106	0.133	0.160	0.186	0.213	0.239	0.266	0.293	0.319	0.346	0.372	0.399	0.425	0.452	0.479	0.505	0.532	0.558	0.585
115.	0.055	0.083	0.111	0.139	0.167	0.195	0.222	0.250	0.278	0.306	0.334	0.361	0.389	0.417	0.445	0.473	0.500	0.528	0.556	0.584	0.612
120.	0.058	0.087	0.116	0.145	0.174	0.203	0.232	0.261	0.290	0.319	0.348	0.377	0.406	0.435	0.464	0.493	0.522	0.551	0.580	0.609	0.638
125.	0.060	0.091	0.121	0.151	0.181	0.212	0.242	0.272	0.302	0.332	0.363	0.393	0.423	0.453	0.484	0.514	0.544	0.574	0.604	0.635	0.665
130.	0.063	0.094	0.126	0.157	0.189	0.220	0.251	0.283	0.314	0.346	0.377	0.409	0.440	0.472	0.503	0.534	0.566	0.597	0.629	0.660	0.692
135.	0.065	0.098	0.131	0.163	0.196	0.229	0.261	0.294	0.327	0.359	0.392	0.425	0.457	0.490	0.523	0.555	0.588	0.620	0.653	0.686	0.718
140.	0.068	0.102	0.135	0.169	0.203	0.237	0.271	0.305	0.339	0.372	0.406	0.440	0.474	0.508	0.542	0.576	0.609	0.643	0.677	0.711	0.745
145.	0.070	0.105	0.140	0.175	0.210	0.245	0.280	0.316	0.351	0.386	0.421	0.456	0.491	0.526	0.561	0.596	0.631	0.666	0.701	0.736	0.771
150.	0.073	0.109	0.145	0.181	0.218	0.254	0.290	0.326	0.363	0.399	0.435	0.472	0.508	0.544	0.580	0.617	0.653	0.689	0.726	0.762	0.798
160.	0.077	0.116	0.155	0.193	0.232	0.271	0.310	0.348	0.387	0.426	0.464	0.503	0.542	0.580	0.619	0.658	0.697	0.735	0.774	0.813	0.851
170.	0.082	0.123	0.164	0.206	0.247	0.288	0.329	0.370	0.411	0.452	0.493	0.535	0.576	0.617	0.658	0.699	0.740	0.781	0.822	0.864	0.905
180.	0.087	0.131	0.174	0.218	0.261	0.305	0.348	0.392	0.435	0.479	0.522	0.566	0.610	0.653	0.697	0.740	0.784	0.827	0.871	0.914	0.958
190.	0.092	0.138	0.184	0.230	0.276	0.322	0.368	0.414	0.460	0.506	0.552	0.598	0.644	0.690	0.736	0.781	0.827	0.873	0.919	0.965	1.011
200.	0.097	0.145	0.194	0.242	0.290	0.339	0.387	0.435	0.484	0.532	0.581	0.629	0.677	0.726	0.774	0.823	0.871	0.919	0.968	1.016	1.065
210.	0.102	0.152	0.203	0.254	0.305	0.356	0.406	0.457	0.508	0.559	0.610	0.661	0.711	0.762	0.813	0.864	0.915	0.965	1.016	1.067	1.118
220.	0.106	0.160	0.213	0.266	0.319	0.373	0.426	0.479	0.532	0.586	0.639	0.692	0.745	0.799	0.852	0.905	0.958	1.011	1.065	1.118	1.171
230.	0.111	0.167	0.223	0.278	0.334	0.390	0.445	0.501	0.557	0.612	0.668	0.724	0.779	0.835	0.891	0.946	1.002	1.058	1.113	1.169	1.224
240.	0.116	0.174	0.232	0.290	0.348	0.407	0.465	0.523	0.581	0.639	0.697	0.755	0.813	0.871	0.929	0.987	1.045	1.104	1.162	1.220	1.278
250.	0.121	0.182	0.242	0.303	0.363	0.424	0.484	0.545	0.605	0.666	0.726	0.787	0.847	0.908	0.968	1.029	1.089	1.150	1.210	1.271	1.331
260.	0.126	0.189	0.252	0.315	0.378	0.441	0.503	0.566	0.629	0.692	0.755	0.818	0.881	0.944	1.007	1.070	1.133	1.196	1.259	1.322	1.385
270.	0.131	0.196	0.261	0.327	0.392	0.458	0.523	0.588	0.654	0.719	0.784	0.850	0.915	0.980	1.046	1.111	1.176	1.242	1.307	1.373	1.438
280.	0.136	0.203	0.271	0.339	0.407	0.474	0.542	0.610	0.678	0.746	0.813	0.881	0.949	1.017	1.085	1.152	1.220	1.288	1.356	1.423	1.491
290.	0.140	0.211	0.281	0.351	0.421	0.491	0.562	0.632	0.702	0.772	0.843	0.913	0.983	1.053	1.123	1.194	1.264	1.334	1.404	1.474	1.545
300.	0.145	0.218	0.291	0.363	0.436	0.508	0.581	0.654	0.726	0.799	0.872	0.944	1.017	1.090	1.162	1.235	1.308	1.380	1.453	1.525	1.598
310.	0.150	0.225	0.300	0.375	0.450	0.525	0.601	0.676	0.751	0.826	0.901	0.976	1.051	1.126	1.201	1.276	1.351	1.426	1.501	1.576	1.651
320.	0.155	0.232	0.310	0.387	0.465	0.542	0.620	0.697	0.775	0.852	0.930	1.007	1.085	1.162	1.240	1.317	1.395	1.472	1.550	1.627	1.705
330.	0.160	0.240	0.320	0.400	0.480	0.559	0.639	0.719	0.799	0.879	0.959	1.039	1.119	1.199	1.279	1.359	1.439	1.519	1.598	1.678	1.758
340.	0.165	0.247	0.330	0.412	0.494	0.576	0.659	0.741	0.824	0.906	0.988	1.071	1.153	1.235	1.318	1.400	1.482	1.565	1.647	1.729	1.812
350.	0.170	0.254	0.339	0.424	0.509	0.593	0.678	0.763	0.848	0.933	1.017	1.102	1.187	1.272	1.357	1.441	1.526	1.611	1.696	1.780	1.865
360.	0.174	0.262	0.349	0.436	0.523	0.610	0.698	0.785	0.872	0.959	1.047	1.134	1.221	1.308	1.395	1.483	1.570	1.657	1.744	1.831	1.919
370.	0.179	0.269	0.359	0.448	0.538	0.627	0.717	0.807	0.896	0.986	1.076	1.165	1.255	1.345	1.434	1.524	1.614	1.703	1.793	1.882	1.972

TABLE 2.7i Ellipsoidal Head Thickness Tables in Inches: Joint Efficiency=0.85

ELLIPSOIDAL HEAD THICKNESS TABLES IN INCHES   MAJOR TO MINOR AXIS RATIO=2:1

JOINT EFF.=0.85   ALLOWABLE STRESS VALUE =13800. PSI

INSIDE DIAMETER, IN.

PSI	12.	18.	24.	30.	36.	42.	48.	54.	60.	66.	72.	78.	84.	90.	96.	102.	108.	114.	120.	126.	132.
15.	0.008	0.012	0.015	0.019	0.023	0.027	0.031	0.035	0.038	0.042	0.046	0.050	0.054	0.058	0.061	0.065	0.069	0.073	0.077	0.081	0.084
20.	0.010	0.015	0.020	0.026	0.031	0.036	0.041	0.046	0.051	0.056	0.061	0.067	0.072	0.077	0.082	0.087	0.092	0.097	0.102	0.107	0.113
25.	0.013	0.019	0.026	0.032	0.038	0.045	0.051	0.058	0.064	0.070	0.077	0.083	0.090	0.096	0.102	0.109	0.115	0.122	0.128	0.134	0.141
30.	0.015	0.023	0.031	0.038	0.046	0.054	0.061	0.069	0.077	0.084	0.092	0.100	0.107	0.115	0.123	0.130	0.138	0.146	0.153	0.161	0.169
35.	0.018	0.027	0.036	0.045	0.054	0.063	0.072	0.081	0.090	0.098	0.107	0.116	0.125	0.134	0.143	0.152	0.161	0.170	0.179	0.188	0.197
40.	0.020	0.031	0.041	0.051	0.061	0.072	0.082	0.092	0.102	0.113	0.123	0.133	0.143	0.154	0.164	0.174	0.184	0.194	0.205	0.215	0.225
45.	0.023	0.035	0.046	0.058	0.069	0.081	0.092	0.104	0.115	0.127	0.138	0.150	0.161	0.173	0.184	0.196	0.207	0.219	0.230	0.242	0.253
50.	0.026	0.038	0.051	0.064	0.077	0.090	0.102	0.115	0.128	0.141	0.154	0.166	0.179	0.192	0.205	0.217	0.230	0.243	0.256	0.269	0.281
55.	0.028	0.042	0.056	0.070	0.084	0.099	0.113	0.127	0.141	0.155	0.169	0.183	0.197	0.211	0.225	0.239	0.253	0.267	0.281	0.296	0.310
60.	0.031	0.046	0.061	0.077	0.092	0.107	0.123	0.138	0.154	0.169	0.184	0.200	0.215	0.230	0.246	0.261	0.276	0.292	0.307	0.322	0.338
65.	0.033	0.050	0.067	0.083	0.100	0.116	0.133	0.150	0.166	0.183	0.200	0.216	0.233	0.249	0.266	0.283	0.299	0.316	0.333	0.349	0.366
70.	0.036	0.054	0.072	0.090	0.107	0.125	0.143	0.161	0.179	0.197	0.215	0.233	0.251	0.269	0.287	0.305	0.322	0.340	0.358	0.376	0.394
75.	0.038	0.058	0.077	0.096	0.115	0.134	0.154	0.173	0.192	0.211	0.230	0.250	0.269	0.288	0.307	0.326	0.345	0.365	0.384	0.403	0.422
80.	0.041	0.061	0.082	0.102	0.123	0.143	0.164	0.184	0.205	0.225	0.246	0.266	0.287	0.307	0.328	0.348	0.369	0.389	0.409	0.430	0.450
85.	0.044	0.065	0.087	0.109	0.131	0.152	0.174	0.196	0.218	0.239	0.261	0.283	0.305	0.326	0.348	0.370	0.392	0.413	0.435	0.457	0.479
90.	0.046	0.069	0.092	0.115	0.138	0.161	0.184	0.207	0.230	0.253	0.276	0.299	0.322	0.346	0.369	0.392	0.415	0.438	0.461	0.484	0.507
95.	0.049	0.073	0.097	0.122	0.146	0.170	0.195	0.219	0.243	0.267	0.292	0.316	0.340	0.365	0.389	0.413	0.438	0.462	0.486	0.511	0.535
100.	0.051	0.077	0.102	0.128	0.154	0.179	0.205	0.230	0.256	0.282	0.307	0.333	0.358	0.384	0.410	0.435	0.461	0.486	0.512	0.538	0.563
105.	0.054	0.081	0.108	0.134	0.161	0.188	0.215	0.242	0.269	0.296	0.323	0.349	0.376	0.403	0.430	0.457	0.484	0.511	0.538	0.564	0.591
110.	0.056	0.084	0.113	0.141	0.169	0.197	0.225	0.253	0.282	0.310	0.338	0.366	0.394	0.422	0.451	0.479	0.507	0.535	0.563	0.591	0.620
115.	0.059	0.088	0.118	0.147	0.177	0.206	0.236	0.265	0.294	0.324	0.353	0.383	0.412	0.442	0.471	0.500	0.530	0.559	0.589	0.618	0.648
120.	0.061	0.092	0.123	0.154	0.184	0.215	0.246	0.276	0.307	0.338	0.369	0.399	0.430	0.461	0.492	0.522	0.553	0.584	0.614	0.645	0.676
125.	0.064	0.096	0.128	0.160	0.192	0.224	0.256	0.288	0.320	0.352	0.384	0.416	0.448	0.480	0.512	0.544	0.576	0.608	0.640	0.672	0.704
130.	0.067	0.100	0.133	0.166	0.200	0.233	0.266	0.300	0.333	0.366	0.399	0.433	0.466	0.499	0.533	0.566	0.599	0.632	0.666	0.699	0.732
135.	0.069	0.104	0.138	0.173	0.207	0.242	0.277	0.311	0.346	0.380	0.415	0.449	0.484	0.518	0.553	0.588	0.622	0.657	0.691	0.726	0.760
140.	0.072	0.108	0.143	0.179	0.215	0.251	0.287	0.323	0.358	0.394	0.430	0.466	0.502	0.538	0.574	0.609	0.645	0.681	0.717	0.753	0.789
145.	0.074	0.112	0.149	0.186	0.223	0.260	0.297	0.334	0.371	0.408	0.446	0.483	0.520	0.557	0.594	0.631	0.668	0.705	0.743	0.780	0.817
150.	0.077	0.115	0.154	0.192	0.231	0.269	0.307	0.346	0.384	0.423	0.461	0.499	0.538	0.576	0.615	0.653	0.691	0.730	0.768	0.807	0.845
160.	0.082	0.123	0.164	0.205	0.246	0.287	0.328	0.369	0.410	0.451	0.492	0.533	0.574	0.615	0.656	0.697	0.738	0.779	0.820	0.861	0.901
170.	0.087	0.131	0.174	0.218	0.261	0.305	0.348	0.392	0.435	0.479	0.522	0.566	0.610	0.653	0.697	0.740	0.784	0.827	0.871	0.914	0.958
180.	0.092	0.138	0.184	0.231	0.277	0.323	0.369	0.415	0.461	0.507	0.553	0.599	0.645	0.692	0.738	0.784	0.830	0.876	0.922	0.968	1.014
190.	0.097	0.146	0.195	0.243	0.292	0.341	0.389	0.438	0.487	0.535	0.584	0.633	0.681	0.730	0.779	0.827	0.876	0.925	0.974	1.022	1.071
200.	0.102	0.154	0.205	0.256	0.307	0.359	0.410	0.461	0.512	0.564	0.615	0.666	0.717	0.769	0.820	0.871	0.922	0.974	1.025	1.076	1.127
210.	0.108	0.161	0.215	0.269	0.323	0.377	0.430	0.484	0.538	0.592	0.646	0.699	0.753	0.807	0.861	0.915	0.969	1.022	1.076	1.130	1.184
220.	0.113	0.169	0.225	0.282	0.338	0.395	0.451	0.507	0.564	0.620	0.677	0.733	0.789	0.846	0.902	0.959	1.015	1.071	1.128	1.184	1.240
230.	0.118	0.177	0.236	0.295	0.354	0.413	0.472	0.530	0.589	0.648	0.707	0.766	0.825	0.884	0.943	1.002	1.061	1.120	1.179	1.238	1.297
240.	0.123	0.185	0.246	0.308	0.369	0.431	0.492	0.554	0.615	0.677	0.738	0.800	0.861	0.923	0.984	1.046	1.107	1.169	1.230	1.292	1.353
250.	0.128	0.192	0.256	0.320	0.384	0.449	0.513	0.577	0.641	0.705	0.769	0.833	0.897	0.961	1.025	1.089	1.153	1.217	1.282	1.346	1.410
260.	0.133	0.200	0.267	0.333	0.400	0.467	0.533	0.600	0.666	0.733	0.800	0.866	0.933	1.000	1.066	1.133	1.200	1.266	1.333	1.400	1.466
270.	0.138	0.208	0.277	0.346	0.415	0.484	0.554	0.623	0.692	0.761	0.831	0.900	0.969	1.038	1.107	1.177	1.246	1.315	1.384	1.453	1.523
280.	0.144	0.215	0.287	0.359	0.431	0.502	0.574	0.646	0.718	0.790	0.861	0.933	1.005	1.077	1.149	1.220	1.292	1.364	1.436	1.507	1.579
290.	0.149	0.223	0.297	0.372	0.446	0.520	0.595	0.669	0.744	0.818	0.892	0.967	1.041	1.115	1.190	1.264	1.338	1.413	1.487	1.561	1.636
300.	0.154	0.231	0.308	0.385	0.462	0.538	0.615	0.692	0.769	0.846	0.923	1.000	1.077	1.154	1.231	1.308	1.385	1.462	1.538	1.615	1.692
310.	0.159	0.238	0.318	0.397	0.477	0.556	0.636	0.715	0.795	0.874	0.954	1.033	1.113	1.192	1.272	1.351	1.431	1.510	1.590	1.669	1.749
320.	0.164	0.246	0.328	0.410	0.492	0.574	0.657	0.739	0.821	0.903	0.985	1.067	1.149	1.231	1.313	1.395	1.477	1.559	1.641	1.723	1.805
330.	0.169	0.254	0.339	0.423	0.508	0.592	0.677	0.762	0.846	0.931	1.016	1.100	1.185	1.270	1.354	1.439	1.523	1.608	1.693	1.777	1.862
340.	0.174	0.262	0.349	0.436	0.523	0.610	0.698	0.785	0.872	0.959	1.047	1.134	1.221	1.308	1.395	1.483	1.570	1.657	1.744	1.831	1.919
350.	0.180	0.269	0.359	0.449	0.539	0.628	0.718	0.808	0.898	0.988	1.077	1.167	1.257	1.347	1.437	1.526	1.616	1.706	1.796	1.885	1.975
360.	0.185	0.277	0.369	0.462	0.554	0.646	0.739	0.831	0.924	1.016	1.108	1.201	1.293	1.385	1.478	1.570	1.662	1.755	1.847	1.939	2.032
370.	0.190	0.285	0.380	0.475	0.570	0.665	0.759	0.854	0.949	1.044	1.139	1.234	1.329	1.424	1.519	1.614	1.709	1.804	1.899	1.994	2.088

TABLE 2.7j Ellipsoidal Head Thickness Tables in Inches: Joint Efficiency=0.80

ELLIPSOIDAL HEAD THICKNESS TABLES IN INCHES   MAJOR TO MINOR AXIS RATIO=2:1

JOINT EFF.=0.80   ALLOWABLE STRESS VALUE =13800. PSI

INSIDE DIAMETER, IN.

PSI	12.	18.	24.	30.	36.	42.	48.	54.	60.	66.	72.	78.	84.	90.	96.	102.	108.	114.	120.	126.	132.
15.	0.008	0.012	0.016	0.020	0.024	0.029	0.033	0.037	0.041	0.045	0.049	0.053	0.057	0.061	0.065	0.069	0.073	0.077	0.082	0.086	0.090
20.	0.011	0.016	0.022	0.027	0.033	0.038	0.043	0.049	0.054	0.060	0.065	0.071	0.076	0.082	0.087	0.092	0.098	0.103	0.109	0.114	0.120
25.	0.014	0.020	0.027	0.034	0.041	0.048	0.054	0.061	0.068	0.075	0.082	0.088	0.095	0.102	0.109	0.116	0.122	0.129	0.136	0.143	0.149
30.	0.016	0.024	0.033	0.041	0.049	0.057	0.065	0.073	0.082	0.090	0.098	0.106	0.114	0.122	0.130	0.139	0.147	0.155	0.163	0.171	0.179
35.	0.019	0.029	0.038	0.048	0.057	0.067	0.076	0.086	0.095	0.105	0.114	0.124	0.133	0.143	0.152	0.162	0.171	0.181	0.190	0.200	0.209
40.	0.022	0.033	0.043	0.054	0.065	0.076	0.087	0.098	0.109	0.120	0.130	0.141	0.152	0.163	0.174	0.185	0.196	0.207	0.217	0.228	0.239
45.	0.024	0.037	0.049	0.061	0.073	0.086	0.098	0.110	0.122	0.135	0.147	0.159	0.171	0.183	0.196	0.208	0.220	0.232	0.245	0.257	0.269
50.	0.027	0.041	0.054	0.068	0.082	0.095	0.109	0.122	0.136	0.150	0.163	0.177	0.190	0.204	0.218	0.231	0.245	0.258	0.272	0.285	0.299
55.	0.030	0.045	0.060	0.075	0.090	0.105	0.120	0.135	0.150	0.165	0.180	0.194	0.209	0.224	0.239	0.254	0.269	0.284	0.299	0.314	0.329
60.	0.033	0.049	0.065	0.082	0.098	0.114	0.131	0.147	0.163	0.180	0.196	0.212	0.229	0.245	0.261	0.277	0.294	0.310	0.326	0.343	0.359
65.	0.035	0.053	0.071	0.088	0.106	0.124	0.141	0.159	0.177	0.194	0.212	0.230	0.247	0.265	0.283	0.300	0.318	0.336	0.353	0.371	0.389
70.	0.038	0.057	0.076	0.095	0.114	0.133	0.152	0.171	0.190	0.209	0.228	0.247	0.266	0.285	0.305	0.324	0.343	0.362	0.381	0.400	0.419
75.	0.041	0.061	0.082	0.102	0.122	0.143	0.163	0.184	0.204	0.224	0.245	0.265	0.286	0.306	0.326	0.347	0.367	0.387	0.408	0.428	0.449
80.	0.044	0.065	0.087	0.109	0.130	0.152	0.174	0.196	0.218	0.239	0.261	0.283	0.305	0.326	0.348	0.370	0.392	0.413	0.435	0.457	0.479
85.	0.046	0.070	0.093	0.116	0.139	0.162	0.185	0.208	0.231	0.254	0.277	0.301	0.324	0.347	0.370	0.393	0.416	0.439	0.462	0.485	0.509
90.	0.049	0.073	0.098	0.122	0.147	0.171	0.196	0.220	0.245	0.269	0.294	0.318	0.343	0.367	0.392	0.416	0.441	0.465	0.490	0.514	0.539
95.	0.052	0.078	0.103	0.129	0.155	0.181	0.207	0.233	0.258	0.284	0.310	0.336	0.362	0.388	0.413	0.439	0.465	0.491	0.517	0.543	0.569
100.	0.054	0.082	0.109	0.136	0.163	0.190	0.218	0.245	0.272	0.299	0.326	0.354	0.381	0.408	0.435	0.462	0.490	0.517	0.544	0.571	0.598
105.	0.057	0.086	0.114	0.143	0.171	0.200	0.228	0.257	0.286	0.314	0.343	0.371	0.400	0.428	0.457	0.486	0.514	0.543	0.571	0.600	0.628
110.	0.060	0.090	0.120	0.150	0.180	0.209	0.239	0.269	0.299	0.329	0.359	0.389	0.419	0.449	0.479	0.509	0.539	0.569	0.598	0.628	0.658
115.	0.063	0.094	0.125	0.156	0.188	0.219	0.250	0.282	0.313	0.344	0.375	0.407	0.438	0.469	0.501	0.532	0.563	0.594	0.626	0.657	0.688
120.	0.065	0.098	0.131	0.163	0.196	0.229	0.261	0.294	0.326	0.359	0.392	0.424	0.457	0.490	0.522	0.555	0.588	0.620	0.653	0.686	0.718
125.	0.068	0.102	0.136	0.170	0.204	0.238	0.272	0.306	0.340	0.374	0.408	0.442	0.476	0.510	0.544	0.578	0.612	0.646	0.680	0.714	0.748
130.	0.071	0.106	0.141	0.177	0.212	0.248	0.283	0.318	0.354	0.389	0.424	0.460	0.495	0.531	0.566	0.601	0.637	0.672	0.707	0.743	0.778
135.	0.073	0.110	0.147	0.184	0.220	0.257	0.294	0.331	0.367	0.404	0.441	0.477	0.514	0.551	0.588	0.624	0.661	0.698	0.735	0.771	0.808
140.	0.076	0.114	0.152	0.190	0.229	0.267	0.305	0.343	0.381	0.419	0.457	0.495	0.533	0.571	0.609	0.648	0.686	0.724	0.762	0.800	0.838
145.	0.079	0.118	0.158	0.197	0.237	0.276	0.316	0.355	0.395	0.434	0.473	0.513	0.552	0.592	0.631	0.671	0.710	0.750	0.789	0.829	0.868
150.	0.082	0.122	0.163	0.204	0.245	0.286	0.327	0.367	0.408	0.449	0.490	0.531	0.571	0.612	0.653	0.694	0.735	0.776	0.816	0.857	0.898
160.	0.087	0.131	0.174	0.218	0.261	0.305	0.348	0.392	0.435	0.479	0.522	0.566	0.610	0.653	0.697	0.740	0.784	0.827	0.871	0.914	0.958
170.	0.093	0.139	0.185	0.231	0.278	0.324	0.370	0.416	0.463	0.509	0.555	0.601	0.648	0.694	0.740	0.787	0.833	0.879	0.925	0.972	1.018
180.	0.098	0.147	0.196	0.245	0.294	0.343	0.392	0.441	0.490	0.539	0.588	0.637	0.686	0.735	0.784	0.833	0.882	0.931	0.980	1.029	1.078
190.	0.103	0.155	0.207	0.259	0.310	0.362	0.414	0.465	0.517	0.569	0.621	0.672	0.724	0.776	0.828	0.879	0.931	0.983	1.034	1.086	1.138
200.	0.109	0.163	0.218	0.272	0.327	0.381	0.436	0.490	0.544	0.599	0.653	0.708	0.762	0.817	0.871	0.926	0.980	1.034	1.089	1.143	1.198
210.	0.114	0.172	0.229	0.286	0.343	0.400	0.457	0.515	0.572	0.629	0.686	0.743	0.800	0.858	0.915	0.972	1.029	1.086	1.143	1.201	1.258
220.	0.120	0.180	0.240	0.300	0.359	0.419	0.479	0.539	0.599	0.659	0.719	0.779	0.839	0.899	0.958	1.018	1.078	1.138	1.198	1.258	1.318
230.	0.125	0.188	0.251	0.313	0.376	0.438	0.501	0.564	0.626	0.689	0.752	0.814	0.877	0.939	1.002	1.065	1.127	1.190	1.253	1.315	1.378
240.	0.131	0.196	0.261	0.327	0.392	0.458	0.523	0.588	0.654	0.719	0.784	0.850	0.915	0.980	1.046	1.111	1.176	1.242	1.307	1.373	1.438
250.	0.136	0.204	0.272	0.340	0.409	0.477	0.545	0.613	0.681	0.749	0.817	0.885	0.953	1.021	1.089	1.158	1.226	1.294	1.362	1.430	1.498
260.	0.142	0.212	0.283	0.354	0.425	0.496	0.567	0.637	0.708	0.779	0.850	0.921	0.991	1.062	1.133	1.204	1.275	1.346	1.416	1.487	1.558
270.	0.147	0.221	0.294	0.368	0.441	0.515	0.588	0.662	0.735	0.809	0.883	0.956	1.030	1.103	1.177	1.250	1.324	1.397	1.471	1.545	1.618
280.	0.153	0.229	0.305	0.381	0.458	0.534	0.610	0.687	0.763	0.839	0.915	0.992	1.068	1.144	1.220	1.297	1.373	1.449	1.526	1.602	1.678
290.	0.158	0.237	0.316	0.395	0.474	0.553	0.632	0.711	0.790	0.869	0.948	1.027	1.106	1.185	1.264	1.343	1.422	1.501	1.580	1.659	1.738
300.	0.163	0.245	0.327	0.409	0.490	0.572	0.654	0.736	0.817	0.899	0.981	1.063	1.144	1.226	1.308	1.390	1.471	1.553	1.635	1.717	1.798
310.	0.169	0.253	0.338	0.422	0.507	0.591	0.676	0.760	0.845	0.929	1.014	1.098	1.183	1.267	1.352	1.436	1.521	1.605	1.690	1.774	1.859
320.	0.174	0.262	0.349	0.436	0.523	0.610	0.698	0.785	0.872	0.959	1.047	1.134	1.221	1.308	1.395	1.483	1.570	1.657	1.744	1.831	1.919
330.	0.180	0.270	0.360	0.450	0.540	0.630	0.720	0.809	0.899	0.989	1.079	1.169	1.259	1.349	1.439	1.529	1.619	1.709	1.799	1.889	1.979
340.	0.185	0.278	0.371	0.463	0.556	0.649	0.741	0.834	0.927	1.019	1.112	1.205	1.297	1.390	1.483	1.576	1.668	1.761	1.854	1.946	2.039
350.	0.191	0.286	0.382	0.477	0.572	0.668	0.763	0.859	0.954	1.050	1.145	1.240	1.336	1.431	1.527	1.622	1.717	1.813	1.908	2.004	2.099
360.	0.196	0.294	0.393	0.491	0.589	0.687	0.785	0.883	0.981	1.080	1.178	1.276	1.374	1.472	1.570	1.668	1.767	1.865	1.963	2.061	2.159
370.	0.202	0.303	0.404	0.504	0.605	0.706	0.807	0.908	1.009	1.110	1.211	1.311	1.412	1.513	1.614	1.715	1.816	1.917	2.018	2.119	2.219

TABLE 2.7k Ellipsoidal Head Thickness Tables in Inches: Joint Efficiency=0.70

ELLIPSOIDAL HEAD THICKNESS TABLES IN INCHES   MAJOR TO MINOR AXIS RATIO=2:1

JOINT EFF.=0.70   ALLOWABLE STRESS VALUE =13800. PSI

INSIDE DIAMETER,IN.

PSI	12.	18.	24.	30.	36.	42.	48.	54.	60.	66.	72.	78.	84.	90.	96.	102.	108.	114.	120.	126.	132.
15.	0.009	0.014	0.019	0.023	0.028	0.033	0.037	0.042	0.047	0.051	0.056	0.061	0.065	0.070	0.075	0.079	0.084	0.089	0.093	0.098	0.103
20.	0.012	0.019	0.025	0.031	0.037	0.043	0.050	0.056	0.062	0.068	0.075	0.081	0.087	0.093	0.099	0.106	0.112	0.118	0.124	0.130	0.137
25.	0.016	0.023	0.031	0.039	0.047	0.054	0.062	0.070	0.078	0.085	0.093	0.101	0.109	0.116	0.124	0.132	0.140	0.148	0.155	0.163	0.171
30.	0.019	0.028	0.037	0.047	0.056	0.065	0.075	0.084	0.093	0.103	0.112	0.121	0.130	0.140	0.149	0.158	0.168	0.177	0.186	0.196	0.205
35.	0.022	0.033	0.043	0.054	0.065	0.076	0.087	0.098	0.109	0.120	0.130	0.141	0.152	0.163	0.174	0.185	0.196	0.207	0.217	0.228	0.239
40.	0.025	0.037	0.050	0.062	0.075	0.087	0.099	0.112	0.124	0.137	0.149	0.162	0.174	0.186	0.199	0.211	0.224	0.236	0.249	0.261	0.273
45.	0.028	0.042	0.056	0.070	0.084	0.098	0.112	0.126	0.140	0.154	0.168	0.182	0.196	0.210	0.224	0.238	0.252	0.266	0.280	0.294	0.308
50.	0.031	0.047	0.062	0.078	0.093	0.109	0.124	0.140	0.155	0.171	0.186	0.202	0.218	0.233	0.249	0.264	0.280	0.295	0.311	0.326	0.342
55.	0.034	0.051	0.068	0.085	0.103	0.120	0.137	0.154	0.171	0.188	0.205	0.222	0.239	0.256	0.273	0.291	0.308	0.325	0.342	0.359	0.376
60.	0.037	0.056	0.075	0.093	0.112	0.131	0.149	0.168	0.187	0.205	0.224	0.243	0.261	0.280	0.299	0.317	0.336	0.355	0.373	0.392	0.411
65.	0.040	0.061	0.081	0.101	0.121	0.141	0.162	0.182	0.202	0.222	0.243	0.263	0.283	0.303	0.323	0.343	0.364	0.384	0.404	0.424	0.444
70.	0.044	0.065	0.087	0.109	0.131	0.152	0.174	0.196	0.218	0.239	0.261	0.283	0.305	0.326	0.348	0.370	0.392	0.413	0.435	0.457	0.479
75.	0.047	0.070	0.093	0.117	0.140	0.163	0.186	0.210	0.233	0.256	0.280	0.303	0.326	0.350	0.373	0.396	0.420	0.443	0.466	0.490	0.513
80.	0.050	0.075	0.099	0.124	0.149	0.174	0.199	0.224	0.249	0.274	0.298	0.323	0.348	0.373	0.398	0.423	0.448	0.472	0.497	0.522	0.547
85.	0.053	0.079	0.106	0.132	0.159	0.185	0.211	0.238	0.264	0.291	0.317	0.344	0.370	0.396	0.423	0.449	0.476	0.502	0.528	0.555	0.581
90.	0.056	0.084	0.112	0.140	0.168	0.196	0.224	0.252	0.280	0.308	0.336	0.364	0.392	0.420	0.448	0.476	0.504	0.532	0.560	0.588	0.615
95.	0.059	0.089	0.118	0.148	0.177	0.207	0.236	0.266	0.295	0.325	0.354	0.384	0.413	0.443	0.473	0.502	0.532	0.561	0.591	0.620	0.650
100.	0.062	0.093	0.124	0.155	0.187	0.218	0.249	0.280	0.311	0.342	0.373	0.404	0.435	0.466	0.497	0.528	0.560	0.591	0.622	0.653	0.684
105.	0.065	0.098	0.131	0.163	0.196	0.229	0.261	0.294	0.326	0.359	0.392	0.424	0.457	0.490	0.522	0.555	0.588	0.620	0.653	0.686	0.718
110.	0.068	0.103	0.137	0.171	0.205	0.239	0.274	0.308	0.342	0.376	0.410	0.445	0.479	0.513	0.547	0.581	0.616	0.650	0.684	0.718	0.752
115.	0.072	0.107	0.143	0.179	0.215	0.250	0.286	0.322	0.358	0.393	0.429	0.465	0.501	0.536	0.572	0.608	0.644	0.679	0.715	0.751	0.787
120.	0.075	0.112	0.149	0.187	0.224	0.261	0.299	0.336	0.373	0.410	0.448	0.485	0.522	0.560	0.597	0.634	0.672	0.709	0.746	0.784	0.821
125.	0.078	0.117	0.155	0.194	0.233	0.272	0.311	0.350	0.389	0.428	0.466	0.505	0.544	0.583	0.622	0.661	0.700	0.739	0.777	0.816	0.855
130.	0.081	0.121	0.162	0.202	0.243	0.283	0.323	0.364	0.404	0.445	0.485	0.526	0.566	0.606	0.647	0.687	0.728	0.768	0.809	0.849	0.889
135.	0.084	0.126	0.168	0.210	0.252	0.294	0.336	0.378	0.420	0.462	0.504	0.546	0.588	0.630	0.672	0.714	0.756	0.798	0.840	0.882	0.924
140.	0.087	0.131	0.174	0.218	0.261	0.305	0.348	0.392	0.435	0.479	0.522	0.566	0.610	0.653	0.697	0.740	0.784	0.827	0.871	0.914	0.958
145.	0.090	0.135	0.180	0.225	0.271	0.316	0.361	0.406	0.451	0.496	0.541	0.586	0.632	0.677	0.722	0.767	0.812	0.857	0.902	0.947	0.992
150.	0.093	0.140	0.187	0.233	0.280	0.327	0.373	0.420	0.467	0.513	0.560	0.607	0.653	0.700	0.747	0.793	0.840	0.886	0.933	0.980	1.026
155.	0.096	0.144	0.193	0.241	0.289	0.338	0.386	0.434	0.482	0.531	0.579	0.627	0.675	0.724	0.772	0.820	0.868	0.916	0.965	1.013	1.061
160.	0.100	0.149	0.199	0.249	0.299	0.348	0.398	0.448	0.498	0.547	0.597	0.647	0.697	0.747	0.796	0.846	0.896	0.946	0.995	1.045	1.095
170.	0.106	0.159	0.212	0.264	0.317	0.370	0.423	0.476	0.529	0.582	0.635	0.688	0.740	0.793	0.846	0.899	0.952	1.005	1.058	1.111	1.164
180.	0.112	0.168	0.224	0.280	0.336	0.392	0.448	0.504	0.560	0.616	0.672	0.728	0.784	0.840	0.896	0.952	1.008	1.064	1.120	1.176	1.232
190.	0.118	0.177	0.236	0.296	0.355	0.414	0.473	0.532	0.591	0.650	0.709	0.769	0.828	0.887	0.946	1.005	1.064	1.123	1.182	1.242	1.301
200.	0.124	0.187	0.249	0.311	0.373	0.436	0.498	0.560	0.622	0.685	0.747	0.809	0.871	0.934	0.996	1.058	1.120	1.183	1.245	1.307	1.369
210.	0.131	0.196	0.261	0.327	0.392	0.458	0.523	0.588	0.654	0.719	0.784	0.850	0.915	0.980	1.046	1.111	1.176	1.242	1.307	1.373	1.438
220.	0.137	0.205	0.274	0.342	0.411	0.479	0.548	0.616	0.685	0.753	0.822	0.890	0.959	1.027	1.096	1.164	1.233	1.301	1.370	1.438	1.507
230.	0.143	0.215	0.286	0.358	0.430	0.501	0.573	0.644	0.716	0.788	0.859	0.931	1.002	1.074	1.146	1.217	1.289	1.360	1.432	1.504	1.575
240.	0.149	0.224	0.299	0.374	0.448	0.523	0.598	0.672	0.747	0.822	0.897	0.971	1.046	1.121	1.196	1.270	1.345	1.420	1.494	1.569	1.644
250.	0.156	0.234	0.311	0.389	0.467	0.545	0.623	0.701	0.778	0.856	0.934	1.012	1.090	1.168	1.245	1.323	1.401	1.479	1.557	1.635	1.713
260.	0.162	0.243	0.324	0.405	0.486	0.567	0.648	0.729	0.810	0.891	0.972	1.053	1.133	1.214	1.295	1.376	1.457	1.538	1.619	1.700	1.781
270.	0.168	0.252	0.336	0.420	0.505	0.589	0.673	0.757	0.841	0.925	1.009	1.093	1.177	1.261	1.345	1.429	1.514	1.598	1.682	1.766	1.850
280.	0.174	0.262	0.349	0.436	0.523	0.610	0.698	0.785	0.872	0.959	1.047	1.134	1.221	1.308	1.395	1.483	1.570	1.657	1.744	1.831	1.919
290.	0.181	0.271	0.361	0.452	0.542	0.632	0.723	0.813	0.903	0.994	1.084	1.174	1.265	1.355	1.445	1.536	1.626	1.716	1.807	1.897	1.987
300.	0.187	0.280	0.374	0.467	0.561	0.654	0.748	0.841	0.935	1.028	1.121	1.215	1.308	1.402	1.495	1.589	1.682	1.776	1.869	1.963	2.056
310.	0.193	0.290	0.386	0.483	0.579	0.676	0.773	0.869	0.966	1.062	1.159	1.256	1.352	1.449	1.545	1.642	1.738	1.835	1.932	2.028	2.125
320.	0.199	0.299	0.399	0.498	0.598	0.698	0.797	0.897	0.997	1.097	1.196	1.296	1.396	1.495	1.595	1.695	1.795	1.894	1.994	2.094	2.194
330.	0.206	0.309	0.411	0.514	0.617	0.720	0.823	0.926	1.028	1.131	1.234	1.337	1.440	1.543	1.645	1.748	1.851	1.954	2.057	2.160	2.262
340.	0.212	0.318	0.424	0.530	0.636	0.742	0.848	0.954	1.060	1.166	1.272	1.378	1.483	1.589	1.695	1.801	1.907	2.013	2.119	2.225	2.331
350.	0.218	0.327	0.436	0.545	0.655	0.764	0.873	0.982	1.091	1.200	1.309	1.418	1.527	1.636	1.745	1.855	1.964	2.073	2.182	2.291	2.400
360.	0.224	0.337	0.449	0.561	0.673	0.786	0.898	1.010	1.122	1.234	1.347	1.459	1.571	1.683	1.795	1.908	2.020	2.132	2.244	2.357	2.469
370.	0.231	0.346	0.461	0.577	0.692	0.807	0.923	1.038	1.153	1.269	1.384	1.500	1.615	1.730	1.846	1.961	2.076	2.192	2.307	2.422	2.538

TABLE 2.7 / Ellipsoidal Head Thickness Tables in Inches: Joint Efficiency=0.65

ELLIPSOIDAL HEAD THICKNESS TABLES IN INCHES   MAJOR TO MINOR AXIS RATIO=2:1

JOINT EFF.=0.65   ALLOWABLE STRESS VALUE =13800. PSI

INSIDE DIAMETER,IN.

PSI	12.	18.	24.	30.	36.	42.	48.	54.	60.	66.	72.	78.	84.	90.	96.	102.	108.	114.	120.	126.	132.
15.	0.010	0.015	0.020	0.025	0.030	0.035	0.040	0.045	0.050	0.055	0.060	0.065	0.070	0.075	0.080	0.085	0.090	0.095	0.100	0.105	0.110
20.	0.013	0.020	0.027	0.033	0.040	0.047	0.054	0.060	0.067	0.074	0.080	0.087	0.094	0.100	0.107	0.114	0.120	0.127	0.134	0.140	0.147
25.	0.017	0.025	0.033	0.042	0.050	0.059	0.067	0.075	0.084	0.092	0.100	0.109	0.117	0.125	0.134	0.142	0.151	0.159	0.167	0.176	0.184
30.	0.020	0.030	0.040	0.050	0.060	0.070	0.080	0.090	0.100	0.110	0.120	0.130	0.141	0.151	0.161	0.171	0.181	0.191	0.201	0.211	0.221
35.	0.023	0.035	0.047	0.059	0.070	0.082	0.094	0.105	0.117	0.129	0.141	0.152	0.164	0.176	0.187	0.199	0.211	0.222	0.234	0.246	0.258
40.	0.027	0.040	0.054	0.067	0.080	0.094	0.107	0.120	0.134	0.147	0.161	0.174	0.187	0.201	0.214	0.228	0.241	0.254	0.268	0.281	0.294
45.	0.030	0.045	0.060	0.075	0.090	0.105	0.120	0.136	0.151	0.166	0.181	0.196	0.211	0.226	0.241	0.256	0.271	0.286	0.301	0.316	0.331
50.	0.033	0.050	0.067	0.084	0.100	0.117	0.134	0.151	0.167	0.184	0.201	0.218	0.234	0.251	0.268	0.284	0.301	0.318	0.335	0.351	0.368
55.	0.037	0.055	0.074	0.092	0.110	0.129	0.147	0.166	0.184	0.202	0.221	0.239	0.258	0.276	0.294	0.313	0.331	0.350	0.368	0.387	0.405
60.	0.040	0.060	0.080	0.100	0.120	0.141	0.161	0.181	0.201	0.221	0.241	0.261	0.281	0.301	0.321	0.341	0.361	0.382	0.402	0.422	0.442
65.	0.044	0.065	0.087	0.109	0.131	0.152	0.174	0.196	0.218	0.239	0.261	0.283	0.305	0.326	0.348	0.370	0.392	0.413	0.435	0.457	0.479
70.	0.047	0.070	0.094	0.117	0.141	0.164	0.187	0.211	0.234	0.258	0.281	0.305	0.328	0.351	0.375	0.398	0.422	0.445	0.469	0.492	0.515
75.	0.050	0.075	0.100	0.126	0.151	0.176	0.201	0.226	0.251	0.276	0.301	0.326	0.351	0.377	0.402	0.427	0.452	0.477	0.502	0.527	0.552
80.	0.054	0.080	0.107	0.134	0.161	0.187	0.214	0.241	0.268	0.295	0.321	0.348	0.375	0.402	0.428	0.455	0.482	0.509	0.536	0.562	0.589
85.	0.057	0.085	0.114	0.142	0.171	0.199	0.228	0.256	0.285	0.313	0.341	0.370	0.398	0.427	0.455	0.484	0.512	0.541	0.569	0.598	0.626
90.	0.060	0.090	0.120	0.151	0.181	0.211	0.241	0.271	0.301	0.331	0.362	0.392	0.422	0.452	0.482	0.512	0.542	0.573	0.603	0.633	0.663
95.	0.064	0.095	0.127	0.159	0.191	0.222	0.254	0.286	0.318	0.350	0.381	0.413	0.445	0.477	0.509	0.540	0.572	0.604	0.636	0.668	0.700
100.	0.067	0.100	0.134	0.167	0.201	0.234	0.268	0.301	0.335	0.369	0.402	0.436	0.469	0.503	0.536	0.570	0.603	0.637	0.670	0.704	0.738
105.	0.071	0.105	0.141	0.176	0.211	0.246	0.281	0.316	0.351	0.387	0.422	0.457	0.492	0.527	0.563	0.598	0.633	0.668	0.703	0.739	0.774
110.	0.074	0.111	0.147	0.184	0.221	0.258	0.295	0.331	0.368	0.405	0.442	0.479	0.516	0.553	0.590	0.626	0.663	0.700	0.737	0.774	0.811
115.	0.077	0.116	0.154	0.193	0.231	0.270	0.308	0.347	0.385	0.424	0.462	0.501	0.539	0.578	0.617	0.655	0.694	0.732	0.771	0.809	0.847
120.	0.080	0.121	0.161	0.201	0.241	0.282	0.322	0.362	0.402	0.442	0.482	0.522	0.563	0.603	0.643	0.683	0.723	0.763	0.803	0.844	0.884
125.	0.084	0.126	0.167	0.209	0.251	0.293	0.335	0.377	0.419	0.461	0.502	0.544	0.586	0.628	0.670	0.712	0.754	0.795	0.837	0.879	0.921
130.	0.087	0.131	0.174	0.218	0.261	0.305	0.348	0.392	0.436	0.479	0.523	0.566	0.610	0.653	0.697	0.740	0.784	0.827	0.871	0.914	0.958
135.	0.090	0.136	0.181	0.226	0.271	0.317	0.362	0.407	0.452	0.497	0.543	0.588	0.633	0.678	0.723	0.769	0.814	0.859	0.904	0.950	0.995
140.	0.094	0.141	0.188	0.235	0.281	0.328	0.375	0.422	0.469	0.516	0.563	0.610	0.657	0.703	0.750	0.797	0.844	0.891	0.938	0.985	1.032
145.	0.097	0.146	0.194	0.243	0.291	0.340	0.389	0.437	0.486	0.534	0.583	0.632	0.680	0.729	0.777	0.826	0.874	0.923	0.971	1.020	1.069
150.	0.101	0.151	0.201	0.251	0.302	0.352	0.402	0.452	0.503	0.553	0.603	0.653	0.704	0.754	0.804	0.854	0.905	0.955	1.005	1.055	1.106
155.	0.104	0.156	0.208	0.260	0.312	0.363	0.415	0.467	0.519	0.571	0.623	0.675	0.727	0.779	0.831	0.883	0.935	0.987	1.039	1.091	1.143
160.	0.107	0.161	0.215	0.268	0.322	0.375	0.429	0.482	0.536	0.590	0.643	0.697	0.751	0.804	0.858	0.911	0.965	1.019	1.072	1.126	1.179
170.	0.114	0.171	0.228	0.285	0.342	0.399	0.456	0.513	0.570	0.627	0.684	0.741	0.797	0.854	0.911	0.968	1.025	1.082	1.139	1.196	1.253
180.	0.121	0.181	0.241	0.302	0.362	0.422	0.483	0.543	0.603	0.664	0.724	0.784	0.845	0.905	0.965	1.025	1.086	1.146	1.206	1.267	1.327
190.	0.127	0.191	0.255	0.318	0.382	0.446	0.509	0.573	0.637	0.700	0.764	0.828	0.892	0.955	1.019	1.083	1.146	1.210	1.274	1.337	1.401
200.	0.134	0.201	0.268	0.335	0.402	0.469	0.536	0.603	0.670	0.737	0.804	0.871	0.939	1.006	1.073	1.140	1.207	1.274	1.341	1.408	1.475
210.	0.141	0.211	0.282	0.352	0.422	0.493	0.563	0.634	0.704	0.774	0.845	0.915	0.986	1.056	1.127	1.197	1.267	1.338	1.408	1.479	1.549
220.	0.148	0.222	0.295	0.369	0.443	0.516	0.590	0.664	0.738	0.811	0.885	0.959	1.033	1.106	1.180	1.254	1.328	1.401	1.475	1.549	1.623
230.	0.154	0.231	0.308	0.386	0.463	0.540	0.617	0.694	0.771	0.848	0.925	1.003	1.080	1.157	1.234	1.311	1.388	1.465	1.542	1.620	1.697
240.	0.161	0.241	0.322	0.402	0.483	0.563	0.644	0.724	0.805	0.885	0.966	1.046	1.127	1.207	1.288	1.368	1.449	1.529	1.610	1.690	1.771
250.	0.168	0.252	0.335	0.419	0.503	0.587	0.671	0.755	0.838	0.922	1.006	1.090	1.174	1.258	1.342	1.425	1.509	1.593	1.677	1.761	1.845
260.	0.174	0.262	0.349	0.436	0.523	0.610	0.698	0.785	0.872	0.959	1.046	1.134	1.221	1.308	1.395	1.483	1.570	1.657	1.744	1.831	1.919
270.	0.181	0.272	0.362	0.453	0.543	0.634	0.725	0.815	0.906	0.996	1.087	1.177	1.268	1.359	1.449	1.540	1.630	1.721	1.811	1.902	1.993
280.	0.188	0.282	0.376	0.470	0.564	0.658	0.752	0.845	0.939	1.033	1.127	1.221	1.315	1.409	1.503	1.597	1.691	1.785	1.879	1.973	2.067
290.	0.195	0.292	0.389	0.487	0.584	0.681	0.778	0.876	0.973	1.070	1.168	1.265	1.362	1.460	1.557	1.654	1.751	1.849	1.946	2.043	2.141
300.	0.201	0.302	0.403	0.503	0.604	0.705	0.805	0.906	1.007	1.107	1.208	1.309	1.409	1.510	1.611	1.711	1.812	1.913	2.013	2.114	2.215
310.	0.208	0.312	0.416	0.520	0.624	0.728	0.832	0.936	1.040	1.144	1.248	1.353	1.457	1.561	1.665	1.769	1.873	1.977	2.081	2.185	2.289
320.	0.215	0.322	0.430	0.537	0.644	0.752	0.859	0.967	1.074	1.181	1.289	1.396	1.504	1.611	1.719	1.826	1.933	2.041	2.148	2.256	2.363
330.	0.222	0.332	0.443	0.554	0.665	0.775	0.886	0.997	1.108	1.219	1.329	1.440	1.551	1.662	1.772	1.883	1.994	2.105	2.216	2.326	2.437
340.	0.228	0.342	0.457	0.571	0.685	0.799	0.913	1.027	1.141	1.256	1.370	1.484	1.598	1.712	1.826	1.940	2.055	2.169	2.283	2.397	2.511
350.	0.235	0.353	0.470	0.588	0.705	0.823	0.940	1.058	1.175	1.293	1.410	1.528	1.645	1.763	1.880	1.998	2.115	2.233	2.350	2.468	2.585
360.	0.242	0.363	0.484	0.604	0.725	0.846	0.967	1.088	1.209	1.330	1.451	1.572	1.692	1.813	1.934	2.055	2.176	2.297	2.418	2.539	2.660
370.	0.249	0.373	0.497	0.621	0.746	0.870	0.994	1.118	1.243	1.367	1.491	1.615	1.740	1.864	1.988	2.112	2.237	2.361	2.485	2.609	2.734

TABLE 2.8 Calculations for Ellipsoidal and Torispherical Flanged and Dished Heads under Internal Pressure

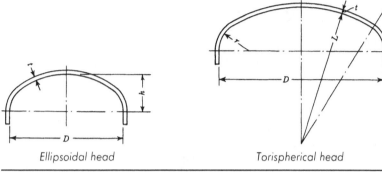

Ellipsoidal head   Torispherical head

Equations	Remarks	Code reference
Ellipsoidal Heads under Internal Pressure		
$t = \dfrac{PD}{2SE - 0.2P}$ or $P = \dfrac{2SEt}{D + 0.2t}$	Equations to be used when ratio of the major axis, D, to the minor axis, h, is 2:1 (see illustration above). For other ratios, see Appendix 1-4	Par. UG-32(d) Appendix 1 (also footnotes)
where t = minimum thickness, in P = internal pressure, psi S = allowable stress, psi E = minimum joint efficiency, percent D = inside diameter of head skirt, in	Minimum thickness (after forming)	Pars. UG-32(a) and (b)
	Allowable percentages of stress values and joint efficiencies for material depend upon type of radiography—full, spot, or no radiography	Pars. UW-11, UW-12 Pars. UW-51, UW-52
	Required skirt length	Pars. UG-32(l) and (m) Fig. UW-13.1
	Attachment of heads to shells	Fig. UW-13
	Tolerances for formed heads	Par. UG-81
	Cold-formed heads of P-1, Group 1 and 2 material may require heat treatment	Par. UCS-79

TABLE 2.8 Calculations for Ellipsoidal and Torispherical Flanged and Dished Heads under Internal Pressure (*Continued*)

Equations	Remarks	Code reference
Torispherical Heads under Internal Pressure		
$t = \dfrac{0.885PL}{SE - 0.1P}$ or $P = \dfrac{SEt}{0.885L + 0.1t}$ in which L equals the inside crown radius (in inches) and where other unknowns are same as above	ASME head is one in which the knuckle radius r is 6 percent of the inside crown radius L	Par. UG-32(e) Appendix 1 (also footnotes)
	Minimum thickness (after forming)	Pars. UG-32(a) and (b); also footnotes
	Allowable percentages of stress values and joint efficiencies for material depend upon type of radiography— full, spot, or no radiography	Pars. UW-11, UW-12, Pars. UW-51 UW-52
	Required skirt length	Pars. UG-32(l) and (m)
	Attachment of heads to shells	Fig. UW-13
	Tolerances for formed heads	Par. UG-81
	Cold-formed heads of P-1, Group 1 and 2 material may require heat treatment	Par. UCS-79

It should be remembered that the *ASME Boiler and Pressure Vessel Code* is a progressive, viable code and that when necessary, changes are made. Therefore the latest edition of the Code should always be used.

Use of Pressure–Thickness Charts for Flat Heads and Bolted Flat Cover Plates (Figs. 2.9 to 2.11)

The required head thickness may be found by the following steps:

1. Locate the required inside diameter (for flat heads) or bolt circle (for flat cover plates) at the bottom of the chart.
2. Read up vertically to the appropriate pressure line.
3. Read horizontally to the required thickness.

TABLE 2.9 Calculation for Ellipsoidal and Torispherical Flanged and Dished Heads under External Pressure

Equations	Remarks	Code references
Compute thicknesses by appropriate procedure in Code Pars. UG-33(a), (d), and (e). Use greater value as indicated	Carbon-steel heads	Pars. UG-33, UCS-33 Appendix L, Code Section II, Part D, Subpart D
	Nonferrous metal heads	Pars. UG-33, UNF-33 Appendix L, Code Section II, Part D, Subpart D
	High-alloy steel heads	Pars. UG-33, UHA-31 Appendix L, Code Section II, Part D, Subpart D
Thickness of cast-iron heads shall not be less than that for plus heads nor less than 1 percent of inside diameter of head skirt	Cast-iron heads	Pars. UCI-32, UCI-33 Appendix L
	Clad-steel heads	Pars. UCL-26, UCS-33
	Cast ductile	Pars. UCD-32, UCD-33
	Iron heads	Appendix L
	Cladding may be included in design calculations	Pars. UCL-23(b), UCL-23(c)
	Stress values of materials (use appropriate table)	Code Section II, Part D, Subpart D
	See Code Appendix L for application of Code formulas and rules	Appendix L

Other problems may be similarly solved. Charts are based on an allowable stress of 13,800 psi. For other stress values, multiply the chart thickness by the appropriate constant in the following table:

Stress, psi	11,000	11,300	12,500	15,000	16,300	17,500	18,800
Constant	1.120	1.104	1.050	0.959	0.920	0.888	0.856

Corrosion allowances are not included in either chart. Thickness equations for flat heads and bolted flat cover plates may be found in Table 2.10. The ASME Code establishes a maximum allowable working

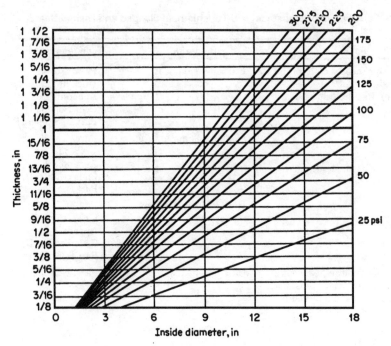

Figure 2.9 Flat head thickness chart, $C=0.50$.

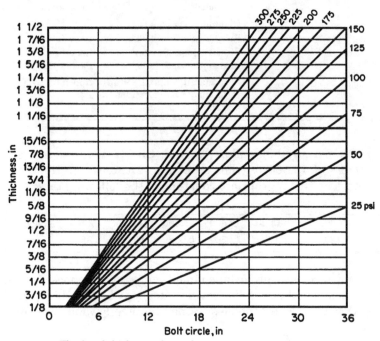

Figure 2.10 Flat head thickness chart, $C=0.162$.

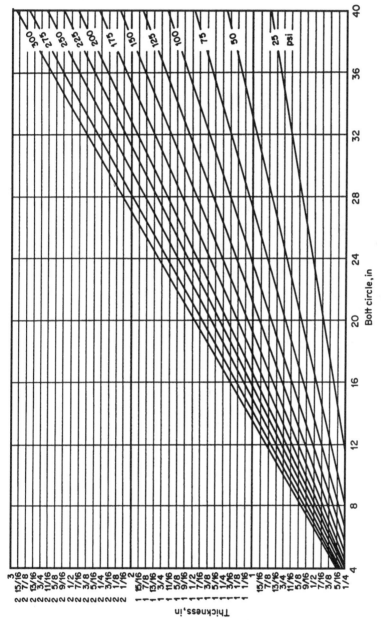

Figure 2.11 Thickness chart for bolted flat cover plates with full-face gasket, $C=0.25$.

TABLE 2.10 Calculations for Unstayed Flat and Bolted Heads and Bolted Flange Connections

Equations	Remarks	Code reference
Unstayed Flat Heads		
Circular: $$t = d\sqrt{\frac{CP}{SE}}$$ or $$P = \frac{St^2E}{d^2C}$$	Values of C and d	Par. UG-34 Fig. UG-34
	Provision if other than full face gasket is used	Par. UG-34
	Details of welded joints	Fig. UG-34
Noncircular: $$t = d\sqrt{\frac{ZCP}{SE}}$$	Noncircular heads	Pars. UG-34, U-2(c)
Nomenclature:	Modify formulas for unstayed flat heads and covers attached by bolts causing edge moments	Par. UG-34, Fn. 21
Bolted Heads (Spherically Dished Covers)		
Use equations in Code Appendix 1		Appendix 1-6 Fig. 1-6
Bolted Flange Connections		
Use equations in Code Appendix 2	General recommendations	Pars. UG-11, UG-44 Appendix 2
	Design considerations	Appendixes 2 and 5
	Required loadings, moments, and stresses	Appendix 2
	Types of flanges	Fig. 2-4
	Optional flanges limited to 300 psi pressure	Appendix 2-4
	Hubbed flanges machined from plate—not allowed	Appendix 2-2(d)
	Blind flanges conforming to an accepted standard and attached by bolting as shown in Figs. UG-34(J) and (K) may be used for the appropriate pressure–temperature rating	Pars. UG-34, UG-11(a)(1), UG-11(a)(2)
	Bolt loads	Appendix 2-5
	Reverse flanges	Appendix 2-13

TABLE 2.11 Maximum Allowable Working Stress per Bolt (Standard Thread and 8-Thread Series)

Bolt diameter, in	Number of threads per inch	Area at bottom of thread, sq. in	Alllowable material stress, psi			
			7000*	16,250	18,750	20,000
			Stress per bolt, psi			
Standard Thread						
½	13	0.126	882	2,047	2,362	2,520
⅝	11	0.202	1,414	3,282	3,787	4,040
¾	10	0.302	2,114	4,907	5,662	6,040
⅞	9	0.419	2,933	6,808	7,856	8,380
1	8	0.551	3,857	8,953	10,331	11,020
1⅛	7	0.693	4,851	11,261	12,993	13,860
1¼	7	0.890	6,230	14,462	16,687	17,800
1⅜	6	1.054	7,378	17,127	19,762	21,080
1½	6	1.294	9,058	21,027	24,262	25,880
1⅝	5½	1.515	10,605	24,618	28,406	30,300
1¾	5	1.744	12,208	28,340	32,700	34,880
1⅞	5	2.049	14,343	33,296	38,418	40,980
2	4½	2.300	16,100	37,375	43,125	46,000
2¼	4½	3.020	21,140	49,075	56,625	60,400
2½	4	3.715	26,005	60,368	69,656	74,300
2¾	4	4.618	32,326	75,042	86,587	92,360
3	4	5.620	39,340	91,325	105,375	112,400
8-Thread Series						
1⅛	8	0.728	5,096	11,830	13,650	14,560
1¼	8	0.929	6,503	15,096	17,418	18,580
1⅜	8	1.155	8.085	18,768	21,656	23,100
1½	8	1.405	9,835	22,831	26,343	28,100
1⅝	8	1.680	11,760	27,300	31,500	33,600
1¾	8	1.980	13,860	32,175	37,125	39,600
1⅞	8	2.304	16,128	37,440	43,200	46,080
2	8	2.652	18,564	43,095	49,725	53,040
2¼	8	3.423	23,961	55,623	64,181	68,460
2½	8	4.292	30,044	69,745	80,475	85,840
2¾	8	5.259	36,813	85,458	98,606	105,180
3	8	6.324	44,268	102,765	118,575	124,480

*Maximum temperature, 450°F.

stress for each bolt which will be used when bolted heads or flanges are to be employed. Table 2.11 sets forth design stress tables and is included as a handy reference guide when the designer wishes to establish the maximum allowable stress applied for each bolt.

Various Other Designs for Heads

The ASME Code sets forth requirements and allows the designer the latitude to design pressure vessels with heads of a "nonstandard"

TABLE 2.12 Calculations for Conical and Toriconical Heads

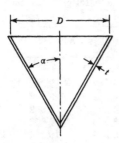
Conical head (no knuckle)

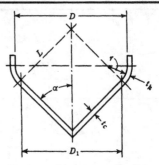
Toriconical head (with knuckle)

Equations	Remarks	Code reference
Conical Heads under Internal Pressure		
$t = \dfrac{PD}{2 \cos \alpha \, (SE - 0.6P)}$	Equations are used when one-half the apex angle (angle α) does not exceed 30°	Par. UG-32(g)
$P = \dfrac{2SEt \cos \alpha}{D + 1.2t \cos \alpha}$	Use toriconical head when angle α exceeds 30°	Par. UG-32(h)
where t = minimum thickness, in P = allowable pressure, psi D = inside diameter, in S = allowable stress, psi E = minimum joint efficiency, percent $\angle \alpha$ = one-half the apex angle of the cone at center line	A stiffening ring shall be provided when required Excess thickness, less corrosion allowance of cones or shells, may be credited to the required area of a stiffening ring	Appendix 1-8 Table 1-8.1
Toriconical Heads under Internal Pressure		
Cone thickness: $t_c = \dfrac{PD_1}{2 \cos \alpha \, (SE - 0.6P)}$ $P = \dfrac{2SEt \cos \alpha}{D_1 + 1.2t \cos \alpha}$	Inside knuckle radius shall not be less than 6 percent of outside diameter of head skirt nor less than 3 times thickness of knuckle	Par. UG-32(h) Appendix 1-4

TABLE 2.12 Calculations for Conical and Toriconical Heads (*Continued*)

Equations	Remarks	Code reference
Toriconical Heads under Internal Pressure (Cont.)		
where D_1 = inside cone diameter at point of tangency to knuckle, in	Value of factor M Required skirt length	Table 1-4.2 Pars. UG-32(l), UG-32(m)
Knuckle thickness: $$t_k = \frac{PLM}{2SE - 0.2P}$$ in which: $$L = \frac{D_1}{2 \cos \alpha}$$ M = factor depending on head proportion, L/r		Par. UG-32

design. This is not to say that the average ASME Code shop will not see such designs; however, they should not rule out the possibility. Such designs include conical, toriconical, and hemispherical. Table 2.12 simplifies the Code design requirements of the conical and toriconical heads under internal pressure and lists the Code design calculation and the Code references.

The hemispherical head is a more common design and is primarily used for gas pressure tanks such as propane tanks. Table 2.13 lists the equations and the Code references for the designs of such heads.

Allowable Pressure, Pitch, and Thickness

Areas of flat surfaces under internal pressure which are not supported have the tendency while under pressure to deform, bulge, and try to form a circle. An example of this phenomenon can be seen by blowing air into a cereal box. As pressure is applied to the box, the box becomes tubular in shape, and if one could blow hard enough, the box would rupture. The Code recognizes this problem and has included requirements for adding bracing or stays between these flat surfaces. Table 2.14 states the equations for braced and stayed surfaces and gives the Code references which apply.

Table 2.15 gives handy and quick thickness, pressure, and pitch requirements for stayed surfaces of stayed vessels built according to the ASME Pressure Vessel Code. To use the table, do the following:

TABLE 2.13 Calculations for Hemispherical Heads

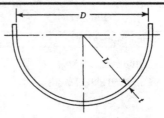

Equations	Remarks	Code reference
Heads under Internal Pressure		
$t = \dfrac{PL}{2SE - 0.2P}$	Equations are used when t is less than $0.356L$ or P is less than $0.655SE$	Par. UG-32(f)
or $P = \dfrac{2SEt}{L + 0.2t}$	Thick hemispherical heads (P exceeds $0.356R$ or P exceeds $0.665SE$)	Appendix 1-3
where t = minimum thickness, in P = allowable pressure, psi L = inside radius, in S = allowable stress, psi E = minimum joint efficiency, percent	Stress values of materials (use appropriate table) Required skirt length and thickness	Code Section II, Part D, Tables UCI-23, UCD-23 Table ULT-23 Par. UG-32(L)
Heads under External Pressure		
	Thickness determined in same manner as for a spherical shell	Pars. UG-33(c), UG-28(d)
	Example of Code calculations	Appendix L-6

1. Locate the desired pressure block in the table.
2. Read up vertically to the appropriate pitch.
3. Read horizontally to the required thicknesses.

Table 2.15 is based on an allowable stress of 16,000 psi. For pressures using other stress values, divide the table pressure by the appropriate constant for the stress desired:

TABLE 2.14 Calculations for Braced and Stayed Surfaces

Equations	Remarks	Code reference
$t = p\sqrt{\dfrac{P}{SC}}$	Measurement of pitch p and values for C	Par. UG-47(a)
$P = \dfrac{t^2 SC}{p^2}$	Minimum thickness of plate that is stayed is ⁵⁄₁₆ in except in welded construction and cylindrical or spherical outer shell plates	Pars. UG-47(b) UW-19
where t = minimum plate thickness, in P = allowable pressure, psi p = maximum pitch, in S = allowable stress, psi C = a factor depending upon the plate thickness and type of stay	If stayed jackets extend completely around a cylindrical shell or spherical vessel or completely cover a formed head, the required thickness shall meet the requirements of par. (a) and the applicable requirements of UG-27(c) and UG-31. For any nozzles one should consult UG-37(d)2.	Par. UG-47(c)
	If two plates are stayed and only one requires staying, the value of C is governed by the thickness of plate requiring staying	Par. UG-47(d)
	Welded stayed construction	Par. UW-19 Fig. UW-19.1
	Maximum pitch for welded and screwed stays with the ends riveted over is 8½ in; for welded-in staybolts, the pitch must not exceed 15 times the diameter of the staybolt	Par. UG-47(f)
	End requirements, location, and dimensions of staybolts	Par. UG-48 to -50
	Hole sizes for screw stays	Par. UG-83
	For allowable pressures and thicknesses see Tables 2.15 and 2.16, this volume	
	Dimpled or embossed assemblies	Appendix 17
	When Appendix 17 is used, state appendix identification and first paragraph number on the data report	

TABLE 2.15 Allowable Pressure, Pitch, and Thickness for Braced and Stayed Surfaces
Table based on allowable stress of 16,000 psi

Thickness, in	\multicolumn{21}{c}{Pitch of stays, in*}

Thickness, in	5	5½	6	6½	7	7½	8	8½	9	9½	10	10½	11	11½	12	12½	13	13½	14	14½	15	Thickness, in†
	\multicolumn{21}{c}{Allowable pressure, psi‡}	
3/16	47	39	33	28	24	21	18	16	...	...	...	...	...	...	...	...	...	...	...	...	...	3/16
1/4	84	69	58	49	43	37	33	29	26	23	21	19	17	15	...	...	...	...	...	...	...	1/4
5/16	131	108	91	78	67	58	51	45	40	36	32	29	27	25	23	21	19	18	17	16	...	5/16
3/8	189	156	131	112	96	84	74	65	58	52	47	43	39	35	32	30	28	26	24	22	21	3/8
7/16	257	212	179	152	131	114	100	89	79	71	64	58	53	48	44	41	38	35	32	30	28	7/16
1/2	352	290	244	208	180	156	138	122	109	98	88	80	73	66	61	56	52	48	45	41	39	1/2
9/16	445	368	309	264	227	198	174	154	137	123	111	101	92	84	77	71	66	61	57	53	49	9/16
5/8	550	454	382	325	281	244	215	190	170	152	138	125	114	104	95	88	81	75	70	65	51	5/8
11/16	666	549	462	394	340	296	260	230	205	180	166	151	138	126	116	106	98	91	85	79	74	11/16
3/4	792	653	550	469	404	352	309	274	244	219	198	180	164	150	138	127	117	109	101	94	88	3/4

*Maximum pitch shall be 8½ in except for welded-in staybolts; the pitch must not exceed 15 times the diameter of the staybolt. (See Code Par. UG-47.)

†Welded stays as shown in Code Fig. UW-19.2 are limited to plate thicknesses not to exceed ½ in.

‡Welded stays as shown on Code Fig. UW-19.2 *ASME Pressure Vessel Code* limited to 300-psi pressure. (See Code Par. UW-19.)

$$t = p\sqrt{\frac{P}{SC}} \qquad P = \frac{t^2 SC}{p^2}$$

t = minimum thickness of plate
P = maximum allowable working pressure
p = maximum pitch between centers of stays
S = maximum allowable stress, 16,000 psi
C = 2.1 for plates not over 7/16 in
C = 2.2 for plates more than 7/16 in

Stress, psi	11,300	12,500	13,800	15,000	16,300	17,500	18,800
Constant	1.416	1.28	1.159	1.066	0.981	0.914	0.851

Table 2.16 is based on an allowable stress of 13,800 psi. For pressures using other stress values, divide the table pressure by the appropriate constant for the stress desired:

Stress, psi	11,300	12,500	15,000	16,300	17,500	18,800
Constant	1.221	1.104	0.920	0.846	0.788	0.734

Simplified Calculations for Reinforcement of Openings (Fig. 2.12)

To determine whether an opening is adequately reinforced, it is first necessary to determine whether the areas of reinforcement available will be sufficient without the use of a pad. Figure 2.12 includes a schematic diagram of an opening showing the involved areas and also all the required calculations. The total cross-sectional area of reinforcement required (in square inches) is indicated by the letter A, which is equal to the diameter (plus corrosion allowance) times the required thickness. The area of reinforcement available without a pad includes

1. The area of excess thickness in the shell or head, A_1
2. The area of excess thickness in the nozzle wall, A_2
3. The area available in the nozzle projecting inward, A_3
4. The cross-sectional area of welds, A_4

If $A_1 + A_2 + A_3 + A_4 \geq A$, the opening is adequately reinforced.

If $A_1 + A_2 + A_3 + A_4 < A$, a pad is needed.

If the reinforcement is found to be inadequate, then the area of pad needed (A_5) may be calculated as follows:

$$A_5 = A - (A_1 + A_2 + A_3 + A_4)$$

If a pad is used, the factor ($2.5t_n$) in the equation for A_2 in Fig. 2.12 is measured from the top surface of the pad and therefore becomes ($2.5t_n + T_p$). The area A_2 must be recalculated on this basis and the smaller value again used. Then:

If $A_1 + A_2 + A_3 + A_4 + A_5 = A$, the opening is adequately reinforced.

TABLE 2.16 Allowable Pressure, Pitch, and Thickness for Braced and Stayed Surfaces

Table based on an allowable stress of 13,800 psi; C = 2.1 for plates 7/16 in thick, and less, and 2.2 for thicker plates (see equations in Table 2.14)

Thickness, in‡	Pitch of stays, in*																				
	5	5½	6	6½	7	7½	8	8½	9	9½	10	10½	11	11½	12	12½	13	13½	14	14½	15
	Allowable pressure, psi†																				
3/16	40	36	28	24	20	18	15														
1/4	72	60	50	42	36	32	28	25	22	20	18	16									
5/16	112	94	79	66	57	50	44	39	34	31	28	25	23	21	19	18	16	15			
3/8	162	133	112	96	83	72	63	56	50	45	40	36	33	30	28	26	24	22	20	19	18
7/16	220	182	153	131	112	98	86	76	68	61	55	50	45	41	38	35	32	30	28	26	24
1/2	302	250	210	179	154	134	118	104	93	83	75	68	62	57	52	48	44	41	38	35	33
9/16	382	313	265	225	194	170	148	131	118	105	95	86	78	72	66	61	56	52	48	45	42
5/8	472	390	327	279	240	210	184	162	145	130	118	106	97	89	82	76	70	65	60	56	52
11/16	572	473	400	340	294	255	225	198	178	159	144	129	118	109	100	92	85	79	73	68	64
3/4	681	568	473	403	347	302	265	235	210	188	171	153	141	128	118	109	100	93	87	81	75

* Maximum pitch shall be 8½ in except for welded-in staybolts, for which the pitch must not exceed 15 times the diameter of the staybolt (Code Par. UG-47).
† Pressures on welded stays shown in Code Fig. UW-19.2 limited to 300 psi (Code Par. UW-19).
‡ Thicknesses of plates for welded stays shown in Code Fig. UW-19.2 shall not exceed ½ inch.

Descriptive Guide to the ASME Code Pressure Vessels

Shell or head data		Nozzle data†	
		P = pressure, psi	
		S = allowable stress, psi	
		R_n = inside radius of nozzle, in	
		E_1 = joint efficiency, percent	
		$t_{rn} = \dfrac{PR_n}{SE_1 - 0.6P}$	
t = actual thickness of shell or head (minus corrosion)		t_n = actual thickness of nozzle (minus corrosion)	
t_r = calculated thickness of shell or head		t_{rn} = calculated thickness of nozzle	
$E_1 t - Ft_r$ = excess thickness in shell or head		$t_n - t_{rn}$ = excess thickness in nozzle	

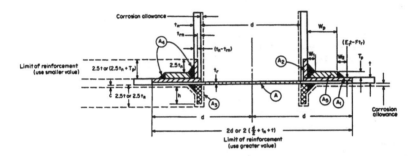

Area of reinforcement required†	$A = dt_r F$	= _____
Area of excess thickness in shell or head (use greater value)	$A_1 = (E_1 t - Ft_r)d$ or $A_1 = 2(E_1 t - Ft_r)(t + t_n)$	= _____
Area available in nozzle projecting outward; use smaller value	$A_2 = 2(2.5t_n)^*(t_n - t_{rn})f_r$ or $A_2 = 2(2.5t)(t_n - t_{rn})f_r$	= _____
Area available in nozzle projecting inward	$A_3 = 2(t_n - c)f_r \times h$	= _____
Cross-sectional area of welds	$A_4 = 2\left[\dfrac{(W_1)^2 + (W_2)^2}{2}\right] \times f_r =$	_____
Area of reinforcement available without pad	$(A_1 + A_2 + A_3 + A_4)$	= _____

If $A_1 + A_2 + A_3 + A_4 \geq A$ Opening is adequately reinforced
If $A_1 + A_2 + A_3 + A_4 < A$ Opening is not adequately reinforced so reinforcing element must be added or thicknesses increased

Area in pad	$A_5 = 2W_p T_p$	= _____
	Total area available	_____

* If reinforcing pad is used, the factor $2.5t_n$ becomes $(2.5t_n + T_p)$.

† For cases when the allowable stress of the nozzle of reinforcing element is less than the allowable stress of the vessel, refer to Code Par. UG-41(a) and Appendix L, latest Addenda, for consideration of this effect.

Figure 2.12 Calculation sheet for reinforcement of openings.

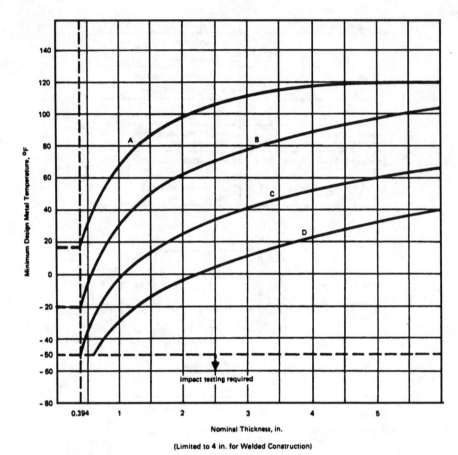

Figure 2.13 Impact-test exemption curve, model 1. (*Courtesy of The American Society of Mechanical Engineers.*)

Other symbols for the area equations in Fig. 2.12 (all values except E_1 and F are in inches) are as follows:

d	diameter in the plane under consideration of the finished opening in its corroded condition
t	nominal thickness of shell or head, less corrosion allowance
t_r	required thickness of shell or head as defined in Code Par. UG-37
t_{rn}	required thickness of a seamless nozzle wall
T_p	thickness of reinforcement pad
W_p	width of reinforcement pad
t_n	nominal thickness of nozzle wall, less corrosion allowance
W_1	cross-sectional area of weld
W_2	cross-sectional area of weld

GENERAL NOTES ON ASSIGNMENT OF MATERIALS TO CURVES:
(a) Curve A applies to:
 (1) all carbon and all low alloy steel plates, structural shapes, and bars not listed in Curves B, C, and D below;
 (2) SA-216 Grades WCB and WCC if normalized and tempered or water-quenched and tempered; SA-217 Grade WC6 if normalized and tempered or water-quenched and tempered.
(b) Curve B applies to:
 (1) SA-216 Grade WCA if normalized and tempered or water-quenched and tempered
   SA-216 Grades WCB and WCC for thicknesses not exceeding 2 in. (51 mm), if produced to fine grain practice and water-quenched and tempered
   SA-217 Grade WC9 if normalized and tempered
   SA-285 Grades A and B
   SA-414 Grade A
   SA-515 Grade 60
   SA-516 Grades 65 and 70 if not normalized
   SA-612 if not normalized
   SA-662 Grade B if not normalized
   SA/EN 10028-2 P295GH as-rolled;
 (2) except for cast steels, all materials of Curve A if produced to fine grain practice and normalized which are not listed in Curves C and D below;
 (3) all pipe, fittings, forgings and tubing not listed for Curves C and D below;
 (4) parts permitted under UG-11 shall be included in Curve B even when fabricated from plate that otherwise would be assigned to a different curve.
(c) Curve C
 (1) SA-182 Grades 21 and 22 if normalized and tempered
   SA-302 Grades C and D
   SA-336 F21 and F22 if normalized and tempered
   SA-387 Grades 21 and 22 if normalized and tempered
   SA-516 Grades 55 and 60 if not normalized
   SA-533 Grades B and C
   SA-662 Grade A;
 (2) all material of Curve B if produced to fine grain practice and normalized and not listed for Curve D below.
(d) Curve D
   SA-203
   SA-508 Grade 1
   SA-516 if normalized
   SA-524 Classes 1 and 2
   SA-537 Classes 1, 2, and 3
   SA-612 if normalized
   SA-662 if normalized
   SA-738 Grade A
   SA-738 Grade A with Cb and V deliberately added in accordance with the provisions of the material specification, not colder than −20°F (−29°C)
   SA-738 Grade B not colder than −20°F (−29°C)
   SA/AS 1548 Grades 7-430, 7-460, and 7-490 if normalized
   SA/EN 10028-2 P295GH if normalized [see Note (g)(3)]
   SA/EN 10028-3 P275NH

General Notes and Notes continue on next page

Figure 2.13 (*Continued*)

E_1 the joint efficiency obtained from Code Table UW-12 when any part of the opening passes through any other welded joint [$E_1 = 1$, when an opening is in the plate or when the opening passes through a circumferential point in a shell or cone (exclusive of head-to-shell joints)]

F a correction factor which compensates for the variation in pressure stresses on different planes with respect to the axis of a vessel. A value of 1.00 is used for F in all cases except when the opening is integrally reinforced. If integrally reinforced, see Code Fig. UG-37.*

h distance the nozzle projects beyond the inner surface of the vessel wall, before corrosion allowance is added.

To correct for corrosion, deduct the specified allowance from shell thickness t and nozzle thickness t_n, but add twice its value to the diameter of the opening, d. [Note: The total load to be carried by attachment

---

*Reinforcement is considered integral when it is inherent in shell plate or nozzle. Reinforcement built up by welding is also considered integral. The installation of a reinforcement pad is not considered integral.

GENERAL NOTES ON ASSIGNMENT OF MATERIALS TO CURVES (CONT'D):

(e) For bolting and nuts, the following impact test exemption temperature shall apply:

Bolting

Spec. No.	Grade	Diameter, in. (mm)	Impact Test Exemption Temperature, °F (°C)
SA-193	B5	Up to 4 (107 mm), incl.	−20 (−29)
	B7	Up to 2½ in. (64 mm), incl.	−55 (−48)
		Over 2½ (64 mm) to 7 (178 mm), incl.	−40 (−40)
	B7M	Up to 2½ (64 mm), incl.	−55 (−48)
	B16	Up to 7 (178 mm), incl.	−20 (−29)
SA-307	B	All	−20 (−29)
SA-320	L7, L7A,	Up to 2½ (64 mm), incl.	See General Note (c) of Fig. UG-84.1
	L7M,	Up to 2½ (64 mm), incl.	See General Note (c) of Fig. UG-84.1
	L43	Up to 1 (25 mm), incl.	See General Note (c) of Fig. UG-84.1
SA-325	1	½ (13 mm) to 1½ (38 mm)	−20 (−29)
SA-354	BC	Up to 4 (102 mm), incl.	0 (−18)
SA-354	BD	Up to 4 (102 mm), incl.	+20 (−7)
SA-437	B4B, B4C	All diameters	See General Note (c) of Fig. UG-84.1
SA-449	...	Up to 3 (76 mm), incl.	−20 (−29)
SA-540	B21 Cl. All	All	Impact test required
SA-540	B22 Cl. 3	Up to 4 (102 mm), incl.	Impact test required
SA-540	B23 Cl. 1, 2	All	Impact test required
SA-540	B23 Cl. 3, 4	Up to 6 (152 mm), incl.	See General Note (c) of Fig. UG-84.1
SA-540	B23 Cl. 3, 4	Over 6 (152 mm) to 9½ (241 mm), incl.	Impact test required
SA-540	B23 Cl. 5	Up to 8 (203 mm), incl.	See General Note (c) of Fig. UG-84.1
SA-540	B24 Cl. 5	Over 8 (203 mm) to 9½ (241 mm), incl.	Impact test required
SA-540	B24 Cl. 1	Up to 6 (152 mm), incl.	See General Note (c) of Fig. UG-84.1
SA-540	B24 Cl. 1	Over 6 (152 mm) to 8 (203 mm), incl.	Impact test required
SA-540	B24 Cl. 2	Up to 7 (178 mm), incl.	See General Note (c) of Fig. UG-84.1
SA-540	B24 Cl. 2	Over 7 (178 mm) to 9½ (241 mm), incl.	Impact test required
SA-540	B24 Cl. 3, 4	Up to 8 (203 mm), incl.	See General Note (c) of Fig. UG-84.1
SA-540	B24 Cl. 3, 4	Over 8 (203 mm) to 9½ (241 mm), incl.	Impact test required
SA-540	B24 Cl. 5	Up to 9½ (241 mm), incl.	See General Note (c) of Fig. UG-84.1
SA-540	B24V Cl. 3	All	See General Note (c) of Fig. UG-84.1

Nuts

Spec. No.	Grade	Impact Test Exemption Temperature, °F (°C)
SA-194	2, 2H, 2HM, 3, 4, 7, 7M, and 16	−55 (−48)
SA-540	B21/B22/B23/B24/B24V	−55 (−48)

(f) When no class or grade is shown, all classes or grades are included.
(g) The following shall apply to all material assignment notes.
   (1) Cooling rates faster than those obtained by cooling in air, followed by tempering, as permitted by the material specification, are considered to be equivalent to normalizing or normalizing and tempering heat treatments.
   (2) Fine grain practice is defined as the procedure necessary to obtain a fine austenitic grain size as described in SA-20.
   (3) Normalized rolling condition is not considered as being equivalent to normalizing.

NOTES:
(1) Tabular values for this Figure are provided in Table UCS-66.
(2) Castings not listed in General Notes (a) and (b) above shall be impact tested.

Figure 2.13 (*Continued*)

welds may be calculated from the equation $W = S(A - A_1)$, in which S equals the allowable stress specified by Code Subsection C and Code Par. UW-15(c). For large openings in cylindrical shells, see Code Appendix 1-7. For openings in flat heads, see Code Par. UG-39. For weld sizes required, see Code Par. UW-16 and Code Fig. UW-16.1.] On all welded vessels built under column C of Code Table UW-12, 80 percent of the allowable stress value must be used in design formulas and calculations. Table 2.17 states the requirements for openings and the reinforcement of these openings as well as detailing the applicable Code paragraphs.

TABLE 2.17 Openings and Reinforced Openings

Remarks	Code reference
For vessels under internal pressure, reinforced area required, A, equals $d \times t_r$ (see Fig. 2.12 for simplified calculations)	Par. UG-37
For vessels under external pressure, reinforced area required need be only one-half that for internal pressure if shell thickness satisfies external-pressure requirements	Par. UG-37
Limits of reinforcement	Par. UG-40
Strength of reinforcement	Par. UG-41
Reinforcement of multiple openings	Par. UG-42
Welded connections	Pars. UW-15, UW-16 Appendix L-7
Openings in or adjacent to welds	Par. UW-14
Large openings in cylindrical shells	Appendix 1-7
Openings in flat heads	Par. UG-39
Single openings not requiring reinforcement	
3-in pipe size, vessel thickness ⅜ in or less	Par. UG-36
2-in pipe size, vessel thickness over ⅜ in	Par. UG-36
Openings in head or girth seam requiring radiography	Par. UW-14 and UG-37
Flued openings in formed heads	Par. UG-38 and Fig. UG-38

Cryogenic and Low-Temperature Vessels

Use of cryogenic and low-temperature pressure vessels is increasing rapidly, but there have been many defects caused by improper materials and poorly designed vessels, particularly those meant for low-temperature service. Many materials exposed to low temperatures lose their toughness and ductility. As the temperature goes down, the metal has a tendency to become brittle. This may cause brittle fracture at full stress.

Most materials used for service at normal temperatures have a marked decrease in impact resistance as temperatures decrease. These materials must have the proven ability to safely resist high-stress changes and shock loads. Only impact-tested materials able to withstand unavoidable stress concentrations should be used for cryogenic vessels. All sharp or abrupt transitions or changes of sections, corners, or notches should be eliminated in the design of the vessel and fabrication of its parts.

Manufacturers, purchasers, and users of low-temperature pressure vessels often misunderstand certain paragraphs in the *ASME Pressure Vessel Code*. Perhaps it might be helpful to clarify various paragraphs applying to low-temperature vessels and to indicate when impact tests are required. If such tests are necessary, each part of the vessel must be considered: shells, heads, nozzles, and reinforcing pads, as well as other pressure parts. Additionally, nonpressure parts welded to pressure parts shall also be considered. Therefore, to impact-test or not can be a large cost factor.

The 1987 Addenda of the *ASME Boiler and Pressure Vessel Code,* Section VIII, Division 1 mandated new toughness requirements and specified new minimum design metal temperature requirements. The −20°F temperature exemption of the pre-1987 Code has been eliminated. The new rules are extensive and largely eliminate the previous exemption from toughness requirements for vessels which may be pressurized at ambient temperatures. These new requirements caused manufacturers to review designs which had been used for many years in order to ascertain the impact of the new requirements on their product standard designs and financial cost factors. Table 2.18 lists the various Code requirements for low-temperature operations.

Impact-Test Exemptions

With cost a consideration in designing and manufacturing carbon and low-alloy steel pressure vessels, new design rules mandate stricter requirements for the notch toughness of certain materials. No longer does the ASME Code Section VIII, Division 1 allow an automatic exemption from impact testing for temperatures above −20°F. Today the designer must evaluate each component material used in the design of the vessel separately. This section will assist the designer to better understand the methods to employ when seeking relief from impact test requirements for carbon-steel vessel materials.

The first step the designer should take is to check which vessel part may be exempted from the impact-test requirement. The method to claim exemptions is to use one of the following three exemption models:

Exemption model 1. The designer should consult Code Par. UG-20(f). When all of the conditions of that paragraph are met, the impact testing requirements of Code Par. UG-84 are no longer mandatory.

It should be noted that there have been several changes to Section VIII impact testing rules since the 1997 edition of the Code, and they have become much more complex. The affected rules include the minimum design metal temperature (MDMT) calculation and the means of allowing a reduction to the MDMT for impact testing requirements.

The fundamental impact testing rule is that when the combination of thickness and MDMT of a material falls below the proper curve shown in Fig. 2.13 (Code Fig. UCS-66), impact testing is required, unless otherwise exempted in the Code.

An important reason for impact testing when it is clearly required is to avoid brittle fracture, a disastrous type of failure. If impact testing is

TABLE 2.18 Requirements for Low-Temperature Service

Remarks	Code reference
Some materials and welds do not require impact test; consult paragraphs in Code Subsection C to which material applies (see Table 2.1 of this volume)	Pars. UG-20, UCS-66, UCS-67, UCS-68 Figs. UCS-66, UCS-66.1 Pars. UHA-51, UCL-27
When impact tests are required weld metal and heat-affected zone must have impact tests.	Pars. UNF-65, UG-84 Appendix NF-6 Par. UHT-5, Part UF
Low-temperature vessels	Part ULT
Welded carbon-steel vessels must be postweld heat-treated unless exempted from impact tests	Pars. UCS-66, UCS-68
All longitudinal seams and girth seams must be double-welded butt joints or the equivalent	Par. UW-2
Remove backing strip on long seams if possible	Table UW-12
Vessels must be stamped according to Code requirements (see Fig. 2.18)	Par. UG-116
Austenitic stainless steels may not require impact tests providing weld procedure specifications also included qualified impact test plates of the welding procedure	Pars. UG-84, UHA-51, UHA-52
All joints of Category A must be Type 1 of Code Table UW-12	Pars. UW-2(b), UW-3
All joints of Categories B must be Type 1 or 2 of Code Table UW-12	Par. UW-2(b)
All joints of Categories C and D must be full penetration welds	Par. UW-2(b)
Brazed vessels	Pars. UB-22, UG-84
High-alloy vessels	Par. UHA-105
Nonferrous vessels	Pars. UNF-65, NF-6

required but barely, one should weigh the cost of testing versus going through the Code procedure to try to lift the material to the not-required position in Fig. 2.13. If the latter seems feasible, one should keep the Code close at hand because there are so many variables involved. The tabular values for Fig. 2.13 included here as Fig. 2.14 are somewhat easier to work with than the graph. If impact testing is not required, there is nothing else to do.

Exemption model 2. In the event that the designer cannot exempt all the remainder of the materials in exemption model 1 above, and depending on whether the stress in tension for the design conditions is less than the maximum allowable design stress, the Code allows one last possible exemption from impact testing. Assuming that the example vessel uses the full allowable stress in its calculated thickness, low-

Thick., in.	Curve A, °F	Curve B, °F	Curve C, °F	Curve D, °F	Thick., in.	Curve A, °F	Curve B, °F	Curve C, °F	Curve D, °F
0.25	18	−20	−55	−55	3.1875	112	80	44	13
0.3125	18	−20	−55	−55	3.25	112	80	44	14
0.375	18	−20	−55	−55	3.3125	113	81	45	15
0.4375	25	−13	−40	−55	3.375	113	82	46	15
0.5	32	−7	−32	−55	3.4375	114	83	46	16
					3.5	114	83	47	17
0.5625	37	−1	−26	−51					
0.625	43	5	−22	−48	3.5625	114	84	48	17
0.6875	48	10	−18	−45	3.625	115	85	49	18
0.75	53	15	−15	−42	3.6875	115	85	49	19
0.8125	57	19	−12	−38	3.75	116	86	50	20
0.875	61	23	−9	−36	3.8125	116	87	51	21
0.9375	65	27	−6	−33	3.875	116	88	51	21
1	68	31	−3	−30	3.9375	117	88	52	22
					4	117	89	52	23
1.0625	72	34	−1	−28					
1.125	75	37	2	−26	4.0625	117	90	53	23
1.1875	77	40	4	−23	4.125	118	90	54	24
1.25	80	43	6	−21	4.1875	118	91	54	25
1.3125	82	45	8	−19	4.25	118	91	55	25
1.375	84	47	10	−18	4.3125	118	92	55	26
1.4375	86	49	12	−16	4.375	119	93	56	27
1.5	88	51	14	−14	4.4375	119	93	56	27
					4.5	119	94	57	28
1.5625	90	53	16	−13					
1.625	92	55	17	−11	4.5625	119	94	57	29
1.6875	93	57	19	−10	4.625	119	95	58	29
1.75	94	58	20	−8	4.6875	119	95	58	30
1.8125	96	59	22	−7	4.75	119	96	59	30
1.875	97	61	23	−6	4.8125	119	96	59	31
1.9375	98	62	24	−5	4.875	119	97	60	31
2	99	63	26	−4	4.9375	119	97	60	32
					5	119	97	60	32
2.0625	100	64	27	−3					
2.125	101	65	28	−2	5.0625	119	98	61	33
2.1875	102	66	29	−1	5.125	119	98	61	33
2.25	102	67	30	0	5.1875	119	98	62	34
2.3125	103	68	31	1	5.25	119	99	62	34
2.375	104	69	32	2	5.3125	119	99	62	35
2.4375	105	70	33	3	5.375	119	100	63	35
2.5	105	71	34	4	5.4375	119	100	63	36
					5.5	119	100	63	36
2.5625	106	71	35	5					
2.625	107	73	36	6	5.5625	119	101	64	36
2.6875	107	73	37	7	5.625	119	101	64	37
2.75	108	74	38	8	5.6875	119	102	64	37
2.8125	108	75	39	8	5.75	120	102	65	38
2.875	109	76	40	9	5.8125	120	103	65	38
2.9375	109	77	40	10	5.875	120	103	66	38
3	110	77	41	11	5.9375	120	104	66	39
					6	120	104	66	39
3.0625	111	78	42	12					
3.125	111	79	43	12					

GENERAL NOTE: See Fig. UCS-66 for SI values.

Figure 2.14 Tabular values for Fig 2.13. (This table appears in the Code as Table UCS-66.)

ering the design stress will increase the vessel thickness. If this is the case, the following sequence will give the reader an idea how to use the chart in Fig. 2.15 to gain additional relief from impact testing. (See also Fig. 2.16.) *Assume for this model an MDMT of 10°F.* Note: For service temperatures below −50°F, impact testing is mandatory.

1. Observe that for the material thickness of 1.5 in, curve C is about 5°F above the 10°F MDMT at the 1.5-in thickness point.
2. The curve of Fig. 2.13 must be lowered by more than 5°F, since the increased vessel thickness will result in a shift on curve C to a higher temperature; try, say, 15°F.

Descriptive Guide to the ASME Code Pressure Vessels   93

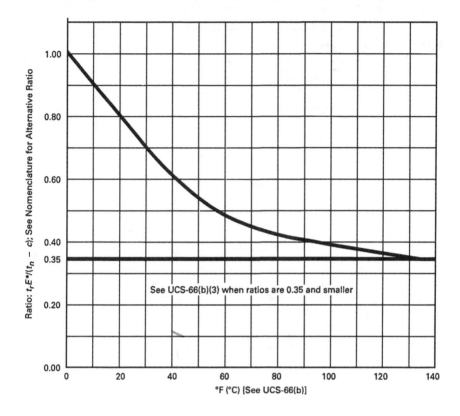

Nomenclature (Note reference to General Notes of Fig.UCS-66-2.)
t_r = required thickness of the component under consideration in the corroded condition for all applicable loadings [General Note (2)], based on the applicable joint efficiency E [General Note (3)], in. (mm)
t_n = nominal thickness of the component under consideration before corrosion allowance is deducted, in. (mm)
c = corrosion allowance, in. (mm)
E^* = as defined in General Note (3).
Alternative Ratio = S^*E^* divided by the product of the maximum allowable stress value from Table UCS-23 times E, where S^* is the applied general primary membrane tensile stress and E and E^* are as defined in General Note (3).

Figure 2.15 Reduction in MDMT without impact testing (Code Fig. UCS-66.1). (*Courtesy of The American Society of Mechanical Engineers.*)

3. On the curved line of Fig. 2.15 enter a "trial" temperature of 15°F on the horizontal temperature line between the 10 and 20°F. Follow this line up vertically until it hits the curved line. Follow the horizontal line to the left and read a design stress to allowable tensile stress ratio of 0.85.

4. Calculate the new design stress by multiplying 0.85 by the allowable stress of 15,000 psi (the stress of the material found in Code Section II, Part D, Table 1A), yielding 12,750 psi.

Step 1 — Establish nominal thickness [General Note (1)] of welded parts, nonwelded parts, and attachments under consideration both before and after corrosion allowance is deducted (t_n and $t_n - c$, respectively), and other pertinent data applicable to the nominal thickness such as:

All applicable loadings [General Note (2)] and coincident minimum design metal temperature (MDMT)
Materials of construction
E = joint efficiency [General Note (3)]
t_n = nominal noncorroded thickness [General Note (1)], in. (mm)
t_r = required thickness in corroded condition for all applicable loadings [General Note (2)], based on the applicable joint efficiency [General Note (3)], in. (mm)
Applicable curve(s) of Fig. UCS-66
c = corrosion allowance, in. (mm)

Step 2 — Select MDMT from Fig. UCS-66 [General Note (4)] for each nominal noncorroded governing thickness [General Note (5)].

Step 3 — Determine Ratio: $\dfrac{t_r E^*}{t_n - c}$
[General Notes (3), (6), (7), and (8)]

Step 4 — Using Ratio from Step 3 to enter ordinate of Fig. UCS-66.1, determine reduction in Step 2 MDMT [General Note (9)].

Step 5 — Determine adjusted MDMT for governing thickness under consideration.

Step 6 — Repeat for all governing thickness [General Note (5)] and take *warmest* value as the lowest allowable MDMT to be marked on nameplate for the zone under consideration [General Note (10)]. See UG-116.
See UG-99(h) for coldest recommended metal temperature during hydrostatic test [General Note (6)].
See UG-100(c) for coldest metal temperature permitted during pneumatic test [General Note (6)].

Legend
☐ Requirement   ┆┄┄┆ Optional

GENERAL NOTES:
(1) For pipe where a mill undertolerance is allowed by the material specification, the thickness after mill undertolerance has been deducted shall be taken as the noncorroded nominal thickness t_n for determination of the MDMT to be stamped on the nameplate. Likewise, for formed heads, the minimum specified thickness after forming shall be used as t_n.
(2) Loadings, including those listed in UG-22, which result in general primary membrane tensile stress at the coincident MDMT.
(3) E is the joint efficiency (Table UW-12) used in the calculation of t_r; E^* has a value equal to E except that E^* shall not be less than 0.80. For castings, use quality factor or joint efficiency E whichever governs design.
(4) The construction of Fig. UCS-66 is such that the MDMT so selected is considered to occur coincidentally with an applied general primary membrane tensile stress at the maximum allowable stress value in tension from Table 1A of Section II Part D, Tabular values for Fig. UCS-66 are shown in Table UCS-66.
(5) See UCS-66(a)(1), (2), and (3) for definitions of governing thickness.
(6) If the basis for calculated test pressure is greater than the design pressure [UG-99(c) test], a Ratio based on the t_r determined from the basis for calculated test pressure and associated appropriate value of $t_n - c$ shall be used to determine the recommended coldest metal temperature during hydrostatic test and the coldest metal temperature permitted during the pneumatic test. See UG-99(h) and UG-100(c).
(7) Alternatively, a Ratio of $S^* E^*$ divided by the product of the maximum allowable stress value in tension from Table 1A of Section II Part D times E may be used, where S^* is the applied general primary membrane tensile stress and E and E^* are as defined in General Note (3).
(8) For UCS-66(b)(1)(b) and (i)(2), a ratio of the maximum design pressure at the MDMT to the maximum allowable pressure (MAP) at the MDMT shall be used. The MAP is defined as the highest permissible pressure as determined by the design formulas for a component using the nominal thickness less corrosion allowance and the maximum allowable stress value from the Table 1A of Section II, Part D at the MDMT. For ferritic steel flanges defined in UCS-66(c), the flange rating at the warmer of the MDMT or 100°F may be used as the MAP.
(9) For reductions in MDMT up to and including 40°F, the reduction can be determined by: reduction in MDMT = (1 − Ratio) 100°F.
(10) A colder MDMT may be obtained by selective use of impact tested materials as appropriate to the need (see UG-84). See also UCS-68(c).

Figure 2.16 Diagram of UCS-66 rules for determining lowest MDMT without impact testing (Code Fig. UCS-66.2).

5. Recalculate the vessel thickness using the design formula and substitute 1.76 in as the new stress value (an approximation for a trial value is 1.50 in/0.85).

6. Look at Fig. 2.13 again. Lower curve C by 15°F at a thickness of 1.76 in to get a curve C MDMT of 15°F. Since the curve is now below the vessel MDMT of 10°F, 1.76-in-thick SA-516 grade 60 may be used without postweld heat treating (PWHT) normalizing, or impact-testing the materials. From the standpoint of cost and economics this method is probably the best choice.

If postweld heat treating is performed when it is not otherwise a requirement of Section VIII, Division 1, a 30°F reduction in impact-testing exemption temperature shall be given to the minimum permissible temperature from Fig. UCS-66 for P-1 material (see Code Par. UCS-68). Utilizing this method would be costly.

Further exemptions shall be determined as follows:

If reinforcing pads are required, such as for the nozzle reinforcement, such reinforcement would be evaluated similarly with one of the exemption models above, using the pad thickness as the governing thickness.

The 6-in SA-106 grade B nozzle neck is a curve B material per Fig. 2.13, note (b)(3). The nozzle neck nominal thickness of 0.280 in and the MDMT of 18°F locate a point above curve B, so no impact testing is required. This is also exempt from impact testing as stated in exemption 1. In this case this material could have been exempt from impact testing utilizing exemption model 1.

Exemption model 3. Code Pars. UCS-66(c), (d), (e), and (f) grant exemptions from impact testing for certain types of special items such as ANSI B16.5 ferritic steel flanges, nuts, or bolts as follows.

Nozzle flange. This is exempt from impact testing per Par. UCS-66(c): "No impact testing is required for ANSI B16.5 ferritic steel flanges used at design metal temperatures no colder than −20°F."

If the body flange is not an ANSI B16.5 flange, but a flange designed by the manufacturer in accordance with Code Appendix 2, the previous exemption does not apply. The governing thickness of the flange is 1.5 in, which is the thickness of the thickest welded joint [see Code Par. UCS-66(a)(1)(a)]. The flange thickness at the flange face thickness which is away from the weld does not govern the thickness. The governing thickness is the same as the shell; hence the body flange material will fall as a curve D–required material. Since no nonimpact-tested carbon steel–forged flange material is included in curve D, impact-tested forging material such as SA-350 must be used. SA-350 grade LF1, with a specified average impact energy of 13 ft · lb at −20°F, is chosen, although grade LF2 is also acceptable. These materials are impact-tested by the mill as part of the material specification requirements.

The cover plate is a nonwelded component part of the vessel. Paragraph UCS-66(a)(3) defines the governing thickness for the cover plate to be the same as that of the shell (and body flange), or 1.5 in in thickness. If the nominal thickness of the plate had exceeded 6 in, impact testing would be required by Par. UCS-66(a)(5). The following are acceptable materials which could be selected for this vessel:

1. An SA-350 impact-tested forging
2. A plate material from exemption curve D, such as normalized SA-516 grade 70, with no impact testing

Since the body flange requires impact testing, weld procedure qualification impact tests and production impact tests are required for the flange weld, per Code Pars. UCS-67 and UG-84. Since impact values for the weld metal and the heat-affected zone must be at least as high as the values for the base metal, the test values must meet the 13 ft · lb of the SA-350 grade LF1 specification.

Welding procedure test plates may be made from the body flange material (SA-350 grade LF1) or from an equivalent material provided the requirements of Code Par. UG-84 are met.

One last important factor when considering impact testing is Form U-1, the Manufacturer's Data Report. It should be noted in the Remarks section of the data report "...Impact test exempt per..." and then written which exemption was used, for example, UG-20(f), UCS-66(a), (b), or (c). In some cases more than one exemption may be listed on the data report.

The important thing to remember is that vessels operating at any temperature must be evaluated for impact-test requirements and demand special attention in design and materials. Specific paragraphs in the *ASME Pressure Vessel Code* prescribe the type of welded joints involved in low-temperature operations. These joints must meet the requirements of Code Par. UW-2, Service Restrictions. Code Par. UW-3, Welded Joint Category, determines the location of the joint. If the material is carbon steel, Code Pars. UG-20(f), UCS-66, and UCS-67 state whether impact tests are required as well as specify the exemptions for impact testing. If tests are required, Code Par. UG-84 describes how they must be made. Code Par. UG-84 also gives instructions on the required size of test specimens and on location of the areas in the specimens, taking impacts, test values, and properties. However, impact testing is not a "must" for welds or materials if certain design requirements are met. For example, a vessel designed for 100 psi at 100°F could not be used for 100 psi at −50°F without impact testing of the vessel's welds and material.

Other materials exempt from impact tests are austenitic stainless steel, with no impact tests mandatory down to a temperature of –325°F, and various stainless steels, such as SA-240, Type 304, Type 304L, and Type 347, with no impact tests necessary down to a temperature of –425°F. Code Par. UHA-51 permits these exemptions, basing them on the fact that these specific materials do not show a transition from ductile to brittle fracture at the decreased temperatures. However, having stated the exemptions for the base materials, welding procedure specifications require additional testing to include impact-test plates of the procedure for these stainless materials (see Code Par. UG-84).

Wrought aluminum alloys do not undergo a marked drop in impact resistance below –20°F. As indicated in Code Par. UNF-65, no impact tests are required down to a temperature of –452°F. Other nonferrous metals for copper, copper alloys, nickel, and nickel alloys can be used to –325°F without impact tests. Code Par. UNF-65 indicates the need for impact testing.

The major concern for metals at low temperature is brittle fracture, which can cause vessel failure. At normal temperatures, metals usually give warning by bulging, stretching, leaking, or other failure. Under low-temperature conditions, some normally ductile metals may fail at low-stress levels without warning of plastic deformation.

Four conditions must be controlled to avoid brittle fractures in low-temperature vessels:

1. Materials should not show a transition from ductile to brittle fracture.
2. Welding materials for low-temperature service must be selected carefully, and a proven welding procedure must be used.
3. Stress raisers from design or fabrication should not be permitted.
4. Localized yielding in the stress-raiser area must not be allowed.

Pressure vessel users should use a material that behaves in a ductile manner at all operating temperatures. This will allow for the possible stress raiser that may be overlooked in fabrication or design. Impact tests could determine the correct materials and welding procedures for these vessels.

Service Restrictions and Welded Joint Category

Many design engineers and pressure vessel manufacturers have been using the wrong type of welded joints in vessels which must meet the requirements in Code Par. UW-2, Service Restrictions, and Par. UW-3, Welded Joint Category, of the *ASME Pressure Vessel Code,* Section VIII, Division 1.

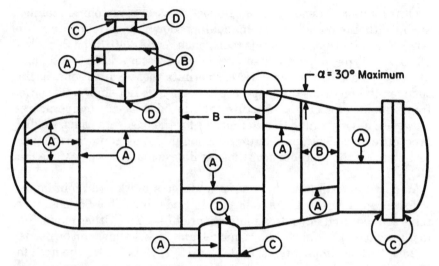

Figure 2.17 Illustration defining locations of welded joint categories. A. Longitudinal joints; circumferential joints connecting hemispherical heads to the main shell; any welded joint in a sphere. B. Circumferential joints; Circumferential joints of torispherical, ellipsoidal heads; angle joints not greater than 30°. C. Flange joints; Van Stone laps; tube sheets; flat heads to main shell. D. Nozzle joints to main shell; heads; spheres; flat sided vessels.

When the paragraph on the welded joint category (see Fig. 2.17) was added to the Code, it was thought that the intent of Code Par. UW-2, Service Restriction, was clarified, but evidently it has not been clear enough to some pressure vessel fabricators. The requirements of Code Par. UW-2 have often been misinterpreted and at times not referred to until too late. For example, if a vessel designed for −140°F which would require impact tests was designed with flanges, the flanges would have to be welded joint category C and would require full penetration welds extending through the entire section of the joint. Also, all longitudinal seams would have to be category A, without backing strips.

We have reviewed drawings of vessels designed for cryogenic service requiring impact tests in which the design on the flanges did not call for full penetration welds and the longitudinal seams had metal backing strips which were to remain in place. This, of course, does not meet the intent of the Code. It is important that design engineers and pressure vessel manufacturers have a clear understanding of the paragraphs in the Code on service restrictions and welded joint category and that the vessels, where required, receive the proper attention during design and in handling in the shop.

What are the vessels which demand special requirements listed in Code Par. UW-2, Service Restrictions? They are vessels intended for lethal service; vessels constructed of carbon- and low-alloy steels which are required to be postweld heat-treated; vessels designed to operate below certain temperatures designated by Code Part UCS (see Code Par. UCS-86); vessels of impact-tested materials, including weld metal, as required by Code Part UHA; unfired steam boilers with design pressures exceeding 50 psi; pressure vessels subject to direct firing; and vessels with special requirements based on material, thickness, and service.

What must be remembered about the welded joint category? Only those joints to which special requirements apply are included in the categories. The term *category* defines only the location of the welded joint in the vessel and never implies the type of welded joint that is required. The *type* of joint is prescribed by other paragraphs in the Code suitable to the design, which may specify special requirements regarding joint type and degree of inspection for certain welded joints.

The purpose of this review is to spell out the requirements of the Code with regard to service restrictions and welded joint category. Designers and manufacturers should examine their practices to ensure that the service restrictions paragraph of the Code is receiving appropriate attention in the engineering department and the shop, for the designers and manufacturers must accept full responsibility for the design and construction of a Code pressure vessel when the vessel is completed and the data sheets are signed.

The ASME Code Form U-1, the Manufacturer's Data Report, in the section for remarks requires a brief description of the purpose of the vessel and also requires that the contents of each vessel be indicated. It will be the Code Inspector's duty to determine whether this item is accurately completed and whether service restrictions were taken into consideration.

Therefore, in designing Code vessels, be sure you have taken the paragraphs on service restrictions and welded joint category into consideration. Also be sure the Authorized Inspector has reviewed the design before fabrication starts so that no difficulties will arise when the vessel is completed.

Noncircular-Cross-Section-Type Pressure Vessels

Code Appendix 13 lists the design, fabrication, and inspection requirements for vessels of noncircular cross section. These rules cover rectangular, stayed, and obround pressure vessels. Table 2.19 lists other code references along with helpful information in the design of such vessels.

TABLE 2.19 Vessels of Noncircular Cross Section

Remarks	Code reference
1. Scope	Appendix 13-1
Types of vessels, descriptions, and figures	Appendix 13-2
Materials	Appendix 13-3
2. Design requirements and stress acceptance criterion	Appendix 13-4
3. Thermal expansion reactions and loads due to attachments	Appendix 13-4
4. Nomenclature and terms used in design formulas	Appendix 13-5 Pars. UG-47, UG-34, UG-101, UW-12
5. End closures	Appendix 13-4
a. If ends are unstayed, flat plate design closure per UG-34 except a C factor of no less than 0.20 shall be used in all cases	
b. All other end closures are special design per U-2, or they may be subject to proof test per UG-101	Pars. U-2, UG-101
Braced and stayed surfaces may be designed per Code Par. UG47	
	Pars. UG-47, UG-48, UG-49, UW-19
	Appendix 13-4
6. Special consideration needs to be given to vessels of aspect ratio less than 4	Appendix 13-4
7. Vessels are designed for membrane and bending stress throughout the structure (for both the short-sided and long-sided plates). Membrane, bending, and total stresses are considered. Formulas for the stresses are given as follows:	Appendix 13-4
a. Vessels of rectangular cross section (nonreinforced)	Appendix 13-7
b. Reinforced vessels of rectangular cross section	Appendix 13-8
c. Stayed vessels of rectangular cross section	Appendix 13-9
d. Vessels having an obround cross section	Appendix 13-10
e. Reinforced vessels of obround cross section	Appendix 13-11
f. Stayed vessels of obround cross section	Appendix 13-12
g. Vessels of circular cross section having a single diametral staying member	Appendix 13-13
h. Vessels of noncircular cross section subject to external pressure	Appendix 13-14
Vessel fabrication	Appendix 13-15
8. Examples illustrating application of formulas	Appendix 13-17
For corner joints	Par. UW-13
9. Special calculations	Appendix 13-18

Jacketed Vessels

Today many manufacturers are manufacturing jacketed vessels. Most jacketed vessels are designed for external pressure on the inner shell and internal pressure on the outer shell. These types may be of the braced and stayed type construction (see Code Appendix 9). Some configurations of such vessels are of the "dimpled" jacketed type (see Code Appendix 17). The outer plate, normally thin gage stainless steel, is formed by presses. Holes are punched in the material at approximately 3-in centers and then a dimple is formed at each hole. This jacket is then welded to an inner shell by welding the hole in which the dimple is formed, thus creating the jacket. This type vessel is normally used in the pharmaceutical industry.

Special design considerations must be accounted for. Table 2.20 lists the Code references for such construction.

TABLE 2.20 Jacketed Vessels

Remarks	Code reference
Minimum requirements	Appendix 9
Any combination of pressures and vacuum in vessel and jacket producing a total pressure over 15 psi on the inner wall should be made to Code requirement	Appendix 9-1
Types of jacketed vessels	Appendix 9-2
Materials	Appendix 9-3
Partial jackets	Figs. 9-2 to 9-7 Appendix 9-7
Dimpled jackets; indicate Appendix 17 identification and paragraph number on data report	Appendix 17
Embossed assemblies	Appendix 17 Pars. UW-19, UG-101
Proof test of dimpled and embossed assemblies, also partial jackets	Par. UG-101
	Appendix 17-7
Design of jackets	Appendix 9-4
Types of jackets	Appendix 9-2
Design of openings through jacket space	Appendix 9-6 Fig. 9-6 Pars. UG-36 to UG-45
Braced and stayed surfaces	Appendix 9-4 Pars. UG-47, UW-19
Inspection openings need not be over 2 in	Appendix 9-4 Par. UG-46
Test pressure glass-lined vessels	Par. UG-99

Pressure Vessels Stamped with the UM Symbol

Vessels stamped with the UM symbol are limited to size and design pressures in accordance with the ASME Code Section VIII, Division 1, Code Par. U-1. The manufacturer should understand that the authority given to the company by the ASME Committee to use the UM symbol is dependent on strict conformance to the ASME Code requirements and is subject to cancellation. The shop must hold an ASME Code U symbol stamp in order to obtain a UM symbol stamp. The UM stamp is renewed annually with a review only by the Authorized Inspection Agency Inspector and his or her supervisor. During the third year for renewal of the U symbol stamp, the joint review team will assess the quality system and verify implementation of both the U and UM stamp. The stamps are then issued by the ASME upon successful completion of the shop review by the inspection agency and/or jurisdiction.

TABLE 2.21 Vessels Exempt from Inspection

Types and sizes exempted	Remarks	Code reference
Vessels not within the Scope of the Code		
Vessels with a nominal capacity of 120 gal or less containing water (or water cushioned with air)	Even though exempted from Code, any vessel meeting all Code requirements may be stamped with the Code symbol	Par. U-1
Vessels having an internal or external operating pressure of 15 psi or less with no limitation on size		
Vessels having inside diameter, width, height, or cross section diagonal of 6 in or less with no vessel length or pressure limitation		
Pressure vessels for human occupancy		Par. U-1 PVH0-1
Vessels within the Scope of the Code		
Vessels not exceeding 5 cu ft volume and 250 psi design pressure	Exempted vessels must comply in all other respects with Code requirements, including stamping with UM symbol by manufacturer	Pars. U-1, UG-91, UG-116 Fig. UG-116(b)
Vessels not exceeding 1.5 cu ft volume and 600 psi design pressure	A certificate for UM vessels must be furnished by manufacturer on Code Form U-3 when requested	Par. UG-120

It is necessary that the UM stamp with the required data be completed as requested by Code Par. UG-116 and Code Fig. UG-116(b). A certificate for UM-stamped vessels, exempted from inspection by an Authorized Inspector, should be furnished on Code Form U-3 when required. Table 2.21 lists the vessels exempt from authorized inspection by an Authorized Inspector. (Some jurisdictions will not recognize the UM stamp and will require inspection by an Authorized Inspector.) The responsibility of the manufacturer is to design, fabricate, inspect, and test all UM vessels to see that they meet Code requirements, with the exception of inspection by an Authorized Inspector. The Authorized Inspector will likely request quarterly visits to verify that the certificate holders are following the approved quality control system.

Stamping Requirements (Fig. 2.18)

If the markings required are stamped directly on the vessel, the minimum letter size is $5/16$ in. If they appear on a nameplate, the minimum letter size is $5/32$ in. If a nameplate is used, only the Code symbol and the manufacturer's serial number are required to be stamped; the other information may be applied by another method, for example, etching.

The stamping should be located in a conspicuous place, preferably near a manhole or handhole opening.

Removable pressure parts should be marked to identify them with the vessel.

For vessels having two or more pressure chambers, see Code Par. UG-116.

```
                    NATIONAL BOARD
                   ┌──────────────┐
                   │ Board Number │  ①
                   └──────────────┘
                      Certified by
                   ┌──────────────────────┐
         ⑥         │ Name of Manufacturer │
       ┌──┐        └──────────────────────┘
       │  │        _____ psi at _____ °F
       └──┘        (Max. allowable working pressure)
         ②
         ③         _____ °F at _____ psi  ④
                   (Min. design metal temperature)

                   _____
                   (Manufacturer's serial number)
                         ┌──────────┐
                         │Year built│
                         └──────────┘

Shell thickness [   ]  Head thickness [   ]  Head radius [   ]  ⑤
```

Key: 1. Stamp *National Board* and the National Board number at top of plate if National Board inspection is required. If not registered with National Board, the words *National Board* shall be removed.
  2. Stamp *Part* under Code symbol on those parts of a vessel for which partial data reports are required.
  3. Stamp type of construction, radiographic, and postweld heat treatment status of main longitudinal and circumferential joints under Code symbol by means of the following letters:

W—arc or gas welded	RT1—completely radiographed 100% when complete vessel satisfies full-RT and when the RT3 requirements have been applied
RES—resistance welded	
B—brazed	
P—pressure welded (except resistance)	
	RT2—when the complete vessel satisfies the requirements of Code Par. UW-11(a)(5) and when the spot requirements of Code Par. UW-11(a)(5)(b) have been applied
HT—completely postweld heat-treated	RT3—spot-radiographed
PHT—partially postweld heat-treated	RT4—part of vessel radiographed where none of RT1, 2 or 3 apply

Vessels intended for special service must also be appropriately stamped separated by a hyphen after lettering of item 3 above.

L—lethal service	WL—Apply to nameplates of layered construction
UB—unfired boiler	ULT—The following markings shall be applied to nameplates for low temperature service
DF—direct firing	
NPV—Nonstationary pressure vessels	UHT—Apply to nameplates when materials of code part UHT are used.

  4. The maximum allowable working pressure at the maximum temperature established by the designer shall be so stamped. Additionally the minimum design metal temperature at the corresponding coincident pressure shall be stamped.
  5. Pennsylvania requires shell and head thickness and head radius.
  6. Stamp *User* above Code symbol if vessel is inspected by the user's inspector.

Note: Nameplates shall be stamped with letters or numbers at least $5/16$ in high. When vessels are designed and fabricated in accordance with Code Part ULT, the following markings shall be used in lieu of item 4 above.

Maximum Allowable Working Pressure: _____ psi at 100°F
Minimum Allowable Temperature Minus _____ °F

Service Restricted to the Following	Operating Temperature
Liquid _____	Minus ____ °F
Liquid _____	Minus ____ °F
Liquid _____	Minus ____ °F
Liquid _____	Minus ____ °F

Figure 2.18 Stamping requirements and applicable form.

Chapter

3

Design for Safety

Basic Causes of Pressure Vessel Accidents

Two new vessels built for equal pressures and temperatures and the same type of service can fare very differently in different locations. In one plant, capable maintenance personnel will keep the vessel in top shape by means of a systematic schedule of inspection and maintenance. In another, an equal number of personnel, through ignorance, incompetence, or neglect, can bring about the failure of the vessel in a comparatively short time. Competent personnel are obviously a key factor in the longevity and safety of a pressure vessel.

Corrosion Failures

Most pressure vessels are subject to deterioration by corrosion or erosion, or both. The effect of corrosion may be pitting or grooving over either localized or large areas. Corrosion over large areas can bring about a general reduction of the plate thickness.

Some of the many causes of corrosion are inherent in the process being used, but others are preventable. Vessels containing corrosive substances, for example, deteriorate more rapidly if the contents are kept in motion by stirring paddles or by the circulation of steam or air. If the corrosion is local, it may not appear serious at first; as the metal thickness is reduced, however, the metal becomes more highly stressed, even under normal pressures. The more stressed the metal becomes, the faster it corrodes, at last reaching the point of sudden rupture. Corrosion in unstressed metal is less likely to cause such sudden failure.

Stress Failures

Many plate materials are not at all suitable for the service required of them. Other materials may be serviceable in an unstressed condition but will fail rapidly in a stressed condition. To give an example, a large chemical company once used stainless steel corrosion buttons to measure corrosion in a large fired cast-steel vessel. When the cast-steel vessel had to be replaced, the company ordered a new vessel made of the same stainless steel used for the buttons because of its proved resistance. After a few days of service, however, the operator noticed leakage through the 1-in shell, immediately reduced the pressure, and shut down the vessel. Examination showed many cracks in the stainless steel. The substance being processed was highly corrosive to the grain boundaries of the metal when under stress (see Code Par. UHA-103). This fact was verified when the metal was tested in both stressed and unstressed conditions in the same solution. The stressed pieces failed within 100 hours.

When metals are subjected to repeated stresses exceeding their limits, they ultimately fail from *fatigue*. It may take thousands of stress applications to produce failure, but a vessel such as an air tank in which there are rapid pressure fluctuations may fail at the knuckle radius. In a large laundry facility, the authors once saw grooving that penetrated halfway into a $\frac{5}{8}$-in head in the knuckle radius of a hot water tank.

Excessive stresses can develop in vessels that are not free to expand and contract with temperature changes. If these stresses are allowed to recur too frequently, eventual failure from fatigue can be expected.

Other Failures

Vessels are often installed in positions that prevent entry through the inspection openings for examination of interior surfaces. Any owner or user of a vessel who is interested in its actual condition would never let this happen.

If a vessel is supported on concrete saddles, the difference in the coefficients of expansion of the concrete and steel will sometimes cause the concrete to crack just beneath the tank, allowing accumulations of moisture and dust that will cause corrosion.

Safety valves or devices that become plugged or otherwise inoperative can lead to accidents. Safety valves may also be adjusted too high for the allowable pressure so that the vessel fails if overpressure occurs.

Defective material not visible at the time of fabrication, such as laminations in the plate or castings with internal defects, may bring on failure. Inferior technique at the time of fabrication, such as poor joint fit-up or bad welding that may appear satisfactory on the surface but suffers from lack of penetration, slag, excessive porosity, or lack of ductility can also cause serious trouble.

It requires consistent effort on the part of plant personnel as well as a well-organized system of maintenance and inspection to safeguard pressure vessel equipment properly.

Design Precautions

The reliable functioning of a pressure vessel depends upon the foresight of the user and the information the user gives to the designer and fabricator. If the process involved is corrosive, for example, these data must include the chemicals to be used and, if possible, the expected rate of corrosion based on a history of vessels in similar service. (Corrosion resistance data are available for most materials and chemicals but actual operating conditions can seldom be duplicated in the laboratory and must be determined by frequent, close examination after the vessel is placed in service.) Table 3.1 lists the requirements of the ASME Code Section VIII, Division 1 corrosion allowances and the applicable Code references.

Usually no one but the user is fully familiar with the operating conditions and the causes of defects and deterioration in the type of vessel being ordered. If the designer is presented with all the facts, however, the materials best suited for the job will be used and all openings placed in the most desirable locations.

Although the Code permits many types of construction, some of them may not be suitable for the service required. Therefore, merely to prescribe an ASME vessel is not always sufficient. For example, the Code permits small openings to be welded from one side only and also provides for offset construction of girth seams and heads [see Code Par. UW-13; also Fig. UW-13.1(K)]. This type of construction may not be suitable for some vessels, however, because it leaves some inside seams and fittings not welded and thus subject to corrosion. The purchaser aware of such facts will be specific in ordering and will state the type of welded joints desired. The drawings will also be checked before fabrication to make sure that the openings are located where desired and that the inspection openings permit easy access to the interior.

The Code inspector must also check the design of the vessel to be assured that the shell, head, and nozzle thicknesses are correct, the knuckle radius sufficient, the openings properly reinforced, the flange thickness sufficient, and the like, but there is no concern with the position of the outlets so long as they comply with the Code.

Code Par. UG-22 and Appendix G state that loadings must be considered when designing a vessel, but do not give clear information about how to take them into account. When vessels and piping systems operate out-of-doors, wind must be considered. Also to be taken into account are external loadings such as snow; the minimum ambient

TABLE 3.1 Corrosion Allowances and Requirements

Remarks	Code reference
General requirements	Par. UG-25
Suggested good practice	Appendix E
For brazed vessels	Par. UB-13
For clad vessels	Pars. UCL-25, UG-26 Appendix F
Vessels subject to corrosion must have suitable drain openings	Par. UG-25
Vessels subject to internal corrosion, erosion, or mechanical abrasion must have inspection openings	Par. UG-46

temperature; the reaction of piping systems; the restricted thermal expansion on attachment of piping systems to the vessel and on its hangers or anchors; the weight of valves; and pumps, blowers, and rotating machinery. It is imperative for the designer to see that all machinery, piping, and equipment are properly balanced and supported. Nonmandatory Code Appendixes D and G also give more useful information to be taken into consideration.

Internal corrosion usually develops on surfaces susceptible to moisture. For this reason, it is important to make sure that the vessel can be properly drained. In vertical vessels equipped with agitating devices, the greatest wear usually occurs on the bottom head. In horizontal vessels with agitators, the most wear will be between the six o'clock and nine o'clock positions for clockwise rotation and between the six o'clock and three o'clock positions for counterclockwise rotation. Horizontal vessels with an agitator should therefore be designed for turning, if possible, so that the thinned-out area can be rotated to the top half. The useful life of the vessel will thus be extended.

Vessels to be subjected to heavy corrosion, erosion, or mechanical abrasion require an extra corrosion allowance and telltale holes. Leakage from these holes will give a warning that the vessel has corroded to its minimum allowable thickness. For example, if the minimum thickness required is $\frac{3}{8}$ in but the plate thickness used (to increase the length of service) is $\frac{7}{8}$ in, the corrosion allowance would thus be $\frac{1}{2}$ in. Telltale holes placed on about 18-in centers would then be drilled $\frac{3}{8}$ in deep from the surface not affected by corrosion.

Code Par. UG-25(e) states: "When telltale holes are drilled, they shall be $\frac{1}{16}$ to $\frac{3}{16}$ in in diameter and shall be drilled to a depth not less than 80 percent of the thickness required for a seamless shell of like dimensions. These holes shall be drilled into the surface opposite to

TABLE 3.2 Minimum Thickness of Plate

The minimum thickness of plate after forming and without allowances for corrosion shall not be less than the minimum thickness allowed for the type of material being used (see Code Pars. UG-16 and UG-25).

Material	Minimum thickness	Code reference
Carbon and low-alloy steel	$3/32$ in for shells and heads used for compressed air, steam, and water service	Par. UG-16
	Minimum thickness of shells and and heads after forming shall be $1/16$ in plus corrosion allowance	
	$1/4$ in for unfired steam boilers	
Heat-treated steel	$1/4$ in for heat-treated steel except as permitted by the notes in Table UHT-23	
Clad vessels	Same as for carbon and low-alloy steel based on total thickness for clad construction and the base plate thickness for applied-lining construction	Par. UCL-20
High-alloy steel	$3/32$ in for corrosive service $1/16$ in for noncorrosive service	Pars. UHA-20, UG-16
Nonferrous materials	$1/16$ in for welded construction in noncorrosive service	Pars. UNF-16, UG-16
	$3/32$ in for welded construction in corrosive service	
Low-temperature vessels: 5, 8, 9 percent nickel steel	$3/16$ in (Max. thickness: 2 in)	Par. ULT-16(a)(2)
Cast-iron dual metal cylinders	5 in (Max. outside diameter: 36 in)	Par. UCI-29
Layered vessels	$1/8$ in min. of each layer	Par. ULW-16

that at which deterioration is expected." For clad or lined vessels, see Code Par. UCL-25(b).

Table 3.2 lists the minimum thickness of plate which may be used when designing Code pressure vessels.

Inspection Openings

Accessibility to the inside of a vessel is essential. Carrying out regular internal inspections is of importance to the safety and extended life of a vessel. Code Par. UG-46 requires inspection openings in all vessels subject to internal corrosion or having internal surfaces subject to erosion or mechanical abrasion so that these surfaces may be examined for defects. Actually, almost all pressure vessels are subject to some kind of interior corrosion, and its seriousness can be determined only by examining the inside of the vessel. The alternative is possible failure and consequent injury to or even loss of human life.

Minimum sizes for the various types of openings have been established by the Code. An elliptical or obround manhole opening must not be less than 11×15 or 10×16 in and a circular manhole's inside diameter not less than 15 in. Handholes must not be less than 2×3 in; however, for vessels with inside diameter exceeding 36 in, the minimum size handhole used in place of a manhole is 4×6 in.

On small vessels it is sometimes difficult to provide extra openings. The Code states that vessels more than 12 but less than 18 in in inside diameter must have not less than two handholes or two threaded plug openings of at least 1½-in pipe size. Vessels 18 to 36 in, inclusive, in inside diameter must have one manhole or two handholes or two threaded plugs of not less than 2-in pipe size. Vessels with inside diameter more than 36 in must have a manhole. Those vessels of a shape or use that make a manhole impracticable must have at least two 4×6-in handholes or two openings of equal area.

Removable heads or cover plates may be used in place of inspection openings provided they are large enough to permit an equivalent view of the interior.

If vessels with inside diameter not more than 16 in are so installed that they must be disconnected from an assembly to permit inspection, removable pipe connections of not less than 1½-in pipe size can be used for inspection openings.

Only vessels 12 in or less in inside diameter are not required to have openings, but to omit them is poor practice. The exact internal condition of a vessel should be known at all times whatever its size.

Openings in small tanks are required to be located in each head, or near each head, so that a light placed in one end will make the interior surfaces visible from the other. Inspection openings may be omitted in vessels covered in Code Par. UG-46(b). When openings are not provided, the manufacturer's data report shall include the notations required under remarks.

When a plant orders a vessel, it should be sure that the design provides for inspection openings that will be accessible after the vessel has been installed. The plant should also have the vessel examined periodically by competent Inspectors to make sure the internal surfaces remain free from deterioration.

Quick-Opening Closures

Code Par. UG-35 cautions users about removable closures on the ends of pressure vessels, utility access manways, and similar openings that are not of the multibolt type (charge openings, dump openings, quick-opening doors, cover plates). It states that these openings must be so arranged that the failure of one holding device will not release all the others.

Because of the serious failures incurred by such closures, fabricators and Inspectors must give them special attention and see that the closures are equipped with a protective device that will ensure their being completely locked before any pressure can enter the vessel and that will prevent their being opened until all pressure has been released.

As there are many types of quick-opening closures, it is not possible to give detailed requirements for protective devices. One example will have to suffice. The most common of these closures is the breech-lock door, which is supported on a hinge or davit and is locked or unlocked by being revolved a few degrees. When locked, the lugs in the door are in a position behind the lugs in the door frame or locking ring. An interlocking device can be easily installed to make sure these lugs are properly engaged before steam is fed to the vessel. If a positive locking device is not possible, some type of warning device should be installed, such as a locking pin that will prevent the door or locking device from turning until an exhaust valve opens. The steam escaping from the opened valve will warn the operator to release all the pressure in the vessel before the door is opened. Likewise, the valve cannot be closed until the door is in the proper closed position. All quick-opening closure vessels must have a pressure gage visible from the operating area.

Figures 3.1 and 3.2 show the result from the loosening of a set screw in the collar on a valve stem which allowed the valve in a safety locking device to be closed when the door was not fully locked. Vibration from the vessel being under pressure caused the locking device to become fully disengaged, allowing the door to be blown away. The 8-ft, 8-in–diameter door traveled about 500 ft through the air before shearing off the roof of a car on a highway. The door continued down the highway about 400 ft, then veered, tore down a chain-link fence, and cut down a 10-in-diameter tree. In reaction, the 129-ft-long cylinder vaulted from the 2-ft-deep pit in which it was located, traveled 200 ft before ramming through a building, and then struck, overturned, and demolished a railroad freight car half loaded with cement. The total force acting on the head of a quick-opening closure when it is under maximum pressure is very large.

Figure 3.1 Drying cylinder explodes when an almost foolproof safety device failed. (*Courtesy of The Hartford Steam Boiler Inspection and Insurance Company.*)

Figure 3.2 Door of cylinder. (*Courtesy of The Hartford Steam Boiler Inspection and Insurance Company.*)

Safety interlocking devices will help prevent many failures, but they are not infallible. The condition of the operating parts of a door should be carefully examined and the door itself should be observed to be in its intended position before pressure is applied. Precautions should be taken to ensure that the closed door cannot be moved while the pressure is on and that the pressure will be entirely released before the door is opened. The importance of maintenance of doors, door frames, and locking rings, especially the lubrication of locking surfaces and the wear on locking wedges of quick-opening closures, must not be overlooked. A precise examination schedule should be carried out.

Pressure Relief Devices

The importance of pressure relief devices in the safe operation of pressure vessels cannot be overemphasized. Improper functioning can result in disaster both to life and equipment. The *ASME Pressure Vessel Code,* in Par. UG-125, states that all pressure vessels must be provided with a means of overpressure protection and that the number, size, and location of pressure relief devices are to be listed on the manufacturer's data report. At times, vessels designed for a system are properly protected against overpressure by one safety valve. If this is the case, the vessel manufacturer may add the statement, "Safety valves elsewhere in the system." The item on the data sheets for safety valves should never be left blank, nor should it contain a statement that no overpressure protection is provided.

It is the manufacturer's responsibility to see that vessels are fabricated according to the ASME Code and that they meet the minimum design requirements. Simply because the customer did not indicate the size and location of the safety valve opening on the purchase order or drawing does not relieve the vessel manufacturer of this responsibility.

The safety devices need not be provided by the vessel manufacturer, but overpressure protection must be installed before the vessel is placed in service or tested for service, and a statement to this effect should be made in the remarks section of the manufacturer's data report.

The user of a pressure vessel should ensure that the requirements of the pressure relief devices, be they pressure relief valves (Code Par. UG-126), nonreclosing pressure relief devices such as rupture disk devices, or liquid relief valves, are designed in accordance with the ASME Code requirements and so stamped in accordance with Code Par. UG-129 as follows:

1. ASME Standard symbols on the valve, rupture disk or nameplate (see Code Figs. UG-129.1 and UG-129.2)
2. Manufacturer's name or identifying trademark
3. Manufacturer's design or type number
4. Size, in in (mm) of the pipe size of the valve inlet
5. Pressure at which the valve is set to blow, in psi (kPa)

6. Capacity, in ft³/min (m³/min) of air; or capacity, in lb/h of saturated steam (see Code Par. UG-131)
7. Year built

Rupture disk devices shall be marked by the manufacturer as required in Code Par. UG-129 in such a way that the marking will not be obliterated in service.

Safety valves must be installed on the pressure vessel or in the system without intervening valves between the vessel and protective device or devices and point of discharge (see also Code Appendix M).

To prevent excessive pressure from building up in the pressure vessel, a safety valve is set at or below the maximum allowable working pressure for the vessel it protects.

Safety valve construction is important in the selection of safety relief valves. By specifying that the construction must conform to ASME requirements, proper fabrication may be ensured.

Before a manufacturer of relief devices is allowed to use the Code stamp on a relief device, the manufacturer must apply to the Boiler and Pressure Vessel Committee in writing for authorization to use the stamp. A manufacturer or assembler of safety valves must demonstrate to the satisfaction of a designated representative of the National Board of Boiler and Pressure Vessel Inspectors that the manufacturing, production, testing facilities, and quality control procedures will ensure close agreement between performance of random production samples and the capacity performance of valves submitted to the National Board for capacity certification. (For capacity conversion formulas, see Code Appendix 11.)

Escape pipe discharge should be so located as to prevent injury to anyone where it is discharging. It is important to have the discharge pipe of the relief device at least equal to the size of the relief device. It must be adequately supported and provisions must be made for expansion of the piping system.

No pressure vessel unit can be considered completed until it is provided with a properly designed safety valve or relief valve of adequate capacity and in good operating condition.

All safety devices should be tested regularly by a person responsible for the safe operation of the pressure vessel.

Chapter

4

Guide to Quality Control Systems for ASME Code Vessels

A Practical Guide to Writing and Implementing a Quality Control Manual

To assure that the manufacturer has the ability and integrity to build vessels according to the Code, the *ASME Boiler and Pressure Vessel Code* requires each manufacturer of Code power boilers, heating boilers, pressure vessels, and fiberglass-reinforced plastic pressure vessels to have and to demonstrate a quality control system. (For pressure vessels, see Code Pars. U-2 and UG-90 and Code Appendix 10.) This system must include a written description or checklist explaining in detail the quality-controlled manufacturing process. The written description, which for clarification we will call a manual, must explain the company's organized and systematic way of specifying procedures involving each level of design, material, fabrication, testing, and inspection. An effective quality program will find errors before fabrication and will reduce or prevent defects during fabrication. To be effective, this system must have managerial direction and technical support. Proper administration is essential in order to provide the quality control and quality assurance necessary for fabrication of ASME Code pressure vessels.

Before issuance or renewal of an ASME Certificate of Authorization to manufacture Code vessels or piping is granted, the manufacturer's facilities and organization are subject to a joint review by an Inspection Agency and the legal jurisdiction concerned. In areas where a jurisdiction does not review a manufacturer's facility, that function may be carried out by a representative of the National Board of Boiler and Pressure Vessel Inspectors.

The quality control manual, therefore, will form the basis for understanding the manufacturer's controlled manufacturing system. It will be reviewed by the inspection agency, the legal jurisdiction, or a representative of the the National Board. The manual will serve as the basis for continuous auditing of the system by the Authorized Inspectors and later by review when the manufacturer's Certificate of Authorization is to be renewed.

The Code emphasizes that the manual must reflect the actual process by which the manufacturer produces a vessel. The Code states just a few specific requirements for the manual. For example, Code Appendix 10 sets out the various details to be described; however, it does not mandate a certain way to do this. Code Par. 10-4 includes the following statement: "The Code does not intend to encroach on the Manufacturer's right to establish and, from time to time, alter whatever form of organization the Manufacturer considers appropriate for its Code work." The suggestions and requirements implied in this book reflect a system that has worked and been accepted by the Code, and therefore can be considered guidelines for the manufacturer intending to develop a quality control system.

The scope and details of the system will depend upon the size, type of vessel being manufactured, and work to be performed. All welded pressure vessels require certain basic provisions which must be included in the manual. An outline of the procedures to be used by the manufacturer in describing a quality control system should include the following items:

1. *Statement of authority and responsibility.* Management's statement of authority, signed by the president of the company or someone else in top management, must specify that the quality control manager shall have the authority and responsibility to establish and maintain a quality assurance program and the organizational freedom to recognize quality control and quality assurance problems and to provide solutions to these problems.
2. *Organization.* Manufacturer's organizational chart giving titles and showing the relationship between management, quality control, purchasing, engineering, fabrication, testing, and inspection must be presented.
3. *Design calculations and specification control.* The quality control program shall include and address those responsible and their responsibilities regarding the design of pressure vessels which shall be produced under the *ASME Boiler and Pressure Vessel Code.* In an approved Code shop an Authorized Inspector's first

step in the shop inspection of Code vessels is to review the design. It is the manufacturer's responsibility to submit drawings and calculations; therefore, in the quality control manual, under design specifications, the following items should be listed.

a. The manufacturer or fabricator must verify that specifications have been reviewed and are available to the Authorized Inspector for his or her review and acceptance. Additionally, a representative of the manufacturer should be responsible to review, accept, and approve any customer-supplied calculations, specifications, and drawings to assure Code compliance prior to ordering of materials or releasing the job to the shop floor for fabrication (see Code Pars. U-2, and UG-90).

b. For those design calculations, drawings, and specifications prepared by the manufacturer, a responsible person shall be assigned to review, accept, and approve these documents. Design drawings, calculations, and specifications for the pressure vessel must be signed by a responsible engineering department person and reviewed with the Authorized Inspector prior to the start of work so that the Authorized Inspector may designate desired hold or inspection points. Figure 4.1 details a typical calculation sheet for ASME Code pressure vessel design. Computerized design programs must be verified for both input and output data. Controls must be described in the quality program regarding security of the system to preclude tampering.

c. The method of handling specification and drawing changes so that any changes in design specifications and calculations can be verified with the Authorized Inspector must be stated.

d. The table of revisions, revision number, and the date of revision should be given.

An important factor in designing a vessel is the accessibility of welding and the position of the joint to be welded. For example, if it is difficult to weld a joint because the welder cannot reach it, see it, and be comfortable at the same time, it is impossible to expect and get good quality work as a result. This is especially important in field erection. Start quality control by reviewing the drawings, for it is here that difficult-to-weld joints may be discovered and possibly corrected before fabrication is begun.

In reviewing a drawing, the Inspector will be looking for the following design information: dimensions, plate material and thickness, weld details, corrosion allowance, design pressure and temperatures, hydrostatic testing pressure, calculations for reinforcement of openings, and the size and type of flange. It is especially important to indicate whether the vessel is to be postweld heat-treated, radio-

Calculation Sheet

CUSTOMER'S NAME							ORDER NO.	
Design pressure		Material specification	Stress	Radiography		Postweld heat treat	Joint efficiency	
Design temperature		Shell		None		Yes	percent	
Corrosion allowance		Heads		Spot		No	percent	
Inside diameter		Bolts		Complete				
Head radius		Nuts						

Shell thickness

$$t = \frac{PR}{SE - 0.6P}$$

- SE
- $0.6P$
- $SE - 0.6P$

Code Par. UG-27

___ Corrosion allowance
___ Thickness required
___ Thickness used

Torispherical head thickness

$$t = \frac{0.885PL}{SE - 0.1P}$$

- SE
- $0.1P$
- $SE - 0.1P$

Code Par. UG-32

___ Corrosion allowance
___ Thickness required after forming
___ Thickness used

Ellipsoidal head thickness

$$t = \frac{PD}{2SE - 0.2P}$$

- $2SE$
- $0.2P$
- $2SE - 0.2P$

Code Par. UG-32

___ Corrosion allowance
___ Thickness required after forming
___ Thickness used

Flat head thickness

$$t = d\sqrt{\frac{CP}{S}}$$

Code Par. UG-34

___ Corrosion allowance
___ Thickness required
___ Thickness used

Maximum Allowable Working Pressure for New and Cold Vessels (see Code Par. UG-99)

Shell
- R
- $0.6t$
- $R + 0.6t$

$$P = \frac{SEt}{R + 0.6t} = \_\_\_\_ \text{ psi}$$

Torispherical head
- $0.885L$
- $0.1t$
- $0.885L + 0.1t$

$$P = \frac{SEt}{0.885L + 0.1t} = \_\_\_\_ \text{ psi}$$

Ellipsoidal head
- D
- $0.2t$
- $D + 0.2t$

$$P = \frac{2SEt}{D + 0.2t} = \_\_\_\_ \text{ psi}$$

Flat head

$$p = \frac{St^2}{d^2C} = \_\_\_\_ \text{ psi}$$

Maximum allowable working pressure = ____ psi (limited by ____)

Hydrostatic test pressure = ____ × ____ psi = ____ psi

Figure 4.1 Calculation sheet for an ASME pressure vessel.

graphed, or spot-radiographed because the Code allows 100 percent efficiency for double-welded butt joints that are fully radiographed, 85 percent for joints that are spot-radiographed, and 70 percent efficiency for joints not radiographed.

Any comments by the Inspector on the design should be brought to the attention of the manufacturer's engineering department immediately. It is easy to make corrections before construction. To make them afterwards is costly and sometimes impossible.

Another important function that the engineer performs in the control of quality is the selection of materials. The method of material procurement control must be shown in the manufacturer's quality control manual.

4. *Material control.* Material control is that quality function which ensures that only acceptable materials are utilized in the fabrication or construction of Section VIII, Division 1 pressure vessels. The quality control program set forth at the shop must address how the program will control the ordering, purchasing, receiving, storage, and material identification during the receipt inspection of materials.

The quality control manual shall address the following:

a. Engineering should have control of all materials which will be ordered for an ASME Code pressure vessel. The purchase requisition, bill of materials, or any other form used to document purchasing requirements shall clearly state the appropriate SA, SB, SFA, or other Code-allowed material specification for the material ordered. A job order number or another method of identifying a particular job order shall be issued (see Code Par. UG-5).

b. The system of material control used in purchasing, to assure identification of all material used in Code vessels and to check compliance with the Code material specification, should be explained (see Code Par. UG-93).

c. The engineering department must supply all quality assurance requirements on its purchase requisitions, bills of materials, or other requisitions before these documents are given to the purchasing department for procurement.

d. In the cold-forming of shell sections and heads or other pressure parts, the purchase order should state that the requirements of Code Pars. UCS-79 and/or UHT-79 must be satisfied.

Suppliers can sometimes cause problems. For example, while examining the material and fit-up of a flange on a carbon-steel vessel, the Inspector noticed the flange was 2.25 percent chromium, 1 percent molybdenum material. The vendor had substituted 2.25 percent chromium material because there was no carbon-steel flange this size in stock. The vendor thought a flange of better material was being shipped. But the 2.25 percent chromium, 1 percent molybdenum material required heat treatment and the carbon steel did not, in addition to the fact that this shop did not have welder qualifications for chrome molybdenum welding or the special welding procedure specification for the welding of that material.

5. *Examination and inspection program.* The examination and inspection program is at the heart of the entire quality control system and is generally not described in enough detail. This section of the quality control manual should describe all functions of inspection, tests, and examinations from the time material is delivered to the shop until the Code vessel is stamped, certified, and shipped from the manufacturing facility to the customer.

In order to stress the importance of performing the correct material-receiving inspection at the time of delivery it was noted that while performing an inspection at a Code manufacturing plant, a welder was attempting to weld a nozzle to a stainless steel type 321 vessel, but the weld metal was not flowing as it should. The welder notified the Inspector, who always carried a small magnet as part of the testing equipment. As the Inspector touched the magnet to the nozzle, there was a magnetic attraction. This easy test proved that the nozzle material was not type 321 stainless steel, since all 300 series stainless steels are nonmagnetic. Somehow a type 400 series stainless steel material, which is magnetic, was shipped and not properly checked when received. These are examples of why material-receiving inspection is so important. In addition to material-receiving inspections, a description of all inspections relative to the completed vessel shall be described as follows:

   a. The quality control department must verify the mill test reports with the material received by checking the heat numbers, material specifications, physical properties, and tensile strength with the mill test reports (see Code Pars. UG-77 and UG-85).

   b. Procedures must be established for receiving inspection and examination of plates, forgings, casting, pipe, nozzles, etc. This inspection shall require looking for laminations, scars, and surface defects to assure that all materials meet the quality assurance requirements of the Code. Formed heads and shells shall be checked for dimensions such as ovality, flange length, and curvature. Table 4.1 lists the Code-required markings and the required identification of materials during receiving inspection and during fabrication. Inspection procedures should be developed to address these areas of inspection.

The thickness of the plates can be gaged at this time to see if it meets Code tolerances, and a thorough examination can be made

Quality Control Systems for ASME Code Vessels 121

TABLE 4.1 Material and Plate Identification and Inspection

Remarks	Code reference
General recommendations.	Pars. UG-4, UG-5
Material must conform to a material specification given in Code Section II.	Pars. UG-6, UG-7, UG-8, UG-9, UG-10, UG-11, UG-12, UG-13, UG-14, UG-15
One set of original identification marks should be visible: name of manufacturer, heat and slab numbers, quality and minimum of the range of tensile strength. If unavoidably cut out, one set must be transferred by vessel manufacturer. Inspector need not witness stamping but must check accuracy of transfer.	Pars. UG-77(a), UG-94
To guard against cracks in steel plate less than ¼ in and in nonferrous plate less than ½ in, any method of transfer acceptable to Inspector to identify material may be used.	Par. UG-77(b)
Plates given full heat treatment by the producing mill shall be stamped with the letters MT.	Pars. UG-85, SA-20 Section II, Part A
When required heat treatments of material are not performed by the mill, they must be made under control of the fabricator who shall place the letter T following the letter G in the mill plate marking. To show that the heat treatments have been completed, the fabricator must show a supplement to the Material Test Report.	Pars. UG-85, UCS-85, UHT-5, UHT-81
Vessel manufacturer shall obtain Material Test Report or Certificate of Compliance for plate material.	Part UG-93
Inspector shall examine Material Test Report and material markings.	Pars. UG-93, UG-94
When stamping ferritic steels enhanced by heat treatment, use low-stress stamps.	Par. UHT-86
Other markings in lieu of stamping.	Pars. UG-77, UG-93
For some product forms, marking of container or bundle is required.	Pars. ULT-86, UG-93
Material must conform to a material specification given in Code Section II except as otherwise permitted.	Pars. UG-10, UG-11, UG-15

for defects such as laminations, pit marks, and surface scars. These dimensions and inspections shall be documented on a Receiving Inspection Report (see Fig. 4.2; see Code Pars. UG-93 and UG-74).

The Inspector's first check on the job itself is making sure that the plate complies with the mill test reports and Code material specifi-

RECEIVING REPORT

PURCHASE ORDER NUMBER: _____   Hold Tag: YES   NO
JOB NUMBER: _____   Circle One
INSPECTED BY: _____ DATE: _____   Released: YES   NO

Plate

T-1	T-2
T-3	T-4

Specification: _____ Grade: ____ Class: _____

Heat Number: _____ Number of sheets: ____
T-1 _____ Size: L _____ by W _____
T-2 _____
T-3 _____ Normalized: YES   NO   NA
T-4 _____   Circle One

Pipe

(cylinder showing T-1 and T-2)

Specification: _____ Grade: ____ Class: _____

Heat Number: _____ Quantity: _____
Length: _____ Diameter: _____
T-1: _____ A.S.M.E. Stamp: Yes   No   NA
T-2: _____ NB or Serial No.: _____
¶ UCS 79 Requirements are met ?   Yes   No   NA
Out of Roundness: Yes   No
                                              Circle one each

Head

T-1 (circle divided into D-1, D-2, K-1, K-2)

Specification: _____ Grade: ____ Class: _____
Hot Formed: _____   or   Cold Formed: _____
D-1: _____   Quantity: _____
D-2: _____   Diameter: _____
K-1: _____   Heat Number if applicable:
K-2: _____   _____
T-1: _____
Head Type: 2:1 Elips.   Hemi   Toris   Semi Ellips
¶ UCS 79 Requirements are met ?   Yes   No   NA
A.S.M.E. Stamp: Yes   No   NA
NB or Serial No. if Codestamped: _____ NA
                                              Circle one each

MISCELLANEOUS:
Couplings: _____

Flanges: _____

Fittings: _____

Welding Rods _____ Type _____

NCR NUMBER: _____
SPECIFICATION: _____

SPECIFICATION: _____

SPECIFICATION: _____

Figure 4.2   Receiving report.

cations. All material should be identifiable and should carry the proper stamping or markings required by the specification and/or Section VIII, Division 1.

c. Nonconforming materials, components, and parts that do not meet the Code requirements or purchase requirements shall be corrected in accordance with Code requirements or shall not be used for construction of Code vessels.

d. Partial data reports when required by the Code must be requested for welded parts that require Code inspection and that are furnished by another shop (see Code Par. UG-120).

The ASME and National Board have always advocated better management procedures and require an orderly pervasive system of quality assurance in all phases of Code vessel engineering and construction. A quality control manual must include the manufacturer's system of process control.

e. The scope of this section is to detail the system established to control quality in the manufacturer's operational sequence. This documented sequence shall be in the form of detailed process flowcharts, inspection records, or traveller sheets.

f. Detailed process flowcharts, inspection records, or traveller sheets should be prepared so that each operation can be examined and signed off by both the manufacturer's quality control representatives and by the Authorized Inspector. The Authorized Inspector shall be given the opportunity to establish inspection points on the inspection records prior to the beginning of the work.

g. The chief engineer must take the responsibility for correct design instructions given to the shop, specifying on the drawings, written specifications, or instructions any special Code requirements which would affect the fabrication or quality control operations.

h. The quality control manager must prepare and maintain procedures for controlling fabrication and testing processes—such as welding, nondestructive examinations, heat-treating, hydrostatic or pneumatic testing, and inspection—in addition to providing an operations process sheet with sign-off space for use by the manufacturer's representatives and the Authorized Inspector, since even the best check-off list may not indicate the hold points an Authorized Inspector will want to see. The list should include a remarks space for the Inspector to indicate hold points of parts that are to be examined (see Fig. 4.3; see also Code Pars. UG-90 to UG-99).

The person (by title) responsible for conducting the final pressure test shall also be stated. Before the hydrostatic test is performed,

Inspection Checklist

Job Number : _____ Drawing Number: _____
Part Number: _____ Prepared by: _____
Q.C.Manager: _____ Date: _____
A.I. Review: _____ Date: _____

Attribute	WPS/NDE NCR's Etc.	Q.C.M. Inspection Accept and Date	A.I. Inspection Accept and Date
1. Certificate of Authorization Number.			
2. Drawing and/or Design Calculation Number.			
3. Material Test Reports / C of C Review.	H		H
4. Fit-up inspection:			
Lower Head to shell			
Upper Head to Shell			
Nozzle # ___ to Shell			
Nozzle # ___ to Shell			
Nozzle # ___ to Shell			
Nozzle # ___ to Shell			
Nozzle # ___ to Shell			
Nozzle # ___ to Shell			
Long. Seam			
Course #1 to Course #2			
Course #2 to Course #3			
5. Verify Material Identification is maintained			
6. Concurrence of A.I. Prior to Weld Repair.	H		H
7. Root Pass Inspection:			
Lower Head to shell			
Upper Head to Shell			
Nozzle # ___ to Shell			
Nozzle # ___ to Shell			
Nozzle # ___ to Shell			
Nozzle # ___ to Shell			
Nozzle # ___ to Shell			
Nozzle # ___ to Shell			
Course #1 to Course #2			
Course #2 to Course #3			

H = Hold Point (Stop Work Notify Q.C.M.)
W = Witness Point (Notify Q.C.M.)

Page 1 of __

Figure 4.3 Inspection checklist.

Inspection Checklist

Job Number : _____ Drawing Number: _____
Part Number: _____ Prepared by: _____
Q.C.Manager: _____ Date: _____
A.I. Review: _____ Date: _____

Attribute	WPS/NDE NCR's Etc.	Q.C.M. Inspection Accept and Date	A.I. Inspection Accept and Date
8. Heat Treatment Procedure Review.			
9. Acceptance of Repair upon completion.		H	H
10. PT of Root Pass.			
11. Final Weld Inspection:			
Lower Head to shell			
Upper Head to Shell			
Nozzle # ___ to Shell			
Nozzle # ___ to Shell			
Nozzle # ___ to Shell			
Nozzle # ___ to Shell			
Nozzle # ___ to Shell			
Nozzle # ___ to Shell			
Long. Seam			
Course #1 to Course #2			
Course #2 to Course #3			

12. NDE Examination:
    RT UT MT PT VT
       Circle one.			

13. RT Film Review		H	H
14. UT, MT, PT, VT Reports			

15. Hydrostatic Test:
    Test Pressure _____ PSI
    at _____ F
		H	H

16. Check of welder I.D. Markings.			
17. Affix Required ASME Code Data Plate.			
18. Certify Manf. Data Report(s).		H	H
19. Affix Code Symbol Stamp.		H	H

H = Hold Point (Stop Work Notify Q.C.M.)
W = Witness Point (Notify Q.C.M.)

Page _2_ of ___

Figure 4.3 (*Continued*)

the vessel should be blocked properly to permit examination of all parts during the test. Large, thin-walled vessels may require extra blocking to guard against undue strains caused by the water load. Adequate venting must be provided to ensure that all parts of the vessel can be filled with water.

The recommended water temperature should be maintained at least 30°F above the minimum design metal temperature to minimize the risk of brittle fracture, but in no case should it exceed 120°F. If the temperature of the test medium is too low, moisture in the air may condense on the surfaces, making proper testing difficult. Two gages should be used to provide a double check on the pressure. If only one gage is used and it is incorrect, the vessel can be damaged by overpressure.

The hydrostatic test must be performed at a pressure equal to at least 1.5 times the maximum design pressure (DP) multiplied by the lowest ratio of the stress value for the test temperature to the stress value for the design temperature.

$$\text{Test pressure} = 1.5 \times \frac{\text{(allowable stress at test temperature)} \times \text{DP}}{\text{allowable stress at design temperature}}$$

Examinations of the vessel must be made at not less than two-thirds of the test pressure. After the vessel has been drained, a final examination should be made of the internal and external surface, the alignment, and the circularity (see Chap. 5, the section titled Hydrostatic Testing, Case 3).

i. Procedures for covering processes such as preheat, postweld heat treatment, hot-forming, cold-forming, etc., should be explained.

j. Material identification should be recorded on an as-built sketch, tabulation of materials, or on the inspection record or traveller sheet. If the plates are to be sheared or cut, the shop is required to transfer the original mill stamping to each section (see Code Par. UG-77). The Authorized Inspector need not witness the stamping but may wish to check the accuracy of the transfer. To guard against cracks caused by stamping in steel plate less than ¼-in thick, and in nonferrous plate less than ½-in thick, the method of transfer must be acceptable to the Inspector (see Code Par. UG-77).

k. This section shall address control of the ASME Code symbol stamp, who has custody of the stamp, who applies it, and when and how it is to be applied to the ASME Code data plate which will be attached to the completed vessel.

l. A description of the person responsible for the preparation of the ASME Code Data Report and the designated representative from the manufacturer who will certify the applicable data sheet form is needed. Additionally, if the vessel is required by the jurisdiction and/or customer to be registered with the National Board, a description of the person responsible for forwarding the forms to the National Board must be given. It should be noted that vessels manufactured outside the United States or the Provinces of Canada are required to be registered with the National Board of Boiler and Pressure Vessel Inspectors in Columbus, Ohio.

m. If the ASME Certificates of Authorization are extended to cover field construction or fabrication, this section should include controls relative to and functions necessary for controlling ASME-field activities, which must include the Authorized Inspector.

Many Code vessels, especially boilers, are fabricated partly in the shop and partly in the field. When construction of a Code vessel is completed in the field, some organization must take responsibility for the entire vessel, including the issuing of data sheets on the parts fabricated in shop and field and the application of Code stamping. Possible arrangements include the following:

(1) The vessel manufacturer assumes full responsibility, supplying the personnel to complete the job.
(2) The vessel manufacturer assumes full responsibility, completing the job with personally supervised local labor.
(3) The vessel manufacturer assumes full responsibility but engages an outside erection company to assemble the vessel by Code procedures specified by the manufacturer. With this arrangement the manufacturer must make inspections to be sure these procedures are being followed.
(4) The vessel may be assembled by an assembler with the required ASME certification and Code symbol stamp, which indicates that the organization is familiar with the Code and that the welding procedures and operators are Code-qualified. Such an assembler, if maintaining an approved quality control system, may assume full responsibility for the completed assembly.

Other arrangements are also possible. If more than one organization is involved, the Authorized Inspector should find out on the first visit which one will be responsible for the completed vessel. If a company other than the shop fabricator or certified assembler is called onto the job, the extent of association must be determined. This is especially important if the vessel is a power boiler and an outside contractor is hired to install the external piping. This piping, often entirely overlooked, lies within the jurisdiction of the Code, must be fabricated in accordance with the applicable rules of the ASME B31.1 Code for pressure piping by a certified manufacturer or contractor, and must be inspected by the Inspector who conducts the final tests before the master data sheets are signed.

Once the responsible company has been identified, the Inspector must contact the person directly responsible for the erection to agree upon a schedule that will not interfere with the erection but will still permit a proper examination of the important fit-ups, chip-outs, and the like. The first inspection will include a review of the company's quality control manual, welding procedures, and test results to make sure they cover all positions and materials used on the job. Performance qualification records of all welders working on parts of the vessel under jurisdiction of the Code must also be examined.

It is the responsibility of the manufacturer, contractor, or assembler to prepare drawings, calculations, and all pertinent design and test data so that the Inspector may make comments on the design or inspection procedures before the field work is started and thereby expedite the inspection schedule. New material received in the field for field fabrication must be acceptable Code material and must carry proper identification markings for checking against the original manufacturer's certified mill test reports. Material may be examined at this time for thickness, tolerances, and possible defects. The manufacturer who completes the vessel or parts of vessels has the responsibility for the structural integrity of the whole or the part fabricated. Included in the responsibilities of the manufacturer are documentation of the quality assurance program to the extent specified in the Code. The type of documentation needed varies in the different sections of the Code.

6. *Correction of nonconformities.* A definition of nonconformities must be given in the quality manual. Generally there are two types of nonconformities: those found in material and/or parts during receiving inspection and those found during fabrication and final testing. Both must be explained as follows:

   a. A nonconformity is any condition in material, manufacturing or fabrication, or quality of the finished product that does not comply with all of the applicable Code rules. Nonconforming materials, components, and parts that do not meet the Code requirements shall be corrected prior to the completion of the vessel in accordance with the quality control system and/or ASME Code requirements or they shall not be used for Code vessels.

   b. When a nonconforming situation is found, a hold tag or other sticker should be placed on the vessel, vessel part, or vessel

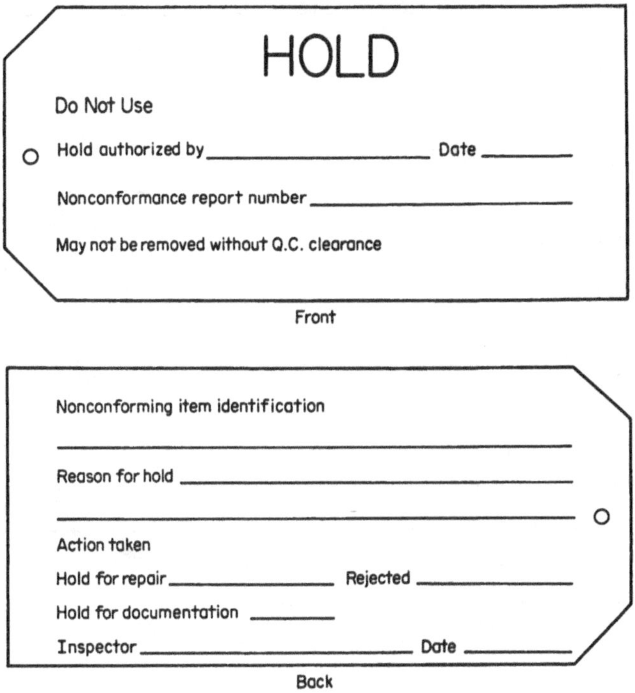

```
┌─────────────────────────────────────────┐
│              HOLD                       │
│  Do Not Use                             │
│ o Hold authorized by_____ Date ____   │
│  Nonconformance report number_____    │
│  May not be removed without Q.C. clearance │
└─────────────────────────────────────────┘
                  Front

┌─────────────────────────────────────────┐
│  Nonconforming item identification      │
│  _____    │
│  Reason for hold _____    │
│  _____  o │
│  Action taken                           │
│  Hold for repair_____ Rejected _____  │
│  Hold for documentation _____         │
│  Inspector_____ Date _____   │
└─────────────────────────────────────────┘
                  Back
```

Figure 4.4 Hold tag used for identifying nonconforming items.

material by the company Inspector describing the nature of the nonconformity and also giving all information possible on the tag and on the nonconformity report (see Figs. 4.4 and 4.5).

c. In order to determine a proper disposition of the nonconformity, it should be brought to the attention of the appropriate company person or nonconformity review board as well as the Authorized Inspector. The method of repair or corrective action should be documented on the nonconformity report and hold tag.

d. When the corrective action has been completed, the report should be signed and dated by the company representatives, normally the quality control manager and any other required company personnel stated in the quality control system, such as engineering or purchasing in conjunction with the Authorized Inspector.

To assure correct use of a nonconformity report, some companies print instructions on the backside of the report giving information

```
                    Nonconformance Report

    Nonconformance report number: _____
                            Date: _____
    Identification and description: _____
    _____
    _____
    _____
    _____


    Corrective action: _____
    _____
    _____
    _____



    Corrective action completed date: _____

    Hold tags removed by: _____ Date: _____

    Quality control manager: _____ Date: _____

    Authorized inspector: _____ Date: _____
```

Figure 4.5 Nonconformance report.

on how the report should be routed to the responsible department or person whose job it is to verify the disposition of the nonconformity and the correct action to take.

7. *Welding.* Welded pressure vessels must be fabricated by welders who are qualified utilizing welding procedures which are also qualified in accordance with ASME Code Section IX. Welding procedures and qualified welders shall be tested, certified, and qualified

prior to welding production joints. The following items must be addressed in the quality control program in order to assure Code requirements are attained:

a. A sound weld is almost impossible without good fit-up. Therefore, the importance of proper edge preparation and fitting of edges on longitudinal seams, girth seams, heads, nozzles, and pipe should be clearly emphasized in the manual. Tack welds, if left in place, shall be feathered at the stops and starts of the weld. If the tack welds are not prepared in such a manner they shall be removed (see Code Par. UW-31).

b. The procedure for fit-up and inspection before welding using either welding procedures or other written instructions should be listed. All tack welds shall be made utilizing qualified welding procedures and qualified welders (see Code Par. UG-90).

c. Sign-off spaces should be initialed by the Inspectors for fit-up inspection before welding. The Inspectors must assure themselves that Code tolerances for the alignment of the abutting plates of the shell courses, heads, and openings are met. A multiple-course vessel may also require staggering of the longitudinal joints. To avoid a flat section along the longitudinal seams, the edges of the rolled plates should be prepared with the proper curvature. The out-of-roundness measurement must be within the amount specified by the Code for the diameter of the vessel and the type of pressure (either internal and/or external) used (see Code Par. UG-96).

Following the fit-up of a part for welding, the quality of the weld is controlled by providing qualified welding procedures and welders' qualifications. Therefore, the quality control manual must show the system for welding quality assurance (see Code Pars. UW-26, UW-28, UW-29, and UW-31; see also Code Section IX, Pars. QW-103 and QW-322).

d. The manufacturer's system for assuring that all welding procedures and welders or welding operators are properly qualified in accordance with the ASME Code Section IX, Welding Qualifications, must be given. A description of who within the company is responsible for certifying and dating the procedure qualification record and welder performance qualification records must be addressed.

e. The engineering department, with the concurrence of the manager of quality control, must determine which welds require new procedure qualification and must provide methods for qualifying the welding procedure, welding operators, and welders. A system for revisions to welding procedures and procedure qualification records must be described.

f. The method of surveillance of the welding during fabrication to assure that only qualified procedures and welders are being used should be given.

g. The method for the qualification of personnel and the assigning of the welders' identification symbols (see Code Pars. UW-28 and UW-29) must be written.

h. The method of identifying the person making welds on a Code vessel, such as stamping the welds every 3 ft or utilizing a weld map which identifies each weld joint and who welded that joint (see Code Par. UW-37), should be described.

i. The method of welding electrode control, i.e., electrode ordering, to an SFA specification, inspection when received, storage of electrodes, how they are issued, and the in-process protection of the electrode (see Code Section II, Part C, Table A-1), should be explained.

j. The title of the person responsible for seeing that item *i* above is carried out must be listed.

k. The method used on the drawing to tell what type and size of weld and electrode are to be used, i.e., whether the drawing shows weld detail in full cross section or uses welding symbols, should be provided.

l. The method of tack welding must be explained. Do the tack welds become part of the weld or are they removed? (See Code Par. UW-31.) The control of tack welding must be part of the welding quality control system.

m. Continuity of welders' qualification records should be maintained and a log kept of all welders' activities to indicate the welders' continuity in specific welding processes. If there is a gap of time in excess of six months in a specific process, the welder shall be requalified. Figure 4.6 gives an example of a welder's qualifications and performance record log.

Tack welds to secure alignment are often under high stress when fit-up is made. Also, the rapid cooling of the tack weld to the parent metal can form brittle crack-sensitive microstructure and can cause cracks in the base material as well as in the tack weld. All tack welds must be made with a qualified welding procedure specification and by qualified welders.

A most important control in welding is actual surveillance on the job. This cannot be minimized. The welding of the vessel must be done in accordance with a qualified procedure. The back-chip of the welded seams must be inspected to see that all defects have been removed and that the chipped or ground groove is smooth and has a

WELDERS QUALIFICATIONS & PERFORMANCE RECORD LOG

Welder's name	Welder's symbol	Welding process qualified	Jan.	Feb.	Mar.	Apr.	May	June	July	Aug.	Sept.	Oct.	Nov.	Dec.

Figure 4.6 Welder's qualification and performance record log.

proper bevel to allow welding without entrapping slag. Each piece of welding must be cleaned of all slag before the next bead is applied; many defects can be caused by improper cleaning.

When a backing strip is used, it is important to see that the strip is closely fitted and that the materials to be welded have sufficient clearance to permit good penetration on the first pass. The first pass is more susceptible to cracking because of the large amount of parent metal involved and the fact that the plate may be cold. As the root pass solidifies quickly, it tends to trap slag or set up stresses that can induce cracking if the fit is not good. The importance of good fit-ups to a sound weld cannot be overemphasized.

The finished butt welds should be free of undercuts and valleys and also should not have too high a crown or too heavy an overlay. Fillet welds should be free of undercuts and overlap. The throat and leg of the welds should be the size specified by the Code.

8. *Nondestructive examination.* A weld may look perfect on the outside and yet reveal cracks, lack of penetration, heavy porosity, or undercutting at the backing strip when the weld is given a nondestructive examination (NDE) by radiography or ultrasonic tests. As with welding, nondestructive examinations may require a procedure, and like the welding procedure NDE must satisfy the requirements of the Code and be acceptable to the Authorized Inspector. Various sections of the ASME Code refer to Section V, Nondestructive Examinations, for nondestructive examinations. These examinations should be made to follow a detailed written procedure according to Section V and the referencing Code section (i.e. Section I, Section VIII, Division 1, etc.). The manufacturer's quality control manual must include the following:

   a. A description of the manufacturer's system for assuring that all nondestructive examination personnel are qualified in accordance with the applicable code is needed. The quality control program must define what NDE will be utilized and whether it will be performed in house or subcontracted.

   b. Written procedures to assure that nondestructive examination requirements are maintained in accordance with the Code shall be addressed.

   c. If nondestructive examination is subcontracted, a statement of who is responsible for checking the subcontractor's site, equipment, and qualification of procedures to assure that they are in conformance with the applicable Code must be given. In a nondestructive examination certification program, the manufacturer has the responsibility for qualification and certification of its NDE personnel. To obtain these goals, personnel may be trained and certified by the manufacturer, or an independent laboratory may be hired. If an independent laboratory is hired, the manufacturer must state in the quality control manual the department and person (by title) responsible for assuring that the necessary quality control requirements are met, because the manufacturer's responsibility is the same as if the parent company had done the work.

    d. If the manufacturer performs radiography testing (RT) or ultrasonic testing (UT), NDE personnel shall be qualified in accordance with the recommended practice known as SNT-TC-1A utilizing the current Code-accepted edition.

    For the pressure vessel fabricator, nondestructive testing provides a means of spotting faulty welds, plates, castings, and forgings before, during, and after fabrication. For the user, it offers a maintenance technique that more or less assures the safety of a vessel until the next scheduled inspection period.

9. *Heat treatment.* Because of certain materials utilized in construction and fabrication, due to material thickness limitations or product form, heat treatment may not be required or desirable. If a company does not have heat-treatment facilities a paragraph should address this fact and simply state that if heat treatment is required it will be subcontracted. The following should be addressed regarding heat treatment when such treatment is required:

    a. An explanation of when preheat is necessary, the welding preheat, and interpass temperature control methods should be provided.

    b. For postweld heat treatment, a description of how the furnace is loaded, how and where the thermocouples are placed, the time–temperature recording system, marking on charts, and how these details are recorded for each thermocouple (see Code Pars. UW-40 and UW-49) should be given.

    c. The methods of heating and cooling, metal temperature, metal temperature uniformity, and temperature control must be explained.

    d. Methods of vessel or parts support during postweld heat treatment should be listed.

10. *Calibration of measurement and test equipment.* A system shall be established documenting the method of calibration of measurement and test equipment as follows:

    a. The system used for assuring that the gage measuring and testing devices used are calibrated and controlled should be described.

    b. An explanation of how records and identification are kept to assure proper calibration must be given. These records shall be available to Authorized Inspectors.

    c. All gages should be calibrated with a calibrated master gage or a standard deadweight tester whenever there is reason to believe that they are in error. It is highly recommended that all gages be tested and if necessary calibrated at least every six months. The method for testing gages and instruments with references to the standards of the National Institute of Standards Testing (NIST) or other recognized national standards should be detailed. In the case of foreign Code manufacturers, for example, the Republic of Korea, records traceable to the Korean Standards (KS) are acceptable. Having accurate and calibrated gages will assure the correct testing pressure when the vessel is undergoing hydrostatic testing or pneumatic testing.

Many fabricators take testing equipment for granted and seldom check the accuracy of gages and measuring equipment. We have often had gages checked that proved to be inaccurate. For example, in testing a vessel 12 ft in diameter with a quick-opening door, the pressure gage was found to be 25 lb off on the low side. On a head 12 ft in diameter its total area is 16,350 in^2. For each pound over the test pressure, the head would carry 16,350 lb. If this gage had been used, the extra load on the head would be 25 × 16,350, or 410,000 lb above the test requirement. This extra pressure would have been enough to cause considerable damage to this type of opening. The quality control manual must state how the measuring and test equipment is controlled.

11. *Record retention.* The ASME Code Section VIII, Division 1 has limited retention requirements relative to records. This topic must be addressed as follows:

    a. The manufacturer must have a system for the maintenance of manufacturer's data reports, radiographs, impact-test results, and the like.

    b. The manager of quality control shall assemble all the manufacturing and inspection documents required by the referencing Code section. These records shall be filed until the data report is signed by the Authorized Inspector and shall include, but not be limited to, the following: design specifications and drawings, design calculations, certified material test reports, nondestructive examination reports and radiographic films, heat-treatment records, hydrostatic and pneumatic test records, impact-test results, copies of signed ASME data reports, nonconformity records, inspection check-off list, and sign-offs. When the data report is signed by the Authorized Inspector, the manufacturer may dispose of these records. If the manufacturer has not registered the vessel with the National Board of Boiler and Pressure Vessel Inspectors, the data reports must be kept in the manufacturer's files for a minimum of five years. UT reports shall also be kept for five years.

    c. All records shall be kept in a safe environment.

    The maintenance of records and revisions in accordance with the

Code should be the quality control manager's responsibility. A system should be maintained for controlling the issuance of the latest quality control manual revisions. Revisions generally are controlled through issuance of the revised section, which is documented on a revisions control sheet or table of contents placed in the front of the manual. Notify the Authorized Inspector of any contemplated revisions for the quality control manual. Verify any revisions with the Authorized Inspector. Revisions and revision record forms could be listed as an appendix of the manual. No revisions are to be issued without prior review and acceptance by the Authorized Inspector.

12. *Sample forms.* The forms used by the company which are generated and which affect and control the quality of the vessel shall be shown by exhibit in the quality control manual as follows:
    a. The type of check-off form used, including those for welding procedures and welding performance qualification, traveller sheets, welder's log records, and other charts used in the completion of records for material purchase, identification, examinations, and tests that were taken before, during, and on completion of the vessel, should be listed.
    b. Examples of all forms should be given and their use, such as the check-off list and data reports which will be presented to the Authorized Inspector for review, should be described. The Inspector will carefully check these documents to make sure that they are properly listed in the data reports, that the vessel complies with the Code, and that the data reports are signed by the manufacturer before being signed by the Inspector. Documentation requires good procedure and organizational review. Experience has proved that with good documentation a manufacturer is able to assess the operation accurately and to improve the efficiency of production.

13. *Authorized Inspector (inspection of vessel and vessel parts)*
    a. The manual should state that the Authorized Inspector is the Code Inspector, and it is the obligation of the company to assist in the performance of these duties.
    b. The Inspector can indicate hold points and discuss with the quality control manager arrangements for additional inspections and audits of personal choosing to establish the company's compliance with the ASME Code.
    c. The manual should state that the checklist of records, drawings, calculations, process sheets, and any other quality control records shall be available for the Authorized Inspector's review.
    d. The final hydrostatic or pneumatic tests required by the Code

shall be witnessed by the Authorized Inspector, and any examinations deemed necessary shall be made before and after these tests.

e. The manufacturer's data reports shall be carefully reviewed by a responsible representative of the manufacturer or assembler. If this representative finds them properly completed, they shall be signed for the company. The Authorized Inspector, after gaining assurance that the requirements of the Code have been met and the data sheets properly completed, may sign these data sheets.

The manufacturer should be carefully acquainted with the provisions in the Code that establish the duties of the Authorized Inspector. This third party in the manufacturer's plant, by virtue of authorization by the state to do Code inspections, is the legal representative. Having this uninterested third party at the shop of fabrication permits the manufacturer to fabricate under state laws. Under the ASME Code rules, the Authorized Inspector has certain duties as specified in the Code.

It is the obligation of the manufacturer to see that the Inspector has the opportunity to perform as required by law. The details of an authorized inspection at the plant or site are to be worked out between the Inspector and the manufacturer. The manufacturer's quality control manual should state provisions for assisting the Authorized Inspector in performing duties (see Code Par. UG-92).

The Authorized Inspector, after witnessing the final hydrostatic tests or pneumatic tests and carrying out the examinations that are necessary, will check the vessel data report sheets to make sure they are properly filled out and signed by the manufacturer and that the material and quality of the work in the vessel are accurately identified. The Authorized Inspector will then allow the Code symbol to be applied in accordance with the Code and any local or state requirements. If everything is in order, the data sheets are signed by the Inspector.

14. *Audits.* Audits are not required by all sections of the Code, but in order to make certain that the quality control system is being properly used during manufacture, a comprehensive system of planned and periodic audits should be carried out by the manufacturer to assure compliance with all aspects of the quality control program. The manufacturer's quality control manual should describe the system of self-auditing.

   a. The manufacturer's system of planned periodic audits to determine the effectiveness of the quality assurance program must be explained.

   b. Audits should be performed in accordance with the manufac-

turer's procedure, using the checklist developed for each audit, to ensure that a consistent approach is used by audit personnel.
   c. Reports of audit results should be given to top management. Positive corrective action to be taken, if necessary, should also be indicated.
   d. Corrective action must be taken by the quality control manager to resolve deficiencies, if any, revealed by the audits.
   e. A re-audit should be made within 30 days to confirm results when corrective action is taken.
   f. Results of the audit must be made available to the Authorized Inspector.
   g. The name and title of the person responsible for the audit system should be specified. Good management will have a self-auditing system that regularly reviews the adequacy of its quality control program and provides for immediate and effective corrective measures when required.
15. *Appendix.* This section of the quality control manual is not mandatory; however, this section may contain the necessary forms, exhibits, and other items or such items may be placed in the applicable sections of the quality control manual. The following are ideas that may be placed here:
   a. The table of revisions, the revision number, and the date of revision can be presented.
   b. All forms can be shown and the form usage can be described.

In our comments, we have given the reader an in-depth outline for inspection and quality control systems that the average-sized manufacturer of boilers and pressure vessels might use in establishing a quality control program and/or quality manual as required by the ASME Code Section I, Power Boilers, Section IV, Heating Boilers, and Section VIII, Divisions 1 and 2, Pressure Vessels. A quality assurance program for nuclear vessels would require a more comprehensive program and manual (see Chap. 10). We have not included Section X, Fiberglass Reinforced Plastic Pressure Vessels. For quality control systems required by these Code vessels, the applicable Code sections must be followed.

It is important to remember that before the issuance or renewal of an ASME Certificate of Authorization to manufacture Code vessels is granted, the manufacturer's facilities and organization will be reviewed by an inspection agency and/or legal jurisdiction, or a representative of the National Board of Boilers and Pressure Vessel Inspectors.

In reviewing the various operations listed in the manufacturer's quality control manual, their objective will be to see the controlled manufacturing system described by the manufacturer's quality control manual in operation. This survey must be jointly made before the survey team can write a report of the manufacturer's system for the American Society of Mechanical Engineers.

The responsibility for establishing the required quality control system is that of the manufacturer. In writing a quality control manual, effective means of planning during all phases of design, fabrication, and inspection should be considered in order to assure a properly constructed Code vessel. Rear Admiral Charles Curtze, in a report to the congressional committee investigating the loss of the USS Thresher, talked about the need for a higher IQ (in this case, integrity and quality) in industry. If your plant has both the ability to build Code vessels and a good IQ, you should have no trouble acquiring and maintaining the ASME Code stamp.

Chapter

5

Inspection and Quality Control of ASME Code Vessels

Shop Inspection

An authorized inspector's first step in the shop inspection of Code vessels is to review the design. It is the manufacturer's responsibility to submit drawings, calculations, and the following design information: dimensions, plate material specifications and thickness, weld detail drawings, calculated corrosion allowance requirements, design pressure and temperature calculations [see Code Pars. UG-20 and UCS-66], impact-test exemptions if allowed [see Code Pars. UG-20(f) and UCS-66], hydrostatic testing pressure (based on the maximum allowable working pressure) and the part of the vessel to which the maximum pressure is limited, calculations for reinforcement of openings, and the design requirements for the size and type of flanges which may be used on the vessel. It is especially important to indicate whether the vessel is to be postweld heat-treated and the degree of radiography because the Code generally allows 100 percent efficiency for double-welded butt joints that are fully radiographed, 85 percent efficiency for joints that are spot-radiographed, and 70 percent efficiency for joints not radiographed. Table 5.1 lists the various inspections, tests, and examinations which are established within the Code. This list is given to assist the reader in determining when certain tests, examinations, and inspections must be made.

Any comment by the Inspector on the design should be brought to the attention of the manufacturer's engineering department immediately. It is easy to make corrections before construction; to make them afterwards is costly and sometimes impossible.

The Inspector's first check on the job itself is making sure that the materials of construction comply with the purchase order require-

TABLE 5.1 ASME Code—Required Tests, Inspections, and Nondestructive Examinations

Remarks	Code reference
Hydrostatic Tests	
Test must be at at least 1.3 times the maximum design pressure multiplied by the lowest ratio of the stress value for the test temperature to that for the design temperature	Par. UG-99, Fn. 35 Par. UG-21, Fn. 8
If the allowable stress at design temperature is less than the allowable stress at test temperature the hydrostatic test pressure must be increased proportionally. Thus $$\text{Test pressure} = 1.3 \times \frac{\text{allowable stress at test temperature} \times \text{design pressure}}{\text{allowable stress at design temperature}}$$	Par. UG-98
Combination units must be so tested that hydrostatic pressure is on one chamber without pressure in the other parts	Pars. UG-21, UG-98
Include corrosion allowance in calculating test pressure	
Inspection must be made at a pressure not less than $2/3$ of test pressure	
Test cast ductile iron at 2 times maximum allowable working pressure	Par. UCD-99
Test cast-iron vessels at 2 times the design working pressure; for design pressure under 30 psi, test at $2\frac{1}{2}$ times design pressure but not to exceed 60 psi	Par. UCI-99
Test pressure shall be at least 1.4 times the design pressure at 100°F	Par. ULT-99
Test for clad-plate vessels	Par. UCL-52
Pneumatic Tests	
Pneumatic test may be used instead of hydrostatic test when: 1. Vessels are so designed and supported that they cannot safely be filled with water 2. Vessels for service in which traces of testing liquid cannot be tolerated are not easily dried	Par. UG-100
Test pressure must not be less than $1\frac{1}{4}$ times maximum allowable pressure	
All attachment welds of a throat thickness greater than $\frac{1}{4}$ in must be given a magnetic particle or liquid penetrant test before pneumatic test	Par. UW-50
Special precautions should be taken when using air or gas for testing	Par. UG-100, Fn. 36

TABLE 5.1 ASME Code—Required Tests, Inspections, and Nondestructive Examinations (*Continued*)

Remarks	Code reference
Magnetic Particle and Liquid Penetrant Tests	
When vessel is to be pneumatically tested, welds around openings and attachment welds of throat thickness greater than ¼ in must be given a magnetic particle or liquid penetrant examination for ferromagnetic materials (liquid penetrant for nonmagnetic materials)	Par. UW-50
Liquid penetrant examination must be made on all austenitic chromium–nickel alloy steel welds in which shell thickness is over ¾ in and on all thicknesses of 36 percent nickel steel welds. Examinations shall be made after heat treatment if heat treatment is given	Par. UHA-34
Heat-treated enhanced materials used for vessels; all welds must be given a magnetic particle examination after hydrostatic testing. Liquid penetrant is an acceptable alternative	Par. UHT-57
Pressure part welded to a plate thicker than ½ in to form a corner joint must be examined by magnetic particle or liquid penetrant method	Pars. UG-93, UW-13 Fig. UW-13.2
Layered vessels	Par. ULW-56
Low-temperature vessel construction; all attachment welds and joints not radiographically examined shall be given a liquid penetrant examination	Par. ULT-57

Qualification of personnel:
  Magnetic particle
  Liquid penetrant
	Code Appendix 6
	Code Appendix 8

ments, mill test reports supplied with the material, and ASME Section II, Material Specifications. All material should be identifiable and carry the proper stamping. The thickness of the plates can be gaged at this time to see if it meets Code tolerances, and a thorough examination can be made for defects such as laminations, pit marks, and surface scars.

If the plates are to be sheared or cut, the shop is required to transfer the original mill stamping to each section. The Inspector need not witness the stamping but must check the accuracy of the transfer. To guard against cracks from stamping in steel plate less than ¼ in thick and in nonferrous plate less than ½ in thick, the method of transfer must be acceptable to the Inspector.

Code tolerances for the alignment of the abutting plates of the shell courses, heads, and openings must be met (see Code Par. UG-33 and

Table UG-33). Table 5.2 instructs the reader on how to inspect these tolerances for strict compliance to the ASME Code. Such inspections are verifying the out-of-roundness measurement of the shell of the vessel. Table 5.2 lists such tolerances and the location of the acceptance criteria in the Code. A multiple-course vessel may also require staggering of the longitudinal joints. To avoid a flat section along the longitudinal seams, the edges of the rolled plates should be prepared with the proper curvature. The out-of-roundness measure must be within the amount specified by the Code for the diameter of the vessel and the type of pressure (internal or external) used (see Code Pars. UG-80 and UG-81 and Code Figs. UG-80.1 and UG-80.2).

The welding of the vessel must be done according to a qualified welding procedure (see Chap. 6). The back-chip of the welded main seams must be inspected to see that all defects have been removed and that the chipped or ground groove is smooth and has a proper bevel to allow welding without entrapping slag. Each bead of welding must be cleaned of all slag before the next bead is applied; many defects can be caused by improper cleaning.

When a backing strip is used, it is important to see that the ring is closely fitted and that the plates have sufficient clearance to permit good penetration on the first pass. The first pass is more susceptible to cracking because of the large amount of parent metal involved and the fact that the plate may be cold. As the root pass solidifies quickly, it tends to trap slag or set up stresses that can induce cracking if the fit is not good. The importance of good fit-up to a sound weld cannot be overemphasized.

The finished butt welds should be free of undercuts and valleys and also should not have too high a crown or too heavy an overlay. Fillet welds should be free of undercuts and overlap. The throat and leg of the weld should be of the size specified by the drawing and in compliance with the Code. Automatic welding properly performed is superior to hand welding, but it requires rigid supervision and inspection. The best check is provided by spot-radiographing. If radiography is unavailable, a test plate can be welded and sectioned (see the section titled Sectioning in this chapter).

Tests

A weld may look perfect on the outside and yet reveal deep cracks, lack of penetration, heavy porosity, or undercutting at the backing strip when it is radiographed.

If the vessel is to be radiographed, all irregularities on the weld plate surfaces must be removed so as not to interfere with the interpretation

TABLE 5.2 Code Tolerance Requirements

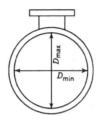

Tolerance	Remarks	Code reference
Out-of-roundness measure in cylindrical shells under internal pressure	$D_{max} - D_{min}$ must not exceed 1 percent of nominal diameter	Par. UG-80(a) Fig. UG-80.2
	At nozzle openings, increase above tolerance by 2 percent of inside diameter of opening	
	Tolerance must be met at all points on shell	
Out-of-roundness measure in cylindrical shells under external pressure	Maximum plus-or-minus deviation from true circle using chord length equal to twice the arc length obtained from Code Fig. UG-29.2 is not to exceed the value e determined by Code Fig. UG-80.1	Par. UG-80(b) Figs. UG-80.1, UG-29.2 Appendix L-4
	Take measurements on unwelded surface of plate	
	For shells with lap joint, increase tolerance by value t	Par. UG-80(b)
	Do not include corrosion allowance in value t	
Permissible variation in plate thickness	No plate shall have an undertolerance exceeding 0.01 in or 6 percent of the ordered thickness, whichever is less	Par. UG-16
Formed heads	Inside surface must not deviate from specified shape by more than 1.25 percent of inside skirt diameter	Pars. UG-81, UW-13
	Minimum machined thickness is 90 percent of that required for a blank head	

TABLE 5.2 Code Tolerance Requirements (*Continued*)

Tolerance	Remarks	Code reference
Pipe under tolerance	Manufacturing undertolerance must be taken into account	Par. UW-16
	Follow pipe specifications	Subsection C and Code Material Section II

Tolerance	Remarks		
	Plate thickness, in	Offset limits	Code reference
Longitudinal joints (category A)	Up to ½ in inclusive Over ½ in to ¾ in Over ¾ in to 1½ in Over 1½ in to 2 in Over 2 in	¼t ⅛ in ⅛ in ⅛ in Lesser of ¹⁄₁₆t or ⅜ in	Par. UW-33 Table UW-33
Circumferential joints (categories B, C, D)	¾ in or less Over ¾ in to 1½ in Over ½ in to 2 in Over 2 in	¼t ³⁄₁₆ in ⅛t Lesser of ⅛t or ¾ in	Par. UW-33 Table UW-33
Offset type joint	⅜ in maximum; longitudinal weld within offset area shall be ground flush with parent metal prior to offset and examined by magnetic particle methodafter offset		Par. UW-13.(c) Fig. UW-13.1(k) plus note
Welding of plates of unequal thickness			Pars. UW-9, UW-42 Fig. UW-9
Spherical vessels and hemispherical heads	Offset must meet longitudinal joint requirements		Par. UW-33
	Offset must be faired at a 3 to 1 taper over width of finished weld		Pars. UW-42, UW-9 Fig. UW-9
	Additional weld metal maybe added to edge of weld if examined by liquid penetrant or magnetic particle method. If the weld buildup is in the area to be radiographed, the buildup shall also be radiographed		Pars. UW-42, UW-9

of subsurface defects. If a vessel is to have only a pneumatic test, all attachment welds with a throat thickness greater than $\frac{1}{4}$ in must be given a magnetic particle examination or liquid penetrant examination before the pneumatic test.

If a vessel is constructed of austenitic chromium–nickel alloy steels over $\frac{3}{4}$ in thick, all welds must be given a liquid penetrant examination (after heat treatment if such treatment is performed).

If the vessel is to be fully postweld heat-treated, the furnace should be checked to make sure that the heat will be applied uniformly and without flame impingement on the vessel.

Vessels with thin walls or large diameters should be protected against deformation by internal bracing. The vessel should be evenly blocked in the furnace to prevent deformation at the blocks and sagging during the stress-relieving operation. Furnace charts should be reviewed to determine if the rate of heating, holding time, and cooling time are correct.

All work should be completed before a vessel is presented for final testing, especially if it is to be postweld heat-treated. A minor repair can assume major proportions because of lack of time to make the repair and repeat postweld heat treatment and testing prior to shipping.

A vessel undergoing a hydrostatic test should be blocked properly to permit examination of all parts during the test. Large, thin-walled vessels may require extra blocking to guard against undue strains caused by the water load. Adequate venting must be provided to ensure that all parts of the vessel can be filled with water. The water should be at approximately room temperature. If the temperature is lower, moisture in the air may condense on the surfaces, making proper testing difficult. Two gages should be used to provide a double check on the pressure. If only one gage is used and it is incorrect, the vessel can be damaged by overpressure. The hydrostatic test must be performed at pressures equal to at least $1\frac{1}{2}$ times the maximum design pressure multiplied by the lowest ratio of the stress value for the test temperature to the stress value for the design temperature. Inspection and examination of the vessel must be made at not less than two-thirds of the test pressure. After the vessel has been drained, a final examination should be made of the internal and external surfaces, the alignment, and the circularity. If everything is satisfactory, the vessel may be stamped in accordance with Code and any local or state or jurisdictional requirements.

The data report sheets are then checked to make sure that they are properly filled out, signed, and certified by the manufacturer and that the material and workmanship of the vessel are accurately stated. If everything is in order, the data sheets are signed by the inspector.

ASME Manufacturer's Data Reports

The importance of having ASME Manufacturer's Data Reports fully and correctly completed, as well as record retention of complete vessel material and fabrication records, must not be underestimated.

When a pressure vessel accident happens, the cause could have been poor operating procedures, insufficient vessel maintenance, inadequate training and indoctrination of the operating personnel, or incorrect sizing and type of relief devices. All of the above have been seen by the authors while doing forensic work involving product liability.

When an employee is injured in an accident, there appears to be no laws under which to sue the employer. The employer only pays for worker's compensation. The employee and hired attorney then go after the manufacturer of the vessel for product liability and try to prove the vessel was improperly designed and fabricated. In the event of litigation, the ASME data reports may be subpoenaed as legal documents. It is here that the vessel manufacturer's properly completed ASME data report and preservation of records even beyond the time required by the ASME Code will be most helpful. *(To assist the manufacturer in completing the data report correctly, refer to Appendix W of Section VIII, Division 1 for guidance.)*

Records should include the customer's purchase order, the user's and vessel manufacturer's design reports, certified mill test reports of materials, records of qualified welding procedures, a list of all qualified welding operators who welded on the vessel, quality control and/or quality assurance checklists, and certified ASME data reports signed by the vessel manufacturer.

The authors have seen situations where a data report was not filled out completely. For example, the data report item on safety relief outlets was left blank. The vessel manufacturer should have tried to find out from the vessel user the type and the size and setting of the relief device. Code Par. UG-125 states that all pressure vessels must be provided with a means of overpressure protection. The number, size, and location of relief devices are to be listed on the Manufacturer's Data Report.

There are times when the vessels are designed for a system and properly protected against overpressure by one safety relief device. If this is the case, the vessel manufacturer may add the statement under the pressure relief devices item *provided by others* or *provided by owner*.

The pressure relief device item should never be left blank, nor should it contain a statement that no overpressure protection is provided.

It is the manufacturers' responsibility to see that the vessels' fabricated items conform to the ASME Code and meet the minimum design requirements. Simply because the customer did not indicate on the purchase order or drawing the size, pressure, and location of the safety relief device does not relieve the manufacturer of this responsibility.

The safety devices need not be provided by the vessel manufacturer, but overpressure protection must be installed before the vessel is placed in service or tested for service, and a statement to this effect should be made in the Remarks section on the Manufacturer's Data Report.

No pressure vessel unit can be considered completed until it is provided with a properly designed safety relief device of adequate capacity and in good operating condition.

Field Assembly Inspection

Many code vessels are fabricated partly in the shop and partly in the field, especially boilers. Although this book deals primarily with pressure vessels, many plants will want to know about the field assembly of boilers as well.

Responsibility for Field Assembly

When construction of a Code vessel is completed in the field, some organization must take responsibility for the entire vessel, including the issuing of data sheets on the parts fabricated in shop and field and the application of Code stamping. Possible arrangements include the following:

1. The vessel manufacturer assumes full responsibility, supplying the personnel to complete the job.
2. The vessel manufacturer assumes full responsibility, completing the job with local labor under the manufacturer's supervision.
3. The vessel manufacturer assumes full responsibility but engages an outside erection company to assemble the vessel by Code procedures specified by the manufacturer. With this arrangement, the manufacturer must make inspections to be sure these procedures are being followed.
4. The vessel may be assembled in the field by an assembler with the required ASME Certificate of Authorization whose scope allows field construction. This certificate along with the applicable ASME Code stamps indicates that the assembler's organization is familiar with the Code and that the welding procedures and operators are Code-qualified. Such an assembler may assume full responsibility for the complete assembly.

Other arrangements are also possible. If more than one organization is involved, the Authorized Inspector should find out on the first visit which one will be responsible for the completed vessel. If a company representative other than the shop fabricator or an authorized assembler is called onto the job, the extent of association must be determined. This is especially important if the vessel is a power boiler and an outside contractor is hired to install the external piping. This piping, often entirely overlooked, lies within the jurisdiction of the Code and must be fabricated in accordance with the applicable rules of the ASME B31.1 Code for pressure piping by a certified manufacturer or contractor and must be inspected by the inspector who conducts the final tests before signing the master data sheets.

Inspection Procedure

Once the responsible company has been identified, the inspector must contact the person directly responsible for the erection to agree upon a schedule that will not interfere with the erection but will still permit a proper examination of the important fit-ups, chip-outs, and the like. The first inspection may include a review of welding procedures (and test results) to make sure they cover all positions and materials used on the job. Performance qualification records of all welders working on parts of the vessel under the jurisdiction of the Code must also be examined.

It is the responsibility of the manufacturer, contractor, or assembler to submit drawings, calculations, and all pertinent design and test data (see the section titled Shop Inspection in this chapter) so that the inspector may make comments on the design or inspection procedures before the field work is started and thereby expedite the inspection schedule.

New material received in the field for field fabrication must be acceptable Code material and must carry proper identification markings for checking against the mill test reports. It may be examined at this time for thicknesses, tolerances, and defects such as laminations, pit marks, and surface scars.

Field examinations of the vessel for fit-up and finished welds, and during and after hydrostatic or pneumatic testing should be conducted in the manner already prescribed for shop inspections.

If all is satisfactory, the Inspector will check the data sheets to see that they are properly filled out and signed by the manufacturers and shop inspectors and that the portion of the vessel that was field-constructed is accurately described and the material and quality of the work are accurately stated. The inspector then signs the certificate of field assembly.

Radiography

Radiography is a tool permitting the examination of the invisible parts of a welded joint. Radiographic equipment requires skilled personnel for safe and efficient handling. Special provisions must be made for their protection, moreover, and unauthorized persons must be kept from entering the surrounding area while the equipment is in use.

The use of radiography is increasing rapidly in the welded pressure vessel industry, especially because of the higher weld joint efficiencies allowed by the Code for vessels that are radiographically examined. Such vessels can be built for the same pressure and diameter with thinner shells and heads and consequently with less weld metal and welding time.

Conditions Requiring Radiography

The ASME Code requires radiographing of pressure vessels for certain design and service conditions, the type of radiography varying with materials, thicknesses, joint efficiency, type of construction, and the like. Conditions for which full or spot radiography is specified are the following:

1. To increase the basic joint efficiency, either full or spot radiography may be required (Code Pars. UW-11 and UW-12; Code Table UW-12).

2. Vessels using lethal substances require full radiography of all longitudinal and circumferential joints (Code Pars. UW-2 and UW-11).

3. If the plate or vessel wall thickness at the weld joint exceeds the thicknesses given in Code Pars. UW-11, UCS-57, Table UCS-57, Pars. UNF-57, UHA-33, UCL-35 or UCL-36, and UHT-57, full radiography is required (see also Table 5.3).

4. Cast ductile iron repair of defects requires full or spot radiography (Code Par. UCD-78).

5. Vessels conforming to type 405 materials that are butt welded with straight chromium electrodes and to types 410, 429, and 430 base materials welded with any electrode require radiography for all thicknesses. If material types 405 or 410 with a carbon content less than 0.08 percent are welded with electrodes that produce an austenitic chromium–nickel weld deposit or a nonhardening nickel–chromium–iron deposit, radiography is required only if the plate thickness at the welded joint exceeds $1\frac{1}{2}$ in (Code Par. UHA-33).

6. Clad and lined vessels may require either full or spot radiography (Code Pars. UCL-35 and UCL-36).

TABLE 5.3 Full Radiographic Thickness Requirements*
Carbon and Low-Allow Steels

P number and group number†—metals		When thickness exceeds
Carbon Steels		
P = 1	Group 1	1.25 in
1	2	
1	3	
Alloy Steels with 0.75 Maximum Chromium and Those with 2.00 Maximum Total Alloy		
P = 3	Group 1	0.75 in
3	2	
3	3	
Alloy Steels with 0.75 to 2.00 Chromium and Those with 2.75 Maximum Total Alloy		
P = 4	Group 1	0.625 in
4	2	
Alloy Steels with 10.00 Maximum Total Alloy		
P = 5	Group 1	0.0 in
5	2	
Nickel Alloy Steels		
P = 9A	Group 1	0.625 in
9B	1	
Other Alloy Steels		
P = 10A	Group 1	0.75 in
10F	6	
P = 10B	Group 2	0.625 in
10C	3	

*These requirements are in addition to those of full or spot radiography needed to increase weld joint efficiency. See Code Pars. UW-11, UW-12, and Table UW-12.

† P numbers and group numbers as given in Code Section III, Part D, Table 1A, Pressure Vessel Code Section VIII, Division 1, and Code Table QW-422, ASME Code Section IX, Welding and Brazing Qualifications.

7. Welded repairs of clad vessels damaged by seepage in the applied lining may require radiography if the inspector so requests (Code Par. UCL-51).

8. If longitudinal joints of adjacent courses are not staggered or separated by a distance at least 5 times the plate thickness of the thinner plate, each side of the welded intersection must be radiographed for a distance of 4 in [Code Par. UW-9(d)].

9. Unreinforced openings located in a circumferential seam must be radiographed for a distance 3 times greater than the diameter of the opening (Code Par. UW-14).

10. Castings for which a higher quality factor is desired require radiography at all critical sections (Code Par. UG-24).

11. Leaking chaplets of cast-iron vessels to be repaired by plugging require radiography of metal around the defect (Code Par. UCI-78).

12. Forgings repaired by welding may require radiographing (Code Par. UF-37).

13. Unfired steam boilers with a design pressure exceeding 50 psi must be fully radiographed [Code Par. U-1(e)].

14. Nozzle butt welds joining the flange or saddle to a vessel requiring radiography must be fully radiographed (Code Par. UW-11 and Code Figs. UW-16.1, f-1, f-2, f-3, and f-4).

15. When weld buildup of the base material is utilized for the purpose of restoring the material thickness to its original thickness or modifying the configuration of welded joints to provide a uniform transition, if radiography is required of the weld joint, the buildup area is required to be radiographed also (Code Par. UW-42).

16. Full or spot radiography must be used in order to increase allowable stress and joint efficiency in the calculations for shell and head thicknesses (Code Pars. UW-11 and UW-12).

Interpretation of Welding Radiographs

Some people falsely assume that because a weld had been radiographed, no defect which has not been revealed by the radiographing process can exist in the weld. On the other hand, some persons interpret each visible mark on the film as a defect in the weld and request chip-out and rewelding. In each of the above cases the radiograph is subjected to false or poor interpretation.

In order for the viewer to interpret the film correctly to find all possible defects in the weld, a proper understanding of the nature and appearance of weld defects, of common processing defects, of the degree of contrast and detail possible in radiography, of the film and film sensitivity, and of the technique of making a radiograph is necessary.

Radiography uses either x rays or gamma rays. X-ray machines operate through a range of 10,000 to 2 million volts. Gamma rays, measured in curies, are produced from radioactive materials of various strengths. (The most common sources of gamma rays in the welding industry are cobalt-60, cesium-136, iridium-192, and radium.)

Both x rays and gamma rays are a form of electromagnetic waves of short wavelength, which allows them to penetrate metals impervious to ordinary light. Some of the radiation is absorbed after penetration, the amount depending on the wavelength and the thickness and density of the material. If a defect such as a crack, slag inclusion, or porosity exists in a material of uniform thickness and density, the thickness and density are less at the point of defect and more radiation will pass through. When this radiation is recorded on sensitive film, the defect will show up as an x-ray shadow. The shadow picture is called a radiograph.

X-ray radiographs show defects more sharply than gamma-ray radiographs, a fact that must be taken into account in the interpretation of the film. When persons experienced with gamma rays are shifted to x-ray work, they tend to see any defect as a cause for rejection, whereas those experienced with x-ray film may tend to underestimate the seriousness of a defect seen on gamma-ray film.

As a check on the radiographic technique, penetrameters are used. The thickness of the penetrameters is given in *ASME Boiler and Pressure Vessel Code,* Section V, Nondestructive Examination, Article 2, Table T-233.1. Penetrameters shall not be more than 2 percent of the thickness of the plate being radiographed, but for thin plates, it need not be less than 0.005 in. As measured by the penetrameter, sensitivity is understood to be the smallest fractional difference in the thickness of the material that can be detected visually on a radiograph. The appearance of the penetrameter image on the film defines vertical sensitivity of 1 to 2 percent. The required appearance of the 2 T-hole or slit in the penetrameter proves a lateral sensitivity of 2 to 4 percent of the thickness being radiographed.

If the weld reinforcement or backing strip of a welded seam is not removed, a shim of the same material as the backing strip must be placed under the penetrameter so that the thickness being radiographed under the penetrameter is the same as the total thickness of the weld, including the reinforcement and backing strip.

Each penetrameter must have three holes with diameters that are usually 1, 2, and 4 times the penetrameter thickness, one of which must be of diameter equal to 2 times the penetrameter thickness, but not less than $\frac{1}{16}$ in. However, smaller holes are permitted. The smallest hole must be distinguishable on the film.

The penetrameters should be placed on the side nearest the source of the radiation. Where it is not possible to do this, they may be placed on the film side of the joint provided that the radiographic technique can be proved adequate, and a lead letter F with a dimension as high as the penetrameter identification numbers is placed adjacent to the penetrameter or placed directly on the penetrameter, provided it does not mask the essential holes; at least one penetrameter must be used for each exposure. The penetrameter should be placed parallel and adjacent to the welded seam except when the weld metal is not similar to the base material, in which case it may be placed over the weld metal with the plane of the penetrameter normal (perpendicular) to the radiation beam. This provides for penetrameter image sharpness and minimum image distortion and allows proper evaluation of radiographic quality.

To assure getting good contrast and seeing a weld defect, the ASME Code gives specific requirements for film density, based on the Hurter and Driffield (H&D) curve.* Density through the radiographic image of the body of the penetrameter shall be 1.8 minimum for single film viewing for radiographs made with an x-ray source and 2.0 for radiographs made with a gamma-ray source. Each radiograph of a composite set shall have a minimum density of 1.3. The maximum density shall be 4.0 for either case. A tolerance of 0.05 in density is allowed for variations between density readings. In the upper limits of density, the viewing equipment is important for correct interpretation of the film. It is essential to have a strong source of light available with flexibility both as to light intensity and area.

In recent years the Code has allowed the inception of alternate methods of determining the acceptable densities of radiographs. One alternative method of measuring film density is by the use of wire penetrameters (Fig. 5.1). Wire penetrameters are just as their name implies. Six wires whose composition is of material similar to the material being radiographed are encapsulated or "sandwiched" between two sheets of 0.030-in clear vinyl plastic. Wire penetrameters are classified by sizes which range from 0.0032 to 0.32 in. These wires are categorized in sets known as sets A, B, C, and D. For set A and B wire penetrameters, the minimum distance between the wires is 3 times the wire diameter and not more than 0.2 in. The length of the wires for sets A and B is a minimum of 1 in. Set C and D wire distance separations are not to be less than 3 times the wire diameter and not more than 0.75 in. The length of set C and D wire penetrameters is a minimum of 2 in. Each plastic case shall have a lead material grade identifying number in the lower left-hand corner of the casing as well as the ASTM set designation.

Much like hole penetrameters, at least one wire penetrameter shall be placed abutting the weld to be radiographed. The radiographic density measured adjacent to the penetrameter must not vary by more than +30 to −15 percent. When the density does vary more than that, two sets shall be used.

---

*The Hurter–Driffield method of defining quantitative blackening of the film expresses the relationship between the exposure applied to a photographic material and the resulting photographic density.

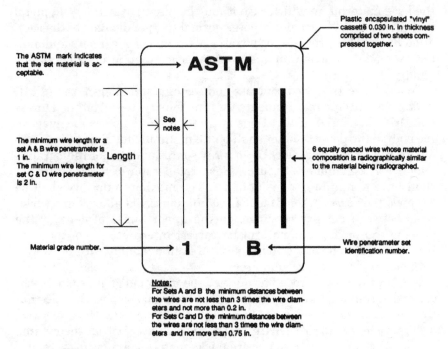

Figure 5.1 Wire penetrameter.

If the source of radiation is placed at the axis of the joint and the complete circumference is radiographed with a single exposure, three uniformly spaced penetrameters are to be used.

Many persons in the welding industry who know how to interpret radiographic film sometimes err in their judgment. Moreover, radiographs often require retakes because of improper techniques or differences of opinion in interpretation. Only the retake can prove whether the technique or the interpretation of the original film was wrong. Basically, the important decisions in a radiographic inspection are the following:

1. Has the radiograph been properly taken?
2. Has complete penetration of the weld been obtained?
3. Are the revealed defects acceptable or should they be repaired?

Proper radiography will show any change in metal thickness greater than 2 percent, the contrast sensitivity required by the Code. Any defect surpassing the allowable minimum—whether slag entrapment, porosity, lack of penetration, undercutting, or cracks—will be recorded on the film as a dark area compared to the normal density of the sound metal. Weld reinforcements, backing strips, splatter, and icicles will show on the film as light areas because they represent metal sections thicker than the average plate. Burn-through shows up as dark, irregular areas, but the drip-through surrounding the burn-through presents a lighter image. In highly absorbent material such as the burn-off from an electrode used in heliarc welding, the burn-off shows up as a light area that is usually spherical in shape.

In heavy wall welds, knowing how deep a defect is from each side is helpful. We have seen chip-outs 2-in deep made in a 3-in weld that could have been only 1-in deep from the opposite side if stereoscopic principles of radiography had been used to make proper depth interpretation. (See Fig. 5.2 for an example.) When available, ultrasonic techniques can and often are utilized to determine the depth of many types of defects. Not every discontinuity in the film is objectionable if vessels are fully radiographed (see Code Par. UW-51 and Appendix 4). The Code allows for a maximum length of slag inclusion according to the plate thickness as follows: $\frac{1}{4}$ in for plates less than $\frac{3}{4}$ in thick; $\frac{1}{3}$ the plate thickness for plates $\frac{3}{4}$ to $2\frac{1}{4}$ in thick; and $\frac{3}{4}$ in for plates more than $2\frac{1}{4}$ in thick.

Any group of slag inclusions in a line with a total length greater than the thickness of plate in a length 12 times the plate thickness is unacceptable, except when the distance between the successive imperfections exceeds 6 times the length of the largest imperfection. With full radiography, moreover, any type of crack or lack of penetration is not

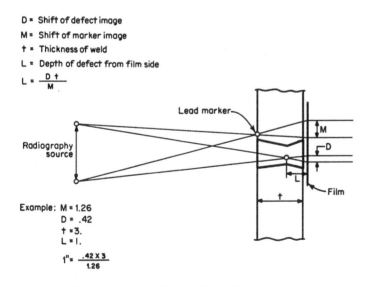

Figure 5.2 Stereoscopic principle in radiography.

acceptable and must be repaired. Porosity is judged by comparison with the standards given in the Code (see Appendix 14).

If only spot radiography is required, the standards are less demanding (see Code Par. UW-52). The length of film need be only 6 in. Porosity, moreover, ceases to be a factor in the acceptability of welds. There is a difference in the other defects allowed as well. Any weld that shows cavities or elongated slag inclusions is unacceptable if these have a length greater than two-thirds of the thickness of the thinner plate of the welded joint.

Any group of imperfections in a line with a total length less than the plate thickness in a length of 6 times the plate thickness is acceptable only if the longest defects are separated by acceptable weld metal equal to 3 times the length of the longest defect. (As the minimum length required for a spot radiograph is 6 in, the total length of imperfections in a line shall be proportional for radiographs shorter than 6 times the plate thickness.) Any defect shorter than ¼ in is acceptable in spot radiography, with the exception of any type of crack or lack of penetration, which is never acceptable and must be repaired. Any defect longer than ¾ in is not acceptable.

If a spot radiograph shows welding that violates Code requirements, two additional spot radiographs must be taken in the same seam at locations chosen by the inspector away from the original spot, preferably one on each side of the defective section. If these additional spots do not meet the acceptable standard, the rejected welding must be removed and the joint rewelded, or the fabricator may choose to reradiograph the entire unit of weld represented, and the defective welding need only be repaired. The rewelded areas must then be spot-radiographed to prove that they meet Code requirements.

For the proper interpretation of films, the view box should be lighted with sufficient intensity to illuminate the darkest area of the film. The view box should be located in a room that can be darkened. Film handling and processing at times will cause marks that give the impression of defects. A fine scratch on the film surface can look like a crack; water marks, stains, and streaks can resemble other defects. If the film surface is viewed in reflected light, these defects can be properly detected. Other apparent but nonexistent flaws in the weld can be caused by a weld overlay that contrasts sharply at the weld edge, by scratches on intensifying screens, or by lines of weld penetration into the backing ring if a backing strip is used. Slag under the backing strip and tack welds on the strip will produce a lighter image and should not be interpreted as defects.

Such conditions may be obvious to the experienced technician, if not to the novice. There are often real defects in a weld that are difficult to determine, however, and it is these one must watch for. Some critical materials are more prone to crack than others, and their film images must be evaluated with extreme care.

A crack that runs perpendicular to the plate surface can be seen easily, but cracks along the fusion line are often at an angle, and the defect image may not be sharp. Defects that are parallel to the plate also may not show up clearly. If a laminated plate is radiographed with the radiographic beam perpendicular to the plate surface, for example, the lamination very probably will not be seen. However, if the plate is cut and radiographed with the beam parallel to the plate as rolled, these laminations will show clearly. (See Figs. 7.1 and 7.2 for examples.)

Most cracks show up longitudinally to the weld, in the weld itself, in the fusion line of the weld, at craters, and occasionally in the plate adjacent to the weld. Transverse cracks are not seen as often but do appear in the weld and plate edges. Slag inclusions are usually sharp and irregular in shape. As stated before, certain small slag inclusions are acceptable to the Code, but elongated slag and too many inclusions in a line are not permitted.

Incomplete penetration, which is not acceptable, is usually sharply defined. Undercuts are usually easily seen at the edge of a weld. Porosity shows up as small, dark spots. These spots may be defined as either fine or coarse porosity; if full radiography is required, their acceptability is measured by the standards of the Code.

In viewing radiographs, it is important to be sure that the penetrameters are properly placed so that the penetrameter holes show and that the H&D density requirements are given in the negative. This will assure better contrast to show any possible defects. The required appearance of the essential hole in the penetrameter is the indication that defects in the weld will be seen in the film. (See also Table 7.1.)

Sectioning

The Code permits the examination of welded joints by sectioning if this method is agreed to by both the user and the manufacturer, but sectioning of a welded joint is not considered a substitute for spot-radiographic examination and permits no increase in joint efficiency; therefore, it is seldom used (see Code Par. UW-41).

Sectioning is accomplished with a trepanning saw, which cuts a round specimen, or with a probe saw, which cuts a boat-shaped specimen. The specimen removed must be large enough to provide a full cross section of the welded joint.

If sections are oxygen-cut from the vessel wall, the opening must not exceed $1\frac{1}{2}$ in or the width of the weld, whichever is greater. A flame-cut section must be sawed across the weld. A saw-cut surface must show the full width of the weld. The specimens must be ground or filed smooth and then etched in a solution that will reveal the defects. Care must be taken to avoid deep etching.

The etching process should be continued just long enough to reveal all defects and clearly define the cross-sectional surfaces of the weld. Several acid solutions are used to etch carbon and low-alloy steels, one of which is hydrochloric acid (commonly known as muriatic acid) mixed with an equal part of water. This solution should be kept at or near its boiling point during the etching process.

Many shops prefer to use ammonium persulfate because the method of its application allows better observation of the specimen and prevents a deep etch. The solution, one part ammonium persulfate to nine parts water, is used at room temperature. A piece of cotton saturated with the solution is rubbed hard on the surface to be etched. The etching process is continued until a clear definition emerges of the cross-sectional surfaces of the weld and any defects present.

Other solutions are used to etch metal specimens, such as iodine, potassium iodide, and nitric acid, but hydrochloric acid and ammonium persulfate are the most common ones. To be acceptable, the specimens should not show any type of crack or lack of penetration. Some slag or porosity is permitted if (1) the porosity is not greater than $\frac{1}{16}$ in and there are no more than six porosity holes of this maximum size per square inch of weld metal, or (2) the combined area of a greater number of porosity holes does not exceed 0.02 in^2 per square inch of weld.

The width of a single slag inclusion lying parallel to the plate must not be greater than one-half the width of the sound metal at that point. The total thickness of all slag inclusions in any plane at approximate right angles to the plate surface must not be greater than 10 percent of the thinner plate thickness.

The plugs or segments should be suitably marked or tagged after removal with a record of their original location, the job number, and the welder or welding operator who did the welding. After etching, they should be oiled and kept in an identifiable container. A record should be made on a drawing of the vessel of all these specimens (with identification marks) after the vessel has been completed and stamped. The plugs may be retained by the purchaser, or they may be discarded.

Holes left by the removal of the trepanned plugs may be closed by any welding procedure approved by the Inspector. The best method is to cut a plug from the same plate material and then bevel the plate around the hole and the plug, insert the plug, weld from one side, backchip the other side, and reweld. When gas welding is used, preheat the area surrounding the plug. If it is possible to weld from only one side, the bottom of the plug should be fitted to permit full penetration of the weld closure. Each layer of weld should be peened to reduce the stresses set up by welding.

If the plate thickness is less than one-third the diameter of the hole, a backing strip may be placed inside the shell over the hole and the hole filled completely with weld metal. If the plate thickness is not less than one-third or greater than two-thirds the diameter of the hole, the hole can be completely filled with weld metal from one side and then chipped clean on the other side. Other methods exist (see Code Appendix K), but the ones described here are the most common.

Hydrostatic Testing

Most fabricating shops take pride in their work and do everything possible to turn out a good job. All this effort can be spoiled by carelessness in preparing for or conducting a hydrostatic test. Several actual incidents may serve as illustrations.

Case 1. A large fabricating shop, one of the best, put all its care and know-how into a critically needed vessel. During hydrostatic testing, however, a 3-in shell plate bulged because of overpressure. This near accident resulted in a costly delay in shipping that should never have occurred. The company had been cautioned by an Authorized Inspector on the first day in its shop, and several times thereafter, that the water pressure on the line to the test floor was 2200 lb and that a reducing valve or some other means of protection was required.

The shop's chief inspector replied that dependable people worked on the test floor and that the two shut-off valves provided for each vessel were sufficient.

The Authorized Inspector told the shop's inspector that, as a precautionary measure, the presence of an Authorized Inspector would be necessary at all hydrostatic tests of Code vessels. The general manager was then informed of the possibility of a vessel's being damaged during these tests. The general manager said that the matter would be taken up with the chief inspector, but the precautionary safety devices recommended were never installed. A few months later, an urgently needed non-Code vessel was badly damaged.

The Code does not specify an upper limit for the hydrostatic test pressure. However, if this pressure exceeds the required test pressure to the degree that a vessel is visibly distorted, the Inspector shall reserve the right to reject the vessel.

Case 2. A thin-shelled vessel was hydrostatically tested at 30 lb. It was filled late in the afternoon and left overnight with the pressure on and

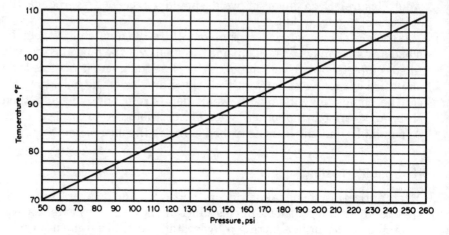

Figure 5.3 Water pressure and temperature graph.

all valves closed. During the night the water in the vessel warmed up, expanded, and bulged out a flat stayed section. This accident occurred despite the cautionary footnote to Code Par. UG-99, which states: "A small liquid relief valve set to 1⅓ times the test pressure is recommended for the pressure test system, in case a vessel, while under test, is likely to be warmed up materially with personnel absent." (For a graph showing the increase in water pressure caused by increase of water temperature, see Fig. 5.3.)

Case 3. A vessel 6 ft in diameter and 15 ft long was damaged while being drained. A tester opened a large opening at the bottom of the tank without first opening a valve or other inlet on top of the vessel to permit air to enter.

The large opening at the bottom drained the water rapidly, creating a partial vacuum in the tank. As the vessel was not designed for external pressure, this vacuum caused the shell to crack between two nozzle outlets.

All three of these mishaps occurred in shops whose work is far above average; they prove that similar accidents are possible in any shop if care is not taken.

Test Gage Requirements

Code Par. UG-102 states: "An indicating gage shall be connected directly to the vessel. If it is not readily visible to the operator controlling the pressure applied, an additional gage shall be provided where it will be visible throughout the duration of the test." It also recommends that a recording gage be used in addition to indicating gages on larger vessels. Not all shops, especially small ones, have recording gages, but good practice indicates the use of two gages on all tests because they are sensitive instruments and can easily be damaged.

The Code requires vents at all the high points of a vessel (in the position in which it is being tested) to purge air pockets while the vessel is filling. During a recent hydrostatic test of a large vessel, the frame of a large quick-opening door failed, and the door dropped to the floor, thereby flooding the shop with several inches of water. Had there been any air in the vessel, serious damage to the building and possible injury to personnel might have occurred.

The Code requires test equipment to be inspected and all low-pressure filling lines and other appurtenances that should not be subjected to test pressure to be disconnected before test pressure is applied. The reason can be seen in the recent test of a vessel for 1500-lb pressure with a 125-lb valve in the line. As this valve was leaking at the bonnet, the inspector immediately had the pressure reduced and the valve replaced. The Inspector had seen the damage in another shop when a bonnet, valve, and stem had been blown through the roof, leaving a hole a foot wide (fortunately, no one was over the valve when it failed).

Not so long ago, a tester was seen placing a cast-iron plug in a vessel that was to be tested to 1400-lb pressure. The tester was stopped by the Inspector, who requested that a steel plug be used. Replying, "Is that why the plug blew out?" the tester then showed the Inspector a dent in the concrete wall.

During a 2500-lb hydrostatic test of a large vessel, a leak was noted at a flanged connection. A worker who was responsible for the hydrostatic test picked up a sledgehammer and tapped on the flange cover trying to stop the leak. The cover then exploded off the vessel, rocketed to the far end of the shop, and struck a fellow worker who was walking across the end of the shop. His leg was severed at the knee. Ignorance of correct procedures can be fatal in testing vessels.

Seven Simplified Steps to Efficient Weld Inspection

A manufacturer's preparation of pressure vessels for examination by a Code Inspector or a customer can be routine and easy, or complicated and time-consuming. The speediest and most efficient inspection takes place in the shop that prepares for inspection in an orderly manner. Such preparation will save not only time but very possibly money.

A job should have advanced to the point at which specifications called for an inspection before one is undertaken; if there is no such specification, the job should have advanced to the point requested by the inspector. Any vessel failing to meet minor requirements should be corrected before being shown to the Inspector. Major defects must be shown to the Authorized Inspector before repairs are made. The Inspector makes an inspection on the basis of a written code or customer specification only. It is the responsibility of the manufacturer, not of the Inspector, to see that these standards are met.

A good Inspector tries to find all defects before the final or hydrostatic tests, especially if the vessel is to be postweld heat-treated. The inspector knows—and certainly the manufacturer knows—that a minor repair can assume major proportions because of a lack of time to make the repair and to repeat postweld heat treatment and testing operations before the scheduled date of delivery.

A careful observation of the following items will help expedite inspection.

1. Inspectors should be provided with drawings that clearly show weld details and dimension, types of material, corrosion allowances, and design pressures and temperatures. If the vessel is to be fully radiographed or spot-radiographed, this fact should be clearly stated because the Code allows different efficiencies for fully radiographed, spot-radiographed, and nonradiographed joints. Hydrostatic test pressure should be based on the maximum working pressure allowed for a new and cold vessel, and the part of the vessel to which the maximum pressure is limited should be defined.

2. A Code Inspector must check and verify all calculations. If they are neatly arranged and include all pertinent design information, the job will be much easier. A well-managed engineering department will have the calculation sheets printed or mimeographed and will fill in pertinent design information.

3. If the vessel is a Code vessel, the data sheet should be checked and all corrections made before it is typed. Data sheets prepared at the start of a job after drawings are made keep the engineering department and the Inspector from having to review the design and calculations a second time.

4. Mill test reports of material should be filed with the Inspector's copies of drawings to save the manufacturer's personnel from being interrupted while performing normal tasks to produce these reports.

5. Stamping information, including the location of stamps, should be readily available. Corrections should be made beforehand.

6. A written copy of the welding procedure (or a note stating that both the procedure and operator are qualified) and of the test results should be available on request. These items should be clipped or filed together.

7. A place for the Inspector to sit and prepare the written part of the inspection should be provided to avoid the inspector feeling like an intruder, which could arise if someone else's desk has to be used. This could even cause the Inspector to do the rest of the work at home or in a hotel room. This inconvenience could conceivably cause important data to be overlooked that might not have been missed had the report been written in the shop.

Chapter

6

Welding, Welding Procedure, and Operator Qualification

Examination of Procedure and Operator

One of the most important functions of an Authorized Inspector is examining welding procedures and operators for their compliance with the intent of Code requirements. The Code does not dictate methods to the manufacturer, but it does require proof that the welding is sound (Code Par. QW-201) and that operators are qualified in accordance with the ASME Code Section IX, Welding and Brazing Qualifications (see Code Par. QW-301). *To avoid confusion, it should be noted that all Code references in this chapter, unlike the rest of this book, refer to ASME Code Section IX, Welding and Brazing Qualifications.*

An Inspector must be familiar with the physical properties of materials and electrodes and should have experience in the evaluation of the written procedures and test plates required by Section IX. An Inspector also must know how to determine the cause of test plate failures. The manufacturer should actually be more interested than the Inspector in finding the cause, for it is far less expensive to prevent failures than to make expensive repairs. This is especially true if the procedure is to be used on a production line where individual items may miss inspection. In a plant where liquified propane gas tanks were being constructed of a high-tensile-strength material, for example, the manufacturer had difficulty in performing satisfactory bend tests on an offset joint for the head and girth seam. After various changes in welding procedures, a test finally satisfied Code requirements. Because of the history the plant had had with this material and type of welded joint, however, the Inspector realized that one successful test might not be sufficient. Knowing that any repairs would be costly and failure could be disastrous, additional bend tests before approving the procedure were requested.

A qualified welding procedure is the most important factor in sound welding. If one has been properly established, welders need only demonstrate their ability to make a weld in the position to be welded as well as other required variables, so that they will pass root and face bend or side bend tests in order to gain qualification. It should be remembered that weld excellence depends not only on good welding electrodes but also on the way in which they are used. Moreover, an Inspector should be satisfied not only that the manufacturer's procedure is acceptable and the operators qualified but that the welding procedure is used and followed during actual fabrication.

Record of Qualifications

Once a welding procedure has been approved, it should be filed for future review by Authorized Inspectors or customers' representatives. Some manufacturers are not as careful as they should be in keeping these records. Not long ago an Inspector asked to see the written welding procedure of a certain shop and also the test results for material similar to that used on the job that was to be inspected. The written procedure could not be located for several days. After the Inspector explained the importance of proper filing and the unnecessary expense of having to requalify procedures, the manufacturer eventually had records brought up to date and sets filed with both the shop superintendent and the engineering office. When the representative of a large refining company made a survey of the plant a few months later, the general manager felt very pleased at being able to show the procedures promptly upon request and felt that this helped the company's business position.

Another company with many approved welding procedures lost the file containing them and was forced into an expenditure of several thousand dollars to replace them. Needless to say, this firm now keeps duplicate copies of all welding qualifications and test results.

Currently there are several commercial programs available that help the manufacturer keep the required records. Besides utilizing the strong database capabilities of computers, they greatly reduce the search time for specific types of welding procedure specifications and associated procedure qualification records and keep up-to-date welder qualifications. The larger the number of welders or procedures the greater the efficiency these programs bring to the manufacturer.

Welding Procedures and Qualification Simplified

Welding procedure specifications (WPS) and *procedure qualification records* (PQR) are the most essential factors contributing to sound welding. A welding procedure specification records the ranges of variables that can be used and gives direction to the welder using the procedure. The procedure qualification record states exactly the variables used in qualifying the WPS, shows test results, and furnishes proof of the weldability of the variables described in the WPS. If these have been properly established, welders need only demonstrate their ability to make a weld that will pass root and face or side bend tests in the position to be welded in order to gain qualification. Other variables do apply, such as thickness limits and diameter and type of electrode used.

In building welded ASME Code vessels, one requirement is to have written welding procedure specifications for the materials to be welded (see Code Par. QW-201), properly completed records of qualification for this procedure (Code Par. QW-200.1), and records of qualification for the welders working on the vessel (Code Par. QW-301.4). These procedures should be used in production by the fabricator. If narrative welding procedures are utilized, the wording of the procedure should be such that the meaning is clear and easily understood by all concerned. If standard ASME Code welding procedure documentation is utilized, then the fabricator need only fill in the welding procedures in writing on the recommended WPS, Code Form QW-482, and make test plates, the results of which are recorded on the PQR, Code Form QW-483. The purpose of the welding procedure is to provide welders with direction for making Code welds, which in turn will assure Code compliance. Once a welding procedure has been qualified, it should be filed for future review by Authorized Inspectors, customers' representatives, and any other interested parties. These records are one of the first things an Authorized Inspector of ASME vessels will examine. Code Par. QW-201 requires each manufacturer or contractor to have a qualified welding procedure specification. Code Par. QW-301 requires each welder or welding operator doing Code welding to be qualified in accordance with a welding procedure specification utilizing a process he or she will be using during actual fabrication. Proven sound welding procedures are a prerequisite to the fabrication and construction of ASME Code vessels. Section IX of the *ASME Boiler and Pressure Vessel Code* currently consists of two parts. Part QW covers requirements for welding, and Part QB contains requirements for brazing; the previous distinction between ferrous and nonferrous metals has been eliminated. For a specific material with a special requirement, such requirements are listed by the appropriate variables. By far the greatest percentage of fabrication involves welding. For this reason, most examples given here apply to Part QW, Welding. For brazing procedures, see Section IX, Part QB, Brazing.

Article V, allowing prequalified standard welding procedure specifications (SWPSs), has been added to Section IX. There are several steps a manufacturer must follow prior to using these SWPSs (see Code Par. QW-510), and Code Form QW-485 is suggested for documenting the steps. This and other forms and records showing compliance and use shall be available to the Authorized Inspector. A listing of the permitted SWPSs is in mandatory Appendix E.

In the case of face, root, and some side bends for nonferrous groupings of base metals for those P numbers listed in Code Table QW-422, coupons may be 1/8 in thick instead of 3/8 in as required by most ferrous materials (see Code Section IX, Table and Code Figs. QW-462.2 and QW 462.3; see also Fig. 6.5). It is therefore clear that each Code welding procedure must be carefully read and separately evaluated. Also note, when welding clad or lined material, the procedure shall be qualified in accordance with Section IX of the Code, with modified provisions as required in the referencing section, Part UCL for clad vessels, in Section VIII, Division 1, of the *ASME Boiler and Pressure Vessel Code*. As stated previously, before a manufacturer can build or repair ASME Code vessels, proof must first be given that welding procedures and welders or welding operators qualify under the provisions of welding qualifications in Section IX (Code Par. QW-201). WPS Code Form QW-482 for recording welding procedure specifications and PQR Code Form QW-483 for procedure qualification records make it easy to fill in the required information. These forms may be purchased from the ASME and The National Board of Boiler and Pressure Vessel Inspectors. The following suggestions should be observed in preparing a WPS Code Form QW-482.

Welding procedure specification forms

The welding procedure should be titled and dated and should have an identifying WPS number and supporting PQR number.

Welding variables (Code Pars. QW-400 and QW-401). The procedure should state the welding processes and other variables as well as the type of equipment—automatic, semiautomatic, or manual. If inert-gas metal-arc welding is used, it should be specified whether a consumable or nonconsumable electrode is used and the material of which the electrode is made. Essential, supplementary, and nonessential variables for procedure qualification and performance qualification records must be addressed.

Joints (Code Par. QW-402). Preparation of base material should include group design, i.e., double butt, single butt with or without a backing strip, or any type of backing material used (e.g., steel, copper, argon, flux, etc.), showing a sketch of groove design on the PQR. The groove design should include the number of passes for each thickness of material and whether the welds are strings or weaving beads. Sketches should be comprehensive and cover the full range of base metal thickness to be welded, the details of the welding groove, the spacing position of pieces to be welded giving maximum fit-up gap, the welding wire or electrode size, the number of weld passes, and the electrical characteristics; if automatic welding is used, the rate of travel and the rate of wire feed should be recorded.

Base metals (Code Par. QW-403). The type of material to be welded, giving the ASME specification number, the P number, and group number under which it is listed, and the thickness range should be specified. On the PQR, show the ASME materials specification, type, and grade number; P numbers to be welded together; and the thickness and diameter of pipes or tubes. If this procedure is for an unlisted material, either the ASTM or other Code designations, or the chemical analysis and physical properties should be specified. This should be done only when you intend to work with a special material permitted by a Code case.

Filler metal (Code Par. QW-404). The SFA specification number and the AWS classification when the type of ASME filler metal is given with the filler metal group F number (Code Table QW-432) and the weld metal analysis A number (QW-404) should be stated. Any manufacturer's product bearing these classifications can be used; if there is no A or F number, the electrode manufacturer's trade name should be listed. In submerged arc welding, the nominal composition of the flux must be stated. For inert-gas arc welding, the shielding gas must be described. The size of the electrode or wire, amperage, voltage, number of passes, rate of travel for automatic welding, and rate of shielding gas flow for inert-gas arc welding must be listed. If a consumable insert is used on single butt welds, it should be stated.

Position (Code Par. QW-405). The position in which the welding will be done should be stated, and if it is done in the vertical position, it should be listed whether uphill or downhill in the progression specified for any pass of a vertical weld. The inclusion of more than one welding position is permitted as long as the intent for each position is clear and shown by attaching sketches to the procedure.

Preheat (Code Par. QW-406). Preheating and temperature control should be stated only if they are to be used, and any preheating and control of interpass temperature during welding that will be done should be described. For example, materials in the P-4 and P-5 groups must be preheated. Carbon steels for pressure vessels in the P-1 group, if preheated to 200°F, can be welded up to 1½ in thickness before postweld heat treatment is required, whereas if not preheated, the limit is 1¼ in thickness [see Section VIII, Division 1, Table UCS-56, Fn. 2(a)].

Postweld heat treatment (Code Par. QW-407) (see also Section VIII, Division 1, Code Table UCS-56). Any postweld heat treatment given, the temperature, time range, and any other treatment should be described.

Gas (Code Par. QW-408). The shielding gas type (e.g., Ar, CO_2, etc.) and the composition of gas mixture used (e.g., 75 percent CO_2, 25 percent Ar), as well as the shielding gas flow rate range should be stated. It should be listed whether gas backing, gas trailing, or shielding gas is used and the gas backing or trailing shielding gas flow rate.

Electrical characteristics (Code Par. QW-409). When describing electrical characteristics, it should be stated whether the current is ac or dc, and if dc current is used, it should be specified whether the base material will be on the negative or positive side of the line. The mean voltage and current for each pass should be given. The mean should be the average of the specified upper or lower limits within 15 percent of the mean; the travel speed should also be given.

Technique (Code Par. QW-410). The joint welding procedure may be shown on the same sketches with the preparation of base material or on a separate sketch. The sketch should show the number of passes for each thickness of material, whether the welds are strings or weaving beads, and single or multiple electrodes. The orifice or gas cup size range, the oscillation parameter ranges covered by WPS (i.e., width, frequency, etc.), and the contact tube to work distance range should be given. The initial and interpass cleaning, including the method and extent of any imperfection removal should be described. The method of back gouging should be explained. It is important that the method of defect removal is stated clearly. For example, flame gouging is detrimental to some materials. If this is so, make sure another method of defect removal is used. Peening, if mentioned at all, should be used only when the desired result cannot be obtained by postweld heat treatment. If peening is used, all details of the peening operation such as not peening first and last passes should be stated clearly (see Section VIII, Division 1, Code Par. UW-39). Treatment of the underside of the welding groove should be clearly indicated, for example, "chipped to clean metal and welding applied as shown in sketch number...."

Variables [Code Pars. QW-400, QW-250, and QW-350 and Tables QW-415 (WPS) and QW-416 (WPQ)]

Any variable which would affect the quality or characteristics of the finished weld may require a new welding procedure. If any such variables exist, they should be clearly indicated. Essential variables will require requalification of the procedure if any are changed from those qualified.

Additional requirements in the Code are known as supplementary essential variables. These variables are to be included on the WPS only when notch toughness is required by the referencing Code section and are then in addition to the essential variables already stated. Changes in nonessential variables as well as editorial changes may be made without setting up new procedures if the changes are recorded and documented as such. These nonessential variables can be changed without requalification, but the changes must be indicated.

When writing a welding procedure specification, it is important to remember that qualification in any position for procedure qualification will qualify for all positions except when notch toughness or stud welding is a requirement (see Code Pars. QW-203 and QW-250). When notch toughness is a factor, qualification in the vertical, uphill position will qualify for all positions (Code Par. QW-405). If the weld must also meet notch toughness requirements as required by another section of the Code, the supplementary essential variable must be included (see Code Tables QW-252 to QW-262 and Code Pars. QW-280 to QW-283 and QW-400). Test plates are required for all procedure qualifications. If the completed WPS Code Form QW-482 and PQR Form QW-483 are clear and the test plates are approved and recorded, the manufacturer will be qualified to weld pressure vessels of the same or similar type of material used in making the test plate and also any other material in the P-number grouping. Starting with the 1974 Code, the P-number listing for base metals has been divided into groups. The purpose of the groupings is to cover qualifications in which notch toughness is a requirement. When notch toughness is required, each group must be separately qualified. Many shops use a simple jig with a hydraulic jack similar to the one shown in Fig. 6.1 to make their own root, face, and side bends on the assumption that if the bend tests satisfy the requirements of Section IX, Code Par. QW-163 with no defect exceeding $\frac{1}{8}$ in (or $\frac{1}{16}$ in for corrosion resistance weld overlay cladding), then the tensile test should present no difficulty. The advantage of conducting your own bend tests is that it keeps a shop from having to go to the expense of sending tensile bars to be machined and pulled if the root, face, or side bends should fail. Moreover, this jig can always be used to qualify

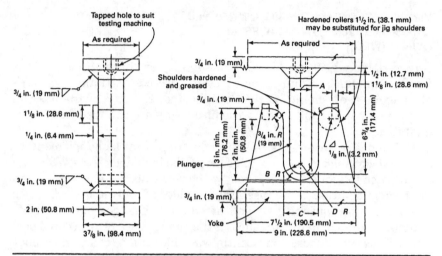

Material	Thickness of Specimen, in.	A, in.	B, in.	C, in.	D, in.
P-No. 23 to P-No. 21 through P-No 25; P-No. 21 through P-No. 25 with F-No. 23; P-No. 35; any P-No. metal with F-No. 33, 36, or 37	$\frac{1}{8}$ $t = \frac{1}{8}$ or less	$2\frac{1}{16}$ $16\frac{1}{2}t$	$1\frac{1}{32}$ $8\frac{1}{4}t$	$2\frac{3}{8}$ $18\frac{1}{2}t + \frac{1}{16}$	$1\frac{3}{16}$ $9\frac{1}{4}t + \frac{1}{32}$
P-No. 11; P-No. 25 to P-No. 21 or P-No. 22 or P-No. 25	$\frac{3}{8}$ $t = \frac{3}{8}$ or less	$2\frac{1}{2}$ $6\frac{2}{3}t$	$1\frac{1}{4}$ $3\frac{1}{3}t$	$3\frac{3}{8}$ $8\frac{2}{3}t + \frac{1}{8}$	$1\frac{11}{16}$ $4\frac{1}{3}t + \frac{1}{16}$
P-No. 51	$\frac{3}{8}$ $t = \frac{3}{8}$ or less	3 $8t$	$1\frac{1}{2}$ $4t$	$3\frac{7}{8}$ $10t + \frac{1}{8}$	$1\frac{15}{16}$ $5t + \frac{1}{16}$
P-No. 52, P-No. 53, P-No. 61, P-No. 62	$\frac{3}{8}$ $t = \frac{3}{8}$ or less	$3\frac{3}{4}$ $10t$	$1\frac{7}{8}$ $5t$	$4\frac{5}{8}$ $12t + \frac{1}{8}$	$2\frac{5}{16}$ $6t + \frac{1}{16}$
All others with greater than or equal to 20% elongation	$\frac{3}{8}$ $t = \frac{3}{8}$ or less	$1\frac{1}{2}$ $4t$	$\frac{3}{4}$ $2t$	$2\frac{3}{8}$ $6t + \frac{1}{8}$	$1\frac{3}{16}$ $3t + \frac{1}{16}$
All others with less than 20% elongation	$t =$ (see Note b)	$32\frac{7}{16}t$, max.	$16\frac{7}{16}t$, max.	$34\frac{7}{8}t + \frac{1}{16}$, max.	$17\frac{7}{16}t + \frac{1}{32}$, max.

(continued)

Figure 6.1 Test jig dimensions (Code Fig. QW-466.1).

welders. The qualification of a procedure is affected by the type of electrode used. If a manufacturer, for example, has been welding Code vessels of SA-285 grade C materials using an E-6020 electrode classed as type F-1 (Section IX, Code Table QW-432, F numbers) but wants to change to an E-6015 or E-6016 electrode classed as type F-4, the manufacturer must have this procedure qualified separately (see Section IX, Code Par. QW-404). The F-number grouping of electrodes and welding rods is based on their usability qualities, which essentially determine the ability of welders to make dependable welds with a given filler metal.

Procedure qualification for both groove welds and fillet welds is made on groove welds (Code Par. QW-202.2).

Material	SI Units Thickness of Specimen, mm	A	B	C	D
P-No. 23 to P-No. 21 through P-No. 25; P-No. 21 through P-No. 25 with F-No. 23; P-No. 35; any P-No. metal with F-No. 33, 36, or 37	3.2 $t = 3.2$ or less	52.4 $16\frac{1}{2}t$	26.2 $8\frac{1}{4}t$	60.4 $18\frac{1}{2}t + 1.6$	30.2 $9\frac{1}{4}t + 0.8$
P-No. 11; P-No. 25 to P-No. 21 or P-No. 22 or P-No. 25	9.5 $t = 9.5$ or less	63.5 $6\frac{2}{3}t$	31.8 $3\frac{1}{3}t$	85.8 $8\frac{2}{3}t + 3.2$	42.9 $4\frac{1}{3}t + 1.6$
P-No. 51	9.5 $t = 9.5$ or less	76.2 $8t$	38.1 $4t$	98.4 $10t + 3.2$	49.2 $5t + 1.6$
P-No. 52; P-No. 53; P-No. 61; P-No. 62	9.5 $t = 9.5$ or less	95.2 $10t$	47.6 $5t$	117.5 $12t + 3.2$	58.7 $6t + 1.6$
All others with greater than or equal to 20% elongation	9.5 $t = 9.5$ or less	38.1 $4t$	19.0 $2t$	60.4 $6t + 3.2$	30.2 $3t + 1.6$
All others with less than 20% elongation	$t =$ (see Note b)	$32\frac{1}{64}t$ max.	$16\frac{1}{64}t$ max.	$34\frac{1}{8}t + 1.6$ max.	$17\frac{1}{16}t + 0.8$ max.

GENERAL NOTES:
(a) For P-Numbers, see QW-422; for F-Numbers, see QW-432.
(b) The dimensions of the test jig shall be such as to give the bend test specimen a calculated percent outer fiber elongation equal to at least that of the base material with the lower minimum elongation as specified in the base material specification.

$$\text{percent outer fiber elongation} = \frac{100t}{A+t}$$

The following formula is provided for convenience in calculating the bend specimen thickness:

$$\text{thickness of specimen } (t) = \frac{A \times \text{percent elongation}}{[100 - (\text{percent elongation})]}$$

(c) For guided-bend jig configuration, see QW-466.2, QW-466.3, and QW-466.4.
(d) The weld and heat-affected zone, in the case of a transverse weld bend specimen, shall be completely within the bend portion of the specimen after testing.

Figure 6.1 (*Continued*)

Many vessels used for cryogenic service require an impact test. Pressure vessels built of austenitic chromium–nickel steels have shown a reliable service record plus consistent good quality as demonstrated by the record of the impact tests required by the Code for vessels in low-temperature service. Therefore Section VIII, Division 1, Code Par. UG-84 does not require impact tests of some vessels fabricated of these steels if impact tests are conducted as part of the welding procedure qualification tests. (See Code Par. UHA-51 and Fig. 6.2 on zone locations for removal of impact-test specimens.)

To summarize, we have listed several suggestions to be followed in the preparation of welding procedures.

1. A WPS Code Form QW-482 should be written by a person with knowledge of production-shop needs and engineering-performance requirements.

2. The PQR Code Form QW-483 should be filed with the WPS Code Form QW-482.

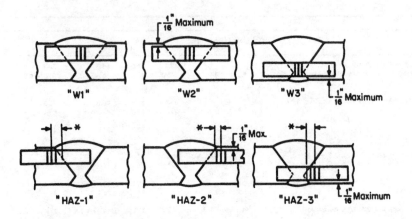

Figure 6.2 Weld metal and heat-affected zone locations for impact-test specimens.

3. Sketches and tables should be shown.
4. The existing Code should be reliably interpreted and the Code or specification translated to shop terminology.
5. Written procedures should be checked by reliable shop personnel who are to use these procedures. Writing a welding procedure specification requires contributions of knowledge and experience from the shop personnel as well as the engineering department. If this is considered, a successful procedure can be used by the designing engineer to assist in designing a better weld at a lower cost.

Welder Qualifications

Unlike the welding procedure specification, which is primarily used to determine that the weldment is capable of providing the required mechanical properties of the welding to be performed, the welder qualification test and subsequent welder qualification record is used to determine the ability of a welder to deposit sound weld metal. Thus the welder qualification does not require a tension test (see Table 6.1).

A welder is automatically qualified for a particular process if the test plates prepared for the welding procedure qualification pass the requirements (see Section IX, Code Pars. QW-301.2 and QW-303). A differentiation must be made here between a welder and a welding operator (Section IX, Code Par. QW-492 contains the definitions). A

TABLE 6.1 Welding Procedure Qualification and Welders' Qualification Records, Welder Identification Markings, and Welding Records

Remarks	Code reference
Manufacturer must assign an identifying number, letter, or symbol to permit identification of welds made by welder or welding operator	Par. UW-29
Manufacturer must maintain a record of welders and welding operators, showing date, test results, and identification mark assigned to each	
Records must be certified by manufacturer and made accessible to Inspector	
Welder must stamp symbol on plate at intervals of 3 ft or less on work done on steel plate ¼ in thick or more; for steel plate less than ¼ in thick or nonferrous plate less than ½ in thick, suitable stencil or other surface marking must be used	Par. UW-37
Records must be kept by manufacturer of all welders and welding operators working on a vessel and the welds made by each so that these data will be available for the Inspector	Par. UW-48
No production work can be done until both the welding procedure and the welders or welding operators have been qualified	Pars. UW-5, UW-26(c), UW-28, UW-29, UW-47, Section IX
Procedure for clad vessels	Pars. UCL-31, UCL-40 to -45
Lowest permissible temperature for welding	Par. UW-30
No welding should be done when temperature of base metal is lower than 0°F	
Heat surface area to above 60°F where a weld is to be started when temperature is between 0 and 32°F	

welder is one who has the ability to perform a manual or semiautomatic welding job. A welding operator is one who operates machine (automatic) welding equipment.

Welders passing the groove weld test need not take the fillet weld test. Welders passing the test for fillet welds are qualified to make fillet welds only [Section IX, Code Par. QW-303.1, Code Figure QW-462.4(b), and Code Par. QW-180]. The welder who is qualified to use type F-4 electrodes (E-6015 and E-6016), however, is also qualified to use type F-1 (E-6020 and E-6030), type F-2 (E-6012, E-6013, and E-6014), and type F-3 (E-6010 and E-6011) (see Section IX, Table QW-432 and Code Par. QW-404).

The welding operator may be qualified with a single test plate or pipe from which bend tests are removed. Although most shop welding is accomplished in the flat position with the weld metal deposited from above, this procedure is often not possible. Therefore welds must be made in the position used in actual fabrication in order to qualify the welder (Section IX, Code Par. QW-303 and Code Fig. QW-461). Some fabricators set up their procedures (Section IX, Code Pars. QW-303 and QW-461) for position 6G, pipe with fixed position and with the axis inclined to 45° to horizontal. Qualification in the inclined fixed position (6G) will also qualify for all positions.

For performance qualification only, all position qualifications for pipe may also be done by welding in the horizontal position (2G), the horizontal fixed position (5G), or the 6G position (see Fig. 6.3). Qualification in these positions will also qualify for the flat, vertical, and overhead positions (Section IX, Code Par. QW-303). The six positions are shown in Fig. 6.3.

In addition to position, Section IX also places diameter restrictions under which a welder may weld in production. For example, if a welder qualifies on a 6-in diameter piece of pipe and wishes to weld 2-in diameter pipe by means of a butt joint, he or she must requalify for the small diameter pipe (see Code Par. QW-452.3). Further restrictions concern the thickness of materials to be welded and P number and F number.

The order of removal and the number and types of the test specimen from the welded test plates are shown in Fig. 6.4, transverse, and Fig. 6.5, longitudinal test specimens. The preparation of the transverse test specimens is shown in Figs. 6.6 and 6.7.

As an alternative to bend tests, radiographic examinations may be employed to prove the performance qualification of welders to make a sound weld, using the welding of groove welds with the manual shielded metal-arc welded process with covered electrodes and the gas tungsten-arc welding (Code Pars. QW-304 and QW-305). A record of the welders' and welding operators' test results should be kept on forms similar to the recommended Code Form QW-484, Section IX. The record of the tests shall be certified by the manufacturer and shall be made accessible to the Authorized Inspector.

It is to the fabricator's advantage to have one welder qualify for position 2G. Should this welder pass the test, this position also brings qualification for welding in the flat position (1G). Another welder could then qualify for the horizontal fixed position (5G). Should the tests be passed, this position also qualifies the welder for the flat, vertical, and overhead positions. It is also better to qualify for a position other than flat because the inclined fixed position (6G), horizontal (2G), vertical (3G), overhead (4G), and horizontal fixed position (5G) also automatically qualify the welder for the flat (1G) position, which, in the long run, is more economical.

Continuity of welders' qualification records should be maintained and a log kept of all welders' activities to indicate their continuity in

Welding, Welding Procedure, and Operator Qualification 179

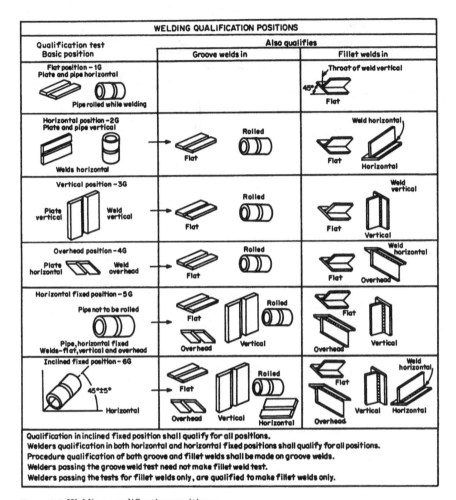

Figure 6.3 Welding qualification positions.

specific welding procedures. If there is a gap of time in excess of six months in a specific process, the welder shall be requalified. See Fig. 4.6 for an example of a welders' qualifications and performance record log. If a manufacturer must qualify a welder on hard-to-obtain or costly material, such as stainless steel or alloy steel, qualification can be obtained by using type F-5 austenitic electrodes on carbon-steel plate.

The Code allows single-welded butt joints without backing strips for some circumferential joints in positions 1G and 2G. If a manufacturer intends to use such joints on shells or pipes, the welders (Section IX, Code Par. QW-303) must be qualified. For this reason, some manufacturers have their welders qualified in metal-arc welding without a

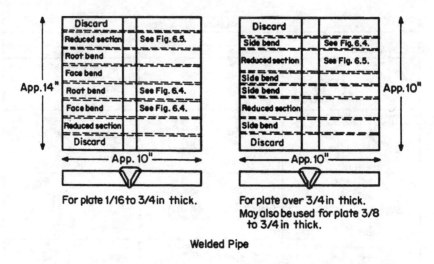

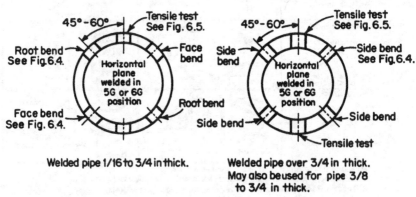

Figure 6.4 Order of removal of welded transverse test specimens.

backing strip (Section IX, Code Fig. QW-469.2) so that they will also be qualified to metal-arc weld with a backing strip.

Welding Details and Symbols

Welding details and symbols are an essential part of every shop drawing used in the design of welded pressure vessels; they guide the welder in the type of weld required, the proper size of weld metal to be deposited, and the size of bevel. Despite the rapid advances in welding proficiency in the past 25 years, many fabricating shops surprisingly still do not use either the details or the symbols.

There is a tendency in some pressure vessel shops toward overwelding, which not only increases production costs but also can result in dis-

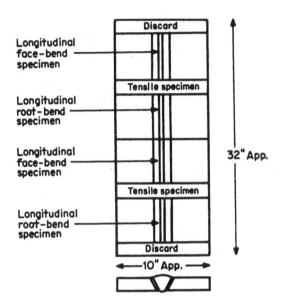

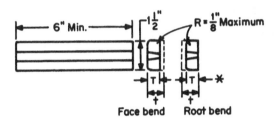

*For T see Code Fig. QW-462.3(b).

Figure 6.5 Longitudinal test specimens for plate materials.

tortion of the part welded, especially on stainless steel and aluminum vessels around nozzles, flanges, and accessory fittings. Other shops, especially those with an incentive rate or piecework rate, have a tendency to underweld. When an Authorized Inspector discovers underwelding on a Code vessel, the proper size of weld will be insisted upon. The extra handling and welding this entails will add to the cost of the vessel. Underwelding that goes undetected, on the other hand, will result in a weakened vessel and possible failure.

Without realizing the possible consequences, many vessel manufacturers leave the amount of weld to be deposited to the discretion of the shop supervisor. As a result, weld sizes do not always meet Code requirements.

Some welders or supervisors given the responsibility to produce a safe weld are inclined to make the weld "heavy enough." Usually, this is too heavy. Others, especially those on an incentive rate, take advan-

182 Chapter Six

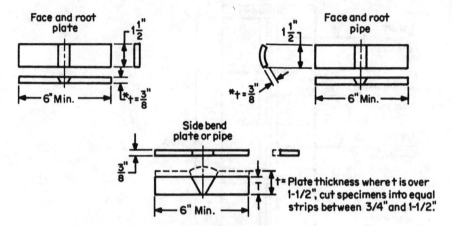

*For material less than 3/8", see Code Fig. QW-462.3a.

Figure 6.6 Preparation of face, root, and side bend specimens.

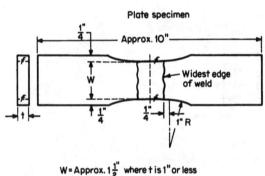

W = Approx. $1\frac{1}{2}"$ where t is 1" or less

W = Approx. 1" where t is more than 1"

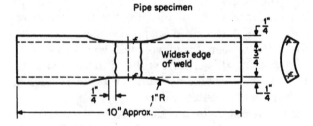

(Note that the pipe specimen and the plate specimen differ in size.)

Figure 6.7 Preparation of reduced section tension test specimens.

tage of the fact and underweld. In the long run, underwelding may be the more costly because of the possibility of failure and the loss of future business.

Responsibility for correct design instructions to the shop must be taken by the engineering department. The designer knows the exact thicknesses required for the shell, heads, nozzles, and flanges and the stress loads to be carried by the welds. The type and size of the welds required must be stated and clearly shown on the shop drawings either as a separate detail or by use of the appropriate symbol. Both underwelding and overwelding may often be attributed to inadequate welding information on shop drawings.

The Code gives fillet weld sizes for nozzles and flanges in terms of throat dimensions (to facilitate load calculations). The effective stress-carrying area is the throat dimension times the length of the weld. The American Welding Society's *Standard Welding Symbols* gives fillet weld sizes in terms of leg dimensions, and it is this measurement that should be given to the shop rather than the throat dimensions. Calculations for leg sizes are illustrated and explained in Fig. 6.8. For example, if the Code-required throat dimension of a nozzle is 0.75 in, the leg size would be 0.75 divided by 0.707, or 1.06 in, and it is this last figure that should appear on the drawing. If this is not done when the designer calls for a throat measurement of 0.75 in, the welder may assume a leg size of 0.75

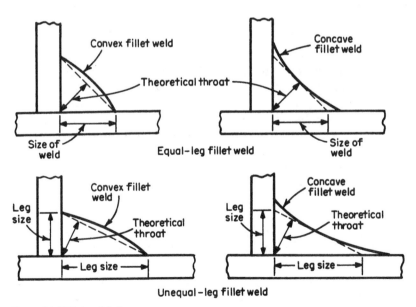

Figure 6.8 Fillet weld size.

in. The person in the shop certainly cannot take the time to calculate the leg sizes of welds and should not be asked to do so.

It can easily be seen why all dimensions given to the shop should be prepared in advance. Some Code welds naturally do not meet Code requirements. Correct weld design will be assured only if weld sizes are given to the shop as leg dimensions (shown either in the weld detail or weld symbol). The American Welding Society's *Standard Welding Symbols* are standard for almost all codes and companies, although some companies make slight variations to meet their particular needs. A chart illustrating their use (Fig. 6.9) can be obtained from the American Welding Society, P. O. Box 351040, Miami FL 33125.

Many of these symbols, which are intended to cover all types of welding, are not used in the fabrication of pressure vessels. In fact, most pressure vessel manufacturers require only a few of them. Figures 6.10 to 6.12—which make no attempt to explain the function of all the AWS symbols—include details of the welded joints used most often in pressure vessel construction and the proper symbols to indicate those joints.

Joint Preparation and Fit-Up

A sound weld is almost impossible without good fit-up. The importance of proper joint preparation of welding edges on longitudinal seams, girth seams, heads, nozzles, and pipe cannot be overemphasized. A good weld also cannot be obtained if the plate adjacent to the welding groove is covered with paint, rust, scale, grease, or oil. The Code states that if flame cutting is used to cut plates, the edges must be uniformly smooth and free of all loose scale and slag accumulations.

Before welding is undertaken, all joint surfaces for a distance of at least $\frac{1}{2}$ in from the edge of the welding groove on ferrous material and at least 2 in on nonferrous material should be free of all paint, scale, rust, grease, and oil. The Code permits a maximum offset, after welding, of one-quarter the plate thickness at the joint, with a maximum allowable offset of the lesser of $\frac{1}{16}$ the plate thickness or $\frac{3}{8}$ in in the longitudinal joints. For circumferential seams, the Code permits a maximum offset, after welding, of one-quarter the plate thickness at the joint, with a maximum allowable offset of the lesser of one-eighth the plate thickness or $\frac{3}{4}$ in (Section VIII, Division 1, Table UW-33).

In pressure vessels of larger diameter in which a weld may be chipped to clean metal from the opposite side and welded the full thickness of the plate, the permitted offsets above are satisfactory. If single-welded butt joints are used, however, special care must be taken in aligning and separating the edges to allow sufficient clearance for complete penetration and fusion at the bottom of the joint.

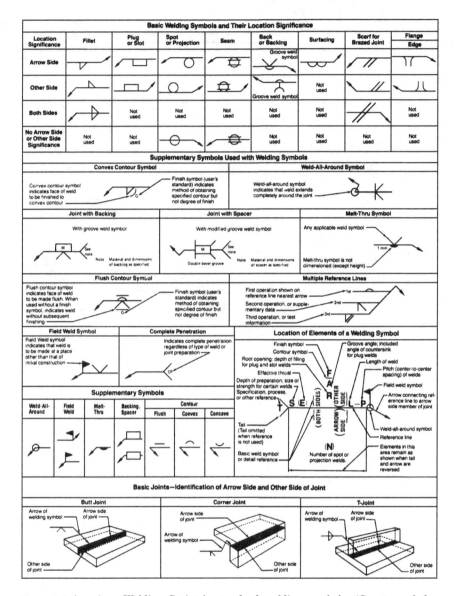

Figure 6.9 American Welding Society's standard welding symbols. (*Courtesy of the American Welding Society.*)

On smaller vessels and pipe that cannot be welded from the inside, special care must be taken when fitting-up the abutting sections. This is especially true in pipe of uneven thickness, in which the proper fit-up of inside diameters is almost impossible to obtain. If there is sufficient thickness in the pipe wall, however, machining can

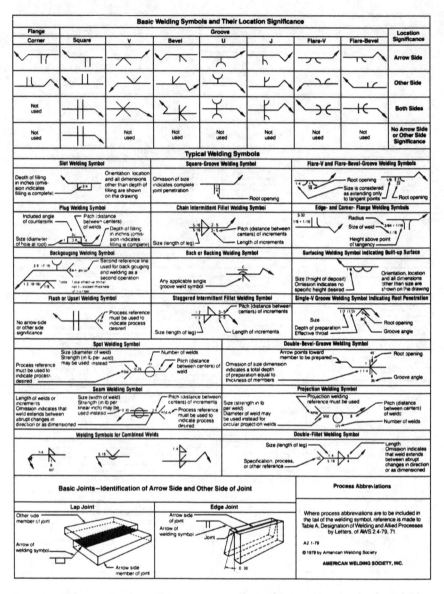

Figure 6.9 (*Continued*)

provide the proper fit-up. If the inside diameters are not equal, it is difficult to achieve full penetration at the root of a weld. As such penetration is needed to prevent root pass defects and cracks, alignment of these diameters is very important. Good joint preparation is especially essential in vessels that are intended for high-pressure and critical service.

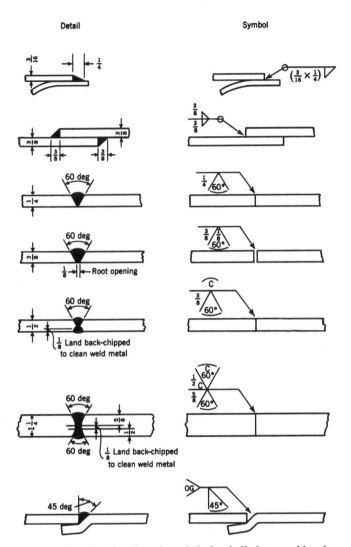

Figure 6.10 Welding details and symbols for shell plates and heads.

Probes of welds that appear sound often reveal cracks that failed to show up in the radiographs made when the joint was first welded. As time went on, the small, unseen crack opened up and became larger—the probe showing the crack to have started at the edge of the plate at the backing ring because of a poor fit or a tack weld that was not properly removed. This crack formation has happened so often that large companies regularly examine pipe that is in service at critical pressures and temperatures. They have also set up more rigid standards of

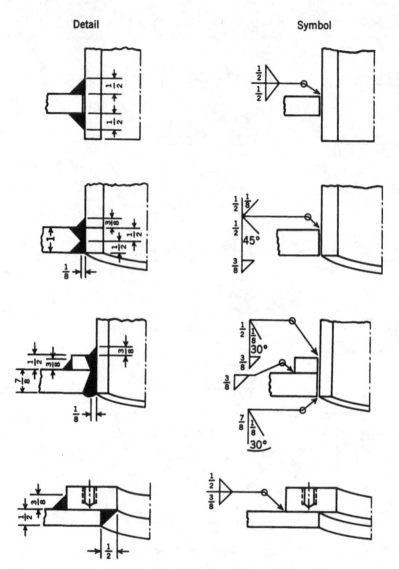

Figure 6.11 Welding details and symbols for nozzles and connections.

inspection for new construction, requiring a fit-up tolerance of 0.005 in and the use of heliarc welding on the first pass to eliminate any incipient cracking that might develop into a major defect. Not all companies have inspection departments to make such field inspections, however, and often the unit cannot be shut down long enough to permit such examinations.

When investigation reveals poor fit-up or a cracked tack weld before welding as the cause of an accident, the manufacturer of the vessel or

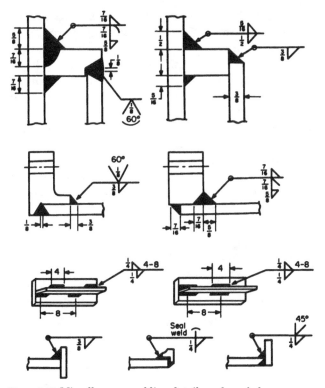

Figure 6.12 Miscellaneous welding details and symbols.

pipe is often unaware that such practices are permitted in the shop. Although care may be taken to prevent like accidents in the future, preventive measures should always be in practice beforehand.

Effects of Welding Heat

Welding metallurgy is the study of the properties of welded metals and the thermal and mechanical effects occurring during the welding process, including the movements of the atoms in the liquid state and the formation of the grain structure of the metal in changing from the liquid to the solid state. Some understanding of the behavior of metal during welding is useful and necessary to the designer and fabricator of pressure vessels in order to help prevent dangerous residual stresses or defects in the welded joints.

Many metallurgists think of metals in terms of a "life cycle," that is, a metal is "born" in the melting operation, "dies" when it fails in service or is no longer used, and is even subject in the interim to "diseases." It is known that heat treatment can change grain structure, which in turn may affect the physical properties. A metal can be hardened or softened, toughened or made more ductile. Welding is an operation in which separate pieces of metal are fused together to produce an integral section. Some metals are more difficult to weld than others. However simple the welding operation may appear, it is noteworthy that certain phenomena that occur in the making of steel also take place in the electric arc welding of steel products (plate, pipe, and the like), but more rapidly and on a smaller scale. Substantially, the same processes and metallurgical changes are involved. The heat of the arc melts a quantity of plate metal and the electrode. Chemical ingredients may be added by the coating on the electrode or by fluxes. The heat of the arc also melts or vaporizes the flux or coating, causing a gaseous shield to surround the arc stream and pool of molten metal and to perform the important job of keeping out all the elements of the air (including nitrogen, hydrogen, and oxygen).

During the cooling and solidification process, the behavior of the molten metal in the weld joint is similar to that of molten metal poured into a mold to form a steel casting. The mold in the welded joint is the solid parent metal; the melt consists of the molten metal of the electrode plus the liquified part of the parent metal, the two combining to form a casting within the solid parent metal. When the metal is in the liquid state, its atoms can move about at random. When it cools to the temperature at which it starts to solidify, the atoms are arranged into a regular pattern (for iron and steel the pattern is a cube). This atomic arrangement and the corresponding effect on grain size are influenced by the rate of cooling from the liquid to the solid state and also by the composition.

Mr. H. S. Blumberg, a New York consulting metallurgist, had prepared two charts on the changes that take place at different temperatures in ferritic (carbon and low-alloy steels) and austenitic (chromium–nickel stainless steels that are nonmagnetic) materials (Figs. 6.13 and 6.14). These charts show some of the simple fundamentals of atom arrangement and the grain size effects that take place when the metals change from the liquid to the solid state. They also show the important grain size changes in the "critical range" (the temperature range, with transition points, in which the structural changes take place) of ferritic carbon and low-alloy steels, including the temperatures used to normalize, anneal, draw, and stress-relieve these steels.

Grain size is influenced by the maximum temperature reached by the metal, the length of time the metal is held at a specific temperature, and the rate of cooling. For example, annealing, which involves

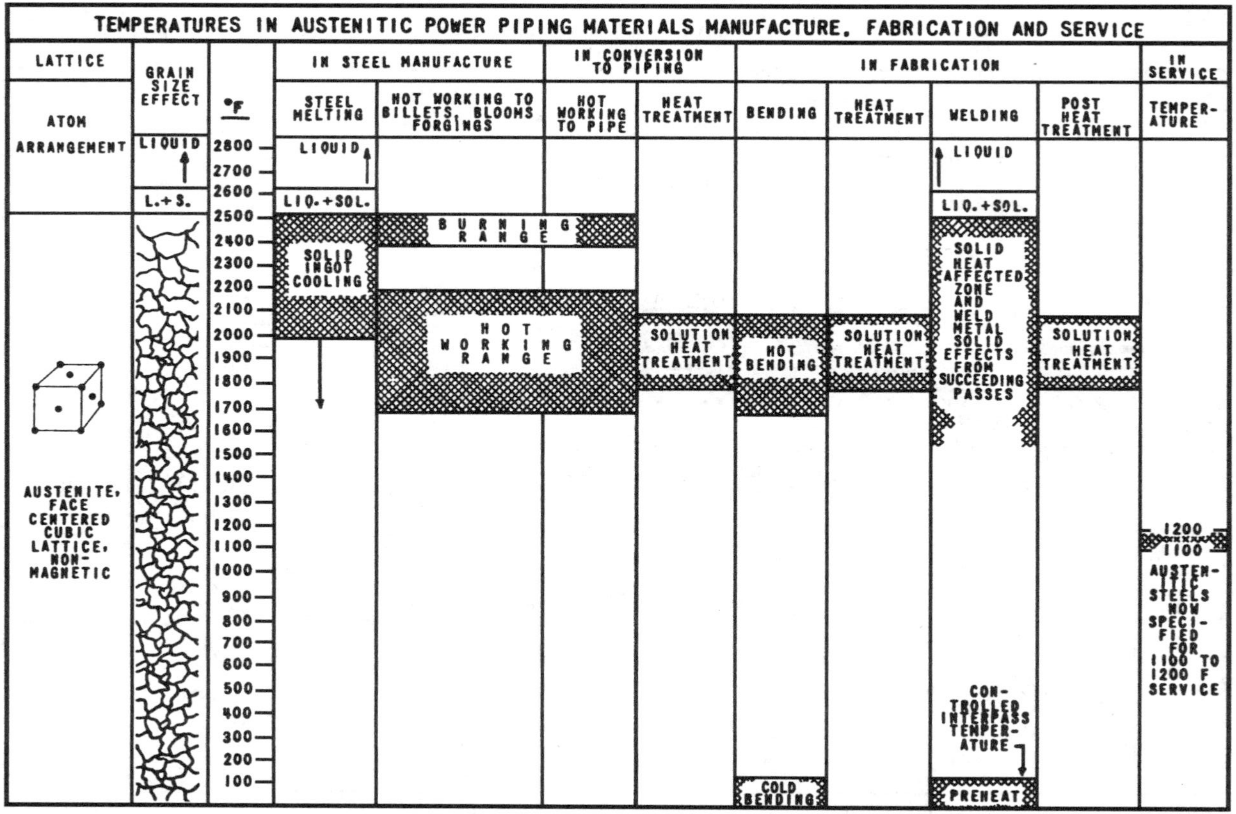

Figure 6.13 Metallurgical changes in austenitic materials at different fabrication and service temperatures.

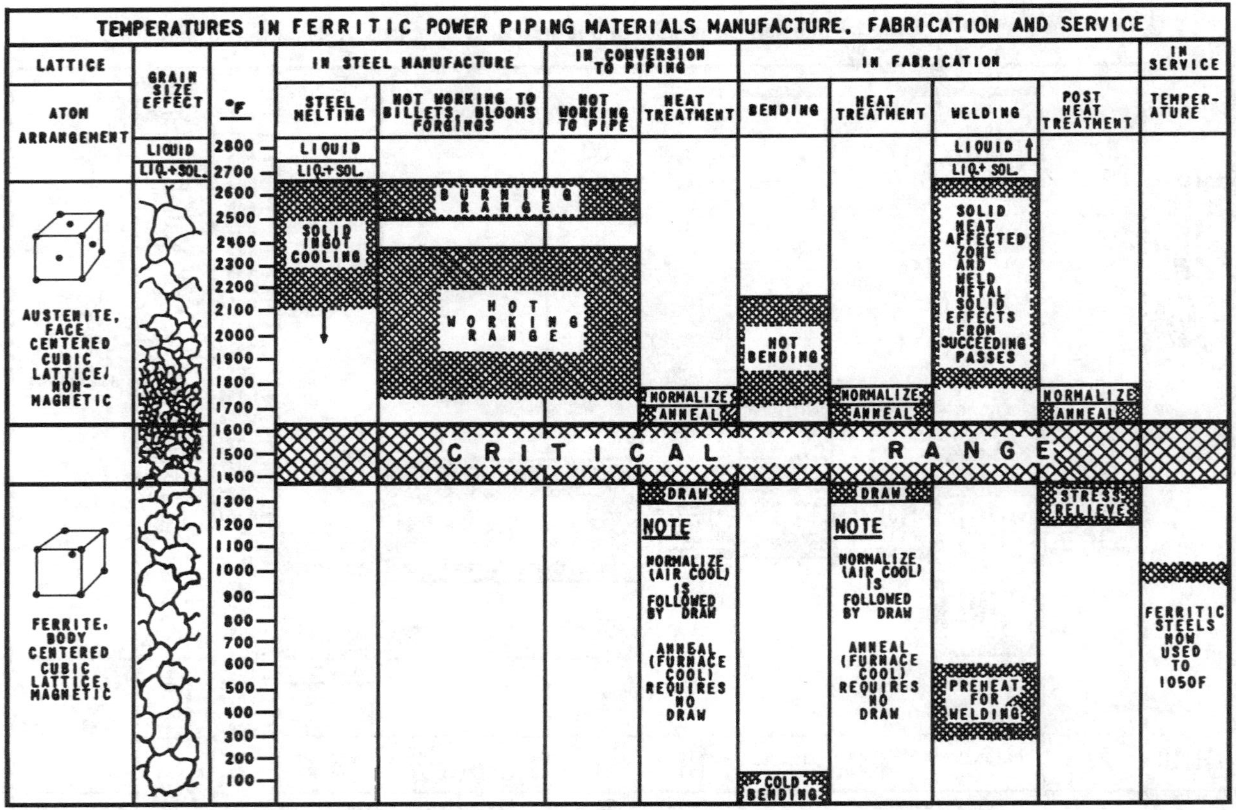

Figure 6.14 Metallurgical changes in ferritic materials at different fabrication and service temperatures.

heating to temperatures above the critical range followed by slow cooling, is used to refine the grain size to provide ductility and softness. Stresses can be relieved by heating beneath the critical range, but unless temperatures exceed that range, full grain refinement does not take place. Normalizing is the process by which the steel is heated to approximately 100°F above the critical range (as in annealing) and then cooled in still air. This heat treatment also refines the grain size and improves certain mechanical properties. Refinement of ferritic steel can be easily achieved by heat treatment because of the movement of the atoms in the solid metal. Note that the austenitic steels do not have a critical range and that consequently their grain size cannot be changed by heat treatment alone.

The production of a fusion weld naturally results in heating of some of the solid metal adjacent to the weld. The temperature reached by the metal in this zone varies from just below the melting point at the line of fusion to just above room temperature at a distance of a few inches from the weld. The effect of the welding heat on the parent metal depends on the temperature reached, the time this temperature is held, and the rate of cooling after welding. It is therefore necessary to understand the changes that take place in the zone where the metal is heated to very high temperatures without ever reaching the molten condition.

Usually, the weld metal zone is stronger than the parent metal but has less ductility. The heat-affected zone is the region where most cracking defects are likely to occur because the grain structure becomes coarse just beyond the fusion line and the ductility is lowest at this point. This reduction of ductility must be taken into consideration and measures taken to minimize the effects of total welding heat, for example:

1. The correct choice of parent material and welding electrode
2. Preheating before welding
3. Use of the proper welding procedure
4. Postweld heat treatment after welding

Preheating

Preheating prior to welding will eliminate the danger of formation of cracks, reduce hardness, reduce distortion, and reduce or prevent shrinkage stresses. Basically, preheating is a means of raising the temperature of a metal to a desired level above the surrounding ambient temperature before welding. It is usually applied between 125 and

500°F but may involve temperatures up to 1200°F depending upon the material to be welded.

Preheating reduces the thermal strains set up during welding that may crack a weld. Preheating is especially necessary in thicker carbon-steel plates in order to guard against the high thermal conductivity of steel when the heat of the welding arc is absorbed rapidly by the metal adjacent to the weld. The rate of cooling is slowed by preheating and the hardness is decreased. This slowing of the cooling rate allows more time for metal transformation to a more favorable structure, minimizing hard zones and cracks; thus preheating retards the formation of undesirable metallurgical structures in metals having high hardness and susceptibility to cracking. Preheating dries up moisture, reducing the incidence of hydrogen and oxygen accumulation in a weld zone; consequently, it minimizes porosity, embrittlement, and cracking in the weld and heat-affected zone. Even though most medium- and low-carbon steels are welded without preheat, the metal adjacent to the start of a weld should be preheated thoroughly.

Preheating is an essential requirement when welding low-alloy steels in the P-No. 3, P-No. 4, and P-No. 5 material groups. Most alloy steels permit higher stress values at high temperatures and have greater strength at low temperatures and allow higher notch toughness. In these steels, the required preheat temperature is stated and may be mandatory in order to comply with Code requirements. The higher yield strength of these metals allows the use of higher stress values. These groups require careful planning of the welding procedure specifications in order to prevent cracking in the heat-affected zone, where the microstructure of the steel has been modified by the heat of welding. This heat-affected zone is comparatively hard and is formed during the welding operation; in some steels it may be necessary to hold the preheat temperature until the weld can be postweld heat-treated to avoid cracking.

Preheating the metal before welding has proved a dependable method of securing a sound weld. Preheating helps equalize and distribute the stresses evenly by distributing the expansion and contraction forces over an enlarged area to which the stress differential is applied due to the heat of welding. (See Fig. 6.15.)

Appendix R, Preheating, in Section VIII, Division 1, refers the reader to other Code paragraphs and to Section IX of the Code, P Numbers and Groups, where preheating is required.

Postweld Heat Treatment

The process of postweld heat treatment consists of uniform heating of a vessel or part of a vessel to a suitable temperature for the material below

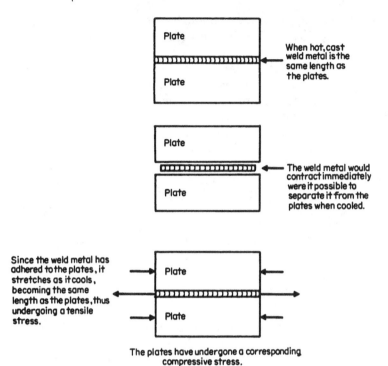

Figure 6.15 Good design provides for expansion and contraction.

the critical range of the base metal, followed by uniform cooling. This process is used to release the locked-up stresses in a structure or weld in order to "stress-relieve" it. Because of the great ductility of steel at high temperatures, usually above 1100°F, heating the material to such a temperature permits the stresses caused by deformation or straining of the metal to be released. Thus postweld heat treatment provides more ductility in the weld metal and a lowering of hardness in the heat-affected zone (HAZ). It also improves the resistance to corrosion and caustic embrittlement. The proper heat treatment before and after fabrication is one that renews the material as near as possible to its original state.

Heat treatment procedures require careful planning and will depend on a number of factors: temperatures required and time control for material; thickness of material and weldment sizes; and contour, or shape, and heating facilities. To be effective, heat treatment must be carefully watched and recorded. Code Par. UW-49 requires that the Inspectors themselves be assured that postweld heat treatment is properly performed.

Furnace charts should be reviewed to determine if the rate of heating, holding time, and cooling time are correct. If a vessel is to be fully postweld heat-treated, the furnace should be checked to make sure that the heat will be applied uniformly and without flame impingement on the vessel.

Vessels with thin walls or large diameters should be protected against deformation by internal bracing. The vessel should be evenly blocked in the furnace to prevent deformation at the blocks and sagging during the postweld heat treatment operation.

During this operation, both the increase and decrease in temperature must be gradual in order to allow uniform temperatures throughout. Where the difference in thickness of materials is greater, the rate of temperature change should be slower. The applicable requirements of Code Pars. UW-10, UW-40, UW-49, UCS-56, UCS-85, Table UCS-56, Par. UHT-56 or Table UHT-56, Pars. UHT-79, UW-80, Table UHA-32, Pars. UHA-32, UHA-105, UCL-34, UNF-56, ULW-26, ULT-56, UF-31, UF-32, and UF-52 must be followed in the postweld heat treatment of a Code vessel.

Conditions requiring postweld heat treatment

The Code requires postweld heat treatment of pressure vessels for certain design and service conditions (see Code Pars. UW-40 and UCS-56 for prescribed procedures). Postweld heat treatment requirements (according to materials, thicknesses, service, etc.) include the following:

1. All carbon-steel weld joints must be postweld heat-treated if their thickness exceeds $1\frac{1}{4}$ in or exceeds $1\frac{1}{2}$ in if the material has been preheated to a minimum temperature of 200°F during welding [Code Par. UCS-56 and Table UCS-56 Fn. 2 (a)].

2. Vessels containing lethal substances must be postweld heat-treated (Code Par. UW-2).

3. Carbon- and low-alloy steel vessels must be postweld heat-treated when the minimum design metal temperatures are below −50°F, unless exempted from impact tests (Code Par. UCS-68).

4. Ferritic steel flanges must be given a normalizing or fully annealing heat treatment if the thickness of the flange section exceeds 3 in (Code Par. UA-46).

5. Unfired steam boilers with a design pressure exceeding 50 psi and fabricated with carbon or low-alloy steels must be postweld heat-treated (Code Par. UW-2).

6. Castings repaired by welding to obtain a 90 or 100 percent casting factor must be postweld heat-treated (Code Par. UG-24).

7. Carbon- and low-alloy steel plates formed by blows at a forging temperature must be postweld heat-treated (Code Par UCS-79).

TABLE 6.2 Postweld Heat Treatment

Remarks	Code reference
All carbon- and low-alloy steel seams must be postweld heat-treated if nominal thickness exceeds 1¼ in or 1½ in if preheated to 200°F before welding	Par. UCS-56
Some materials must be postweld heat-treated at lower thickness	Table UCS-56
Vessels containing lethal substances must be postweld heat-treated	Par. UW-2
Unfired steam boilers	Par. UW-2
Carbon-steel vessels for service at lowered temperature must be postweld heat-treated unless exempted from impact test	Pars. UCS-66, UCS-67, UCS-68
Welded vessels	Pars. UW-10, UW-40
Carbon- and low-alloy steel vessels	Pars. UCS-56, UCS-66, UCS-67, UCS-79, UCS-85
High-alloy steel vessels	Par. UHA-32
Cast-iron vessels	Par. UCI-3
Clad-plate vessels	Par. UCL-34
Bolted flange connections	Appendix 2
Castings	Par. UG-24
Forgings	Par. UF-31
When only part of vessel is postweld heat-treated, extent of postweld heat treatment must be stated on Manufacturer's Data Report Form U-1	Par. UG-116
The letters HT must be stamped under the Code symbol when the entire vessel has been postweld heat-treated	Par. UG-116
The letters PHT must be stamped under the Code symbol when only part of the vessel has been postweld heat-treated	Par. UG-116
Low-temperature vessels	Par. ULT-56
Ferritic steel vessels	Pars. UHT-80, UHT-81
Nonferrous vessels	Par. UNF-56
Layered vessels	Par. ULW-26
Vessels or parts subject to direct firing, when thickness exceeds ⅝ in	Par. UW-2

8. Some vessels of integrally clad or applied corrosion-resistance lining material must be postweld heat-treated when the base plate is required to be postweld heat-treated (Code Par. UCL-34).

9. Some high-alloy chromium steel vessels must be postweld heat-treated (Code Pars. UHA-32 and UHA-105 and Table UHA-32).

10. Some welded repairs on forgings require postweld heat treatment (Code Par. UF-37). (See Table 6.2.)

11. When the thickness of welded joints for vessel parts or pressure vessels constructed with carbon (P-No. 1) steels exceeds 5/8 in and for all thicknesses of low-alloy steels (other than P-No. 1 steels) which are subject to direct firing, postweld heat treatment must be performed (Code Par. UW-2).

Chapter 7

Nondestructive Examination

Nondestructive examinations allow trained personnel to evaluate the integrity of pressure vessel parts without damaging them. Various sections of the *ASME Boiler and Pressure Vessel Code* now refer to the ASME Code Section V, Nondestructive Examinations, for nondestructive examination methods to detect internal and surface discontinuities in welds, materials, components, and fabricated parts.

Subsection A of Section V describes the methods of nondestructive examination to be used if referenced by other Code sections. They include the following:

General requirements

Radiographic examination (RT)

Ultrasonic examination (UT)

Liquid penetrant examination (PT)

Magnetic particle examination (MT)

Eddy current examination (ET)

Visual examination (VT)

Leak testing (LT)

Acoustic emission examination (AE)

Subsection B lists American Society of Testing Material Standards covering nondestructive examination methods which have been accepted as Code standards. These are included for use or for reference, or as sources of technique details which may be used in preparation of manufacturer's procedures.

The acceptance standards for these methods and procedures are stated in the referencing Code sections. Subsection B includes radiographic, ultrasonic, liquid penetrant, magnetic particle, eddy current, leak testing, acoustic emission, and visual examination standards.

Subsections A and B of Section V, Nondestructive Examination, provide rules and instruction in the selection of nondestructive test methods, development of inspection procedures, and evaluation of test indications. With the growing emphasis upon quality, nondestructive evaluation is becoming increasingly important in the construction of pressure vessels.

These examinations should be made according to a procedure that satisfies the requirements of the referencing Code and should be acceptable to the Authorized Inspector.

For the pressure vessel fabricator, nondestructive examinations provide a means of spotting faulty welds, plates, castings, and forgings before, during, and after fabrication. For the user, it offers a maintenance technique that more or less assures the safety of a vessel until the next scheduled inspection period. A serious flaw in a welded seam or a dangerously diminished plate thickness can cause a costly shutdown, or worse, a serious accident. Such defects must be located as soon as possible. Visual inspection is not enough. A weld may look perfect on the outside and yet reveal cracks, lack of penetration, heavy porosity, or undercutting at the backing strip when the weld is given a nondestructive examination by radiographic or ultrasonic tests. The internal condition of a material or weld can be accurately appraised only by use of the proper testing techniques.

Accident prevention requires many skills. Nondestructive testing is one that should be used more often. Since no weld or material is perfect, nondestructive testing is an excellent way to examine the internal conditions of a material. Nondestructive testing is often used on materials supplied to the fabricator of pressure vessels. Additional tests are made during fabrication and are also used during a vessel's operating life to assure safe operation and to gage plate thicknesses, thus enabling accurate prediction of the life of the vessel. Flaws are not uncommon in plate material and welds during any stage of fabricating a vessel. The point is to find them as soon as possible in order to prevent difficulties later. For example, a serious flaw in material or a welded seam may cause an accident. Many such accidents can be prevented with nondestructive tests.

In this chapter, some nondestructive test procedures are explained with regard to Code pressure vessel fabrication. The understanding of their use in addition to visual inspection is essential to safe pressure vessel fabrication and operation.*

---

*For more complete instructions on the methods of nondestructive examinations, refer to Section V of the *ASME Boiler and Pressure Code.*

Table 7.1 and Figs. 7.1 to 7.5 analyze and illustrate some nondestructive testing procedures commonly used in Code pressure vessel fabrication. An authorized Code vessel manufacturer's quality control manual must describe the system that is employed for assuring that all nondestructive examination personnel are qualified in accordance with the applicable Code.

There should be written procedures to assure that nondestructive examination (NDE) requirements are maintained in accordance with the Code. If the nondestructive examination is subcontracted, the name of the person responsible for checking the subcontractor's site, equipment, and qualification of procedures to assure their conformance with the applicable Code should be given. In a nondestructive examination certification program, the manufacturer has the responsibility for qualification and certification of NDE personnel. To obtain these goals, the manufacturer may train and certify personnel or may hire an independent laboratory that has qualified NDE procedures and personnel in accordance with the reference Code section and Section V to perform these duties. If others do the work, it must be stated in the quality control manual which department and person (by name and title) are responsible for ensuring that the necessary quality control requirements are met, because the manufacturer's responsibility is the same as if the parent company had done the work.

When required by the referencing section of the pressure vessel Code, personnel performing nondestructive examinations must be qualified in accordance with publication SNT-TC-1A of the American Society for Nondestructive Testing. SNT-TC-1A, *Recommended Practice for Nondestructive Examining Personnel Qualification and Certification,* is published by the American Society for Nondestructive Testing, Inc., 1711 Arlingate Lane, P.O. Box 28518, Columbus, OH 43228.

It is the responsibility of the manufacturer to conduct training and qualification tests of nondestructive examination personnel in accordance with the requirements of SNT-TC-1A, to arrange for eye examinations for the personnel interpreting nondestructive examinations, and to have audits made of the system in order to assure that qualified procedures and personnel are being used. The manufacturer must also prepare and maintain test examinations as well as test procedures which comply with Section V of the Code and SNT-TC-1A. Only personnel qualified to perform examinations should be assigned this task.

Adequate nondestructive examining procedures are an essential part of any well-managed operation. Procedure format and content must be appropriate to the specific requirements of the shop. Without a written procedure, there is no guarantee as to the effectiveness or repeatability of the test. Nondestructive testing should be performed

TABLE 7.1 Analysis of Nondestructive Testing Techniques

Applications	Advantages	Remarks
X-ray radiographic inspection*		
For examination of internal soundness of welds, castings, forgings, and plate material (see Fig. 7.1)	1. Gives sharper definition and greater contrast for thicknesses up to 3 in 2. Gives a graphic and permanent record indicating size and nature of defects 3. Offers established standards of interpretation for guidance 4. Radiation source can be turned off when not in use	1. Protective precautions necessary to protect personnel in surrounding area 2. Trained technicians required to take and interpret film
Gamma-ray radiographic inspection†		
For examination of internal soundness of welds, castings, forgings, and plate material (see Fig. 7.2)	1. More suitable for heavy thicknesses 2. Portable equipment 3. Source holder permits use through small openings 4. Lower initial cost 5. Requires no cooling mechanism 6. Gives a graphic and permanent record 7. No tube replacement	1. Government license is required for possession and use of isotopes 2. Protective precautions necessary 3. Sensitivity is not as high as x-ray testing, and persons not properly trained in interpreting film may underestimate seriousness of a defect 4. Energy cannot be adjusted, so isotope must be chosen to meet sensitivity requirements and thickness of material

*See Code Pars. UW-51 and UW-52.
†See Code Section V, Article 2.

in accordance with a written procedure. Each manufacturer should follow the procedure in accordance with the requirements of the referencing section of the Code and ASME Code Section V, Nondestructive Examinations.

The manufacturer is responsible for seeing that all personnel performing nondestructive examinations for Code vessels are competent and knowledgeable in applicable nondestructive examining requirements. They must be qualified in accordance with the requirements of SNT-TC-1A. The intent of SNT-TC-1A is to serve as a guide for the qualification and certification of nondestructive test personnel. This document establishes the general framework for a qualification and certification program. It explains the various levels of qualification. It gives the requirements and shows the type of examination required. Certification may take any one of several forms.

TABLE 7.1 Analysis of Nondestructive Testing Techniques (*Continued*)

Applications	Advantages	Remarks
Magnetic particle inspection*		
For locating surface and subsurface defects that are not too deep (see Figs. 7.3 and 7.4) Technique employs either electric coils wound around the part or prods to create a magnetic field. A magnetic powder applied to the surface will show defects as local magnetic fields. The nature of the defects will be revealed by the way the powder is attracted.	1. Useful for the inspection of nozzle and manhole welds for which radiography would be difficult-at best and most often impossible 2. Reveals small surface defects that cannot be detected radiographically 3. Useful in weld repair to assure removal of defect before rewelding 4. Can be used to detect laminations at plate edges	1. Useful only for magnetic material 2. Not suitable for defects parallel to magnetic field 3. Training needed to evaluate visual indications of defects 4. Prod methods of testing may mar the surface of a machined or smooth part 5. Magnetic particle examination methods (see Code Appendix 6)
Liquid penetrant inspection†		
For locating surface defects A liquid dye penetrant is applied to a dry, clean surface and allowed to soak long enough to penetrate any surface defects. After a time interval up to an hour, the excess penetrant is wiped off and a thin coating of developer applied. The penetrant entrapped in a defect will be drawn to the surface by the developer, and the defect will be indicated by the contrast between the color of the penetrant and that of the developer.	1. Especially applicable for nonmagnetic material 2. Useful for the inspection of nozzle and manhole welds when radiography is impossible 3. Easy to apply, accurate, fast, and low cost	1. Detects only defects that are open to the surface 2. Not practical on rough surface 3. Liquid penetrant examination methods (see Code Appendix 8) 4. Take care in applying dye penetrant to ensure that excessive amounts of vapors are not inhaled

*See Code Appendix 6.
†See Code Appendix 8.

The manufacturer may employ persons previously trained and certify these people. If this is done, the responsibility would be the same as if the manufacturer had trained and qualified these persons. The manufacturer may conduct a training program to meet the requirements of SNT-TC-1A. This responsibility would be carried out through a designated employee who meets Level III requirements. This employee will train the persons, examine them, and certify them as to

TABLE 7.1 Analysis of Nondestructive Testing Techniques (*Continued*)

Applications	Advantages	Remarks
Fluorescent penetrant inspection*		
For locating defects that run through to surface. Penetrant is applied by spraying, dipping, or brushing. Excess penetrant is washed from surface with a water spray and the surface allowed to dry. Dry powder or water-suspension developer is applied to part to draw penetrant to surface. Penetrant glows under black (ultraviolet) light. For leak test, penetrant is applied to one side and the other side is examined under black light for indications of glow	1. Easy to perform 2. Defects show clearly under black light 3. Especially applicable for nonmagnetic material 4. Can be used on rough surfaces 5. Detects porosity and defects that would otherwise be hard to find	1. Detects only defects that are open to the surface 2. Black light requires a source of electricity 3. Not very effective as leak test for plates more than $\frac{1}{4}$ in thick

*See Code Appendix 8.

their ability to do nondestructive examinations. Through the examinations, this employee will be able to judge the qualifications of these persons for Level I and/or Level II certification.

The American Society of Nondestructive Testing (ASNT) has training course programs to help a manufacturer write a training program to conduct the courses, which are usually given by a qualified Level III individual who uses questions derived from the ASNT course.

In addition to the required general examination, a specific written examination is necessary. The specific examination should reflect the procedures and equipment used in the plant.

In the practical examination, test objects should be similar to those fabricated by the manufacturer, applying the written procedures and the same equipment used in production. The examinee must be able to use the written procedures and equipment.

In summarizing, a manufacturer must have qualified personnel. In addition, the manufacturer should have written procedures if the shop does nondestructive examinations. The Code requires that nondestructive examining personnel be trained and qualified, also that their qualifications be documented as required in the referencing Code and SNT-TC-1A with its supplements and appendixes. It cannot be emphasized too strongly that nondestructive examination of ASME Code pressure vessels must be performed by qualified personnel.

TABLE 7.1 Analysis of Nondestructive Testing Techniques (*Continued*)

Applications	Advantages	Remarks
Ultrasonic inspection*		
For detecting defects in welds and plate material and for gaging thickness of plate (see Fig. 7.5) High-frequency sound impulses are transmitted through a search unit, usually a quartz crystal. This search unit is held in intimate contact with the part being tested, using an intermediary such as oil or glycerine to exclude air. The sound waves pass through the part being tested and are reflected from the opposite side or from a defect. The time of travel shows the defect on a cathode-ray tube or scope. This scope can also measure depth of crack or defect. Although the impulse instrument used to detect defects will also give thickness measurements, the resonance instrument is actually an electronic micrometer designed to measure thickness from one side of the material within tolerances of ±0.002 in.	1. Portable equipment 2. Access from one side of part being tested is possible 3. Reveals small root cracks and defects not indicated by radiographic film, especially in thick-walled vessels 4. Thickness measurements are rapidly made 5. Good for detection of laminated plates 6. Can be used in nuclear vessels, where induced radiation eliminates the use of radiography 7. Remote inspection can be made in hostile environment 8. Can be used to determine whether use or corrosion has affected the thickness of vessel walls and piping	1. Training required for interpretation of visual indications of defects 2. Rough surfaces must be made smooth if crystal is to make contact with part 3. Photographs must be taken to provide permanent records 4. Ultrasonic test methods when required (see Code Appendix 12) 5. Not very effective on welds with backing rings

*See Code Appendix 12.

Leak Testing

Several leak-detecting methods are used in testing pressure vessels. One is the halogen diode detector testing method, which is generally performed by pressurizing the vessel being tested with an inert gas, usually Freon or a trace of Freon, and examining the areas to be tested with a device that is sensitive to Freon. The presence of a leak will give an audible alarm, will be indicated on a meter, or may light up an indicator light.

TABLE 7.1 Analysis of Nondestructive Testing Techniques (*Continued*)

Applications	Advantages	Remarks
\multicolumn{3}{c}{Eddy current testing}		

Applications	Advantages	Remarks
Eddy current testing is based upon the principles involving circulating currents into an electrically conductive article and observing the interaction between the article and the currents	1. Detects small discontinuities 2. High-speed testing 3. Accurate measurement of conductivity 4. Checks variation in wall thickness 5. Noncontacting	1. Limited to use with conductive materials or conductive-base materials 2. Depth of penetration restricts testing to depths of less than ¼ in in most cases 3. The presence of strong magnetic fields will cause erroneous readings 4. Testing of ferromagnetic metals is sometimes difficult 5. Training required for interpretations of visual defects and indications

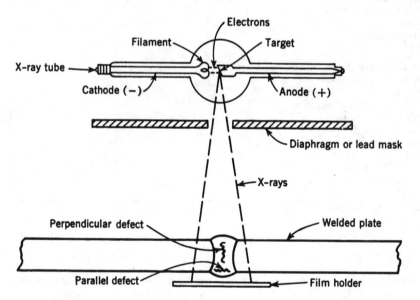

Figure 7.1 X-ray inspection of welds.

Nondestructive Examination   207

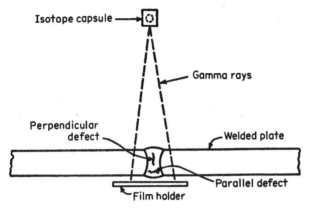
Figure 7.2 Gamma-ray inspection.

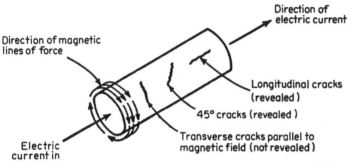
Figure 7.3 Magnetic particle inspection (circumferential magnetic field).

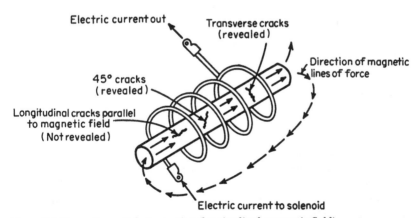
Figure 7.4 Magnetic particle inspection (longitudinal magnetic field).

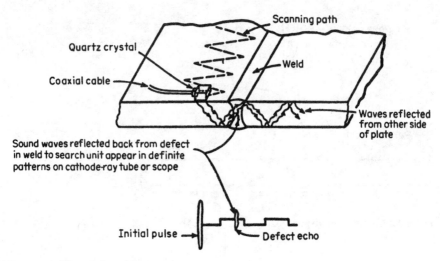

Figure 7.5 Ultrasonic weld inspection.

Another method consists of pressurizing the vessel to be tested with air containing a Freon gas. A halide torch, whose flame changes color when a leak appears, is employed to search for leaks.

Often a helium mass spectrometer leak detector is used. It measures a tracer medium, usually helium gas. If a leak is present, the leak detector detects the helium gas and gives an electrical signal. In the helium mass spectrometer test, the tracer medium helium gas passes through a leak from the pressurized side to the nonpressurized side. The test can be made under a vacuum, pressure test, or a combination of both.

A common method in pressure vessel testing is the gas and bubble formation test. Its purpose is to detect leaks in a vessel by applying a solution which will form bubbles at the leak area when gas passes through it.

Acoustic Emission Testing

Acoustic emission testing is a recent NDE method that has been successfully used in industry with hydrostatic tests and in-service examination of pressure vessels and piping. Acoustic emission systems use piezoelectric transducers attached directly to the structure being evaluated to detect the stress waves generated by discontinuities in materials under stress. The acoustic emission device measures the emission of these minute pulses of elastic energy; when properly detected and analyzed, the test provides information concerning the discontinuities contained in the structure under stress. Once detected, follow-up analysis of these signals verifies the location and meaning of a particular discontinuity.

The number of transducers needed to test a structure is determined by its complexity and not its size. Twelve to fifteen transducers may be required to test a storage vessel. Sixty to ninety may be required to test the pressure containment system of a nuclear plant, which includes the reactor vessel, pumps, steam generator, pressurizer, main piping, and valves. Pipeline testing needs at the least 1 transducer every 1000 ft.

Acoustic emission technology is most beneficial where accessibility to a pressure containment system is limited, because of its ability to conduct a complete structural integrity analysis with a minimum of shutdown time by sensing the presence of potential problems.

Chapter

8

ASME Code Section VIII, Division 2, Alternative Rules, and Division 3, Alternative Rules for High Pressure Vessels

The *ASME Boiler and Pressure Vessel Code,* Section VIII, Pressure Vessels, is published in three parts, Division 1; Division 2, Alternative Rules; and Division 3, Alternative Rules for High Pressure Vessels. In comparison to Division 1, Division 2 vessels require more complex design procedures as well as more restrictive requirements on design, materials, and nondestructive examinations; however, higher design stress values are permitted in Division 2, and it is possible to design pressure vessels with a safety factor of 3. [At the present time Division 2 is being extensively rewritten, and it might take another year or more to finish. For information on the new volume see the paper "An Overview of the New ASME Section VIII, Division 2 Pressure Vessel Code" by D. Osage (published in *Pressure Vessels and Piping Codes and Standards—2003*, PVP Vol. 453). This chapter discusses the current version.] It should be noted that in this chapter, all Code paragraph references are to specific parts of Division 2. The alternative rules apply only to vessels installed in a fixed (stationary) location (see Code Par. AG-100).

In Division 2 vessels, the requirements of Section VIII have been upgraded. A very detailed stress analysis of the vessel will be required. There are many restrictions in the choice of materials, more exacting designs, and stress analyses. Closer inspection of the required fabrication details, material inspection, welding procedures and welding, and more nondestructive examinations will be necessary since design stresses are larger than those permitted in Division 1. There will, in reality, be a theoretical design safety factor of 3, based on the ultimate strength of materials. With larger allowable stresses, a vessel can be built for the same pressure and diameter using thinner shells and heads, resulting in less weld metal and lighter vessels. However, the cost of engineering and quality control will be much greater and may exceed the savings made in material cost and fabrication.

Also, vessels which will be fabricated under Division 1 may not meet the requirements of Division 2. For example, only one quality factor for a vessel is provided for in Division 2, with all longitudinal and circumferential butt welds 100 percent radiographed. In another example, the allowable stress values are higher than those in Division 1: the lowest of (1) one-third of minimum tensile strength at room temperature, (2) two-thirds of minimum yield strength at room temperature, (3) two-thirds of minimum yield strength at temperature, (4) one-third of tensile strength at temperature.

An important feature of Division 2 is that the user of the vessel, or a representative agent of the user, must provide the manufacturer with design specifications, giving detailed information about the intended operating conditions (see Code Pars. AG-301 and AG-100). It is the user's responsibility to give sufficient design information to evaluate whether or not a fatigue analysis of the vessel must be made for cyclic service (see Code Par. AD-160). If the vessel has to be evaluated for cyclic service, it should be designed as required in the Appendix. It is also the user's responsibility to state whether or not a corrosion and/or erosion allowance must be provided for, and if so, the amount required. The user's design specifications must be certified by a registered professional engineer experienced in pressure vessel design (see Code Par. AG-301.2).

It is the responsibility of the vessel manufacturer to prepare a manufacturer's design report at the completion of the job, establishing conformance with the rules of Division 2 to meet the user's design specifications. This report must specify and reconcile any changes made during the fabrication of the vessel. The manufacturer's design report must be certified by a registered professional engineer experienced in pressure vessel design (see Code Par. AG-302.3).

Division 2 of the Code does not hold the Authorized Inspector responsible for approving design calculations. The Inspector verifies the existence of the user's design specifications and the manufacturer's design reports and that these reports have been certified by a registered professional engineer (see Code Par. AG-303).

It is the Authorized Inspector's duty to review the design, with respect to materials used, to see that the manufacturer has qualified welding procedures and welders to weld the material and to review the vessel geometry, weld details, nozzle attachment details, and nondestructive testing requirements. The manufacturer should obtain the Inspector's review of these details before ordering materials and starting fabrication so that the Inspector can point out the parts that may not meet the intent of the Code to avoid later rejection of the material or vessel.

Material controls will be stricter for Division 2 vessels. For example, Division 1, under Par. UG-10, allows tests of materials not identified, but Division 2 requires identification of all materials and is more restrictive than Division 1 in the choice of materials that can be used. Because of these requirements, Division 2 permits higher allowable stress values; therefore more precise design procedures are required. Special inspection and testing requirements of the materials have added to the reliability of the material. For example, plates and forgings over 4 in thick require ultrasonic inspection (see Code Par. AM-203). Cut edges of base materials over 1½ in thick also must be inspected for defects by liquid penetrant or magnetic particle methods. In addition, welding requires specific examination (see Code Table AF-241.1).

Steel-casting requirements and inspection in Division 2 are more demanding than those in Division 1. Division 2 requires inspection and tests for welded joints that are not covered in several welded pipe specifications. Therefore, welded pipe is not covered in the new division (see Code Par. AG-120).

In Division 2, because of the higher stress allowed for materials and for protection against brittle fracture of some steels like SA-285, thicknesses over 1 in will be limited to 65°F or more unless they are impact-tested (see Code Par. AM-211 and Fig. AM-218.1).

The tighter restrictions can be seen in plate steels, such as SA-516, which will permit a plate thickness of 1 in at −25°F when normalized or a temperature of 50°F in the as-rolled condition.

Heat treatment must be tightly controlled (see Code Table AF-402.1 and Code Pars. AF-635 and AF-636 for quenched and tempered materials; Code Pars. AF-730 and AF-776 for forged vessels; and Code Par. AL-820 for layered vessels).

Division 2 requires more nondestructive examinations. See Code Pars. AF-220 and AF-240 and Code Table AF-241.1, which summarize the required examination methods for the type of weld joints. The table indicates the applicable Code paragraphs and Code figures, gives helpful remarks, and shows when NDE examination methods may be substituted. Magnetic particle and liquid penetrant examiners need only be certified by the manufacturer (Code Par. AI-311 and Appendixes 9-1 and 9-2). Radiographers and ultrasonic examiners must be qualified (Code

Par. AI-501 and Appendix 9-3), their qualifications documented as specified when reference is made by Division 2 of the ASME Code Section V and publication SNT-TC-lA, which is the *Recommended Practice for Nondestructive Examining Personnel Qualification and Certification,* published by the American Society for Nondestructive Testing. Code Par. AI-511 gives unacceptable defects and repair requirements.

Hydrostatic testing requires a minimum of 1.25 times the design pressure, corrected for temperature. Division 2 differs from Division 1 in that it establishes an upper limit on test pressure. This upper limit must be specified by the design engineer (Code Pars. AT-302 and AD-151). When pneumatic tests are used, the minimum test pressure is 1.15 times the design pressure.

The Manufacturer's Data Report, required by Division 2, has a space provided for recording the name and identifying number of the registered professional engineer who certified the user's design specifications and manufacturer's design report.

The manufacturer must also sign the ASME data report before submitting it to the Authorized Inspector for the Inspector's review and signature. This data report must be retained for the expected life of the vessel or 10 years (see Code Par. AS-301), whichever is longer.

The manufacturer is also required to have a system for the maintenance of records for five years of specifications, drawings, materials used, fabrication details, repairs, nondestructive examinations and inspections, heat treatments, and testing.

It may be seen, with the above examples, that a Division 1 vessel, although thicker and heavier, may offset the cost of special design engineering and quality control that are required in Division 2, especially in smaller vessels.

Division 2 will allow more flexibility in design, which will enable the use of thinner materials. This will be helpful to the user in the manufacture of large-diameter, high-pressure vessels in which the additional cost of design and inspection is secondary to the weight of the vessel and in which the additional cost of engineering and quality control can be offset by the use of these thinner materials, thereby saving in the weight of the vessel and in handling.

It must also be remembered that the quality control system required by Code Par. AG-304 and mandatory Appendix 18 will be more rigorous than the quality control system for Division 1 vessels. The quality control system for Division 2 vessels should be established more on the methods used in that part of Section III on nuclear vessel quality assurance systems (see Chaps. 4 and 10 for a more detailed explanation of a quality control system for pressure vessels and a quality assurance system for nuclear vessels, respectively), because vessels designed under ASME Code Section VIII, Division 2 have special requirements. Some of these requirements detail how changes are made in either specification or design and their reconciliation via the Registered Professional Engineer control program, for example.

The quality control manual should also state that it is the duty of the Authorized Inspector to see that the certified documents are on file (Code Par. AG-303).

In addition, with respect to design, drawings, and specification control, the quality control manual should state that, before engineering and fabrication starts, the user's design specifications must be certified by a Registered Professional Engineer experienced in pressure vessel design who must assume the responsibility for its use.

The user's design specification must interface with the manufacturer's design report.

The quality control manual with regard to user's design specifications should cover, as a minimum, those conditions set forth in Code Par. AG-301.

The manufacturer's design report, including calculations and drawings certified by a Registered Professional Engineer experienced in pressure vessel design, must be furnished by the vessel manufacturer, stating that the user's design specifications have been satisfied (Code Par. AG-302).

A copy of the manufacturer's report must be given to the user and a copy, along with a copy of the user's design specifications, shall be kept on file for five years (Code Par. AG-302.3).

With the growing use of increased pressures in the operation of vessels it became apparent that a third division limiting itself to higher-pressure vessels, generally above 10,000 psi, would be helpful. Section VIII does not establish minimum pressure limits for Division 3 vessels or maximum pressure limits for Division 1 or 2 vessels.

Division 3, like Division 2, is limited to fixed vessels (Code Par. KG-104). Vessels fabricated to Division 3 may be transported from work site to work site between pressurizations. Division 3 explicitly excludes cargo tanks mounted on transport vehicles (see Chapter 11). If a direct-fired vessel is not within the scope of Section I it may be constructed to the rules of Division 3.

As is required for Division 2, a Professional Engineer registered in the United States or Canada must certify the design, as well as the final design report, of a Division 3 vessel. In general Division 3 provides very sophisticated methods of analysis, manufacture, and control and therefore any more than a cursory discussion falls outside the boundaries of this book's simplified approach to the Code. If the reader is contemplating constructing or designing a Division 3 vessel, he or she should use the Code itself as a source.

Chapter

9

Power Boilers: Guide to the ASME Code Section I, Power Boilers

The *ASME Boiler and Pressure Vessel Code* Section I, Power Boilers, includes rules and general requirements for all methods of construction of power, electric, and miniature boilers and high-temperature water boilers used in stationary service. It also includes power boilers used in locomotive, portable, and traction service.

The rules of Section I are applicable to boilers in which steam or other vapor is generated at a pressure more than 15 psig and high-temperature water boilers intended for operation at pressures exceeding 160 psig and/or temperatures exceeding 250°F. Superheaters, economizers, and other pressure parts connected directly to the boiler without intervening valves are considered as part of the scope of Section I (see Code Section I, Power Boilers Preamble).

References in this chapter are for Code Section I, Power Boilers.

The 1914 ASME Boiler Code was the original pressure vessel code, which now includes 11 sections. Section VIII, Division I, Pressure Vessel Code, is used to build more vessels than all the others combined. This book was written mainly for users of Section VIII, Division I. However, many engineers, fabricators, and inspectors must work with all sections of the Code.

Those who use the ASME Code, Section I, Power Boilers, regularly in construction and inspection may not have difficulty in finding the information needed. It is the engineer, fabricator, or inspector who consults it infrequently who may have difficulty.

A Quick Reference Guide (Fig. 9.1) illustrates some of the types of boiler construction which are provided for under Section I of the ASME

Code and furnishes direct reference to the applicable rule in the Code. In the event of a discrepancy, the rules in the current edition of the Code shall govern. This chart should be used only as a quick reference. The current edition of the Code should always be referenced.

There are also many conditions in the boiler Code for which all boiler manufacturers, assemblers, and Authorized Inspectors have to take special precautions, especially when large boilers are fabricated partly in the shop and partly in the field by one manufacturer and the piping and components are manufactured by others. This is particularly important if all the parts are being assembled in the field by the boiler manufacturer or an assembler who is certified by the ASME to assemble boilers. When construction of a boiler is to be completed in the field, either the manufacturer of the boiler or the assembler has the responsibility of designing the boiler unit, fabricating or obtaining all the parts that make up the entire boiler unit including all piping and piping components within the scope of the boiler code, and including these components in the master data report to be completed for the boiler unit (see Code Par. PG-107).

Possible arrangements include the following:

1. The boiler manufacturer assumes full responsibility by supplying the personnel to complete the job.

2. The boiler manufacturer assumes full responsibility by completing the job with personally supervised local labor.

3. The boiler manufacturer assumes full responsibility but engages an outside erection company to assemble the boiler to Code procedures specified by the manufacturer. With this arrangement, the manufacturer must make inspections to be sure these procedures are being followed.

4. The boiler may be assembled by an assembler with the required ASME certification and Code symbol stamp and quality control system. This indicates that the organization is familiar with the boiler Code and that the welding procedures and operators are Code-qualified. Such an assembler, if maintaining an approved quality control system, may assume full responsibility for the completed boiler assembly.

Other arrangements are also possible. If more than one organization is involved, the Authorized Inspector should find out on the first visit which one will be responsible for the completed boiler. If a company other than the shop fabricator or certified assembler is called to the job, the extent of association must be determined. This is especially important with a power boiler when an outside contractor is hired to install the external piping. This piping, often entirely overlooked, lies within the jurisdiction of the Code and must be fabricated in accordance with the applicable rules of the ASME B31.1 Code for pressure piping by a certi-

ASME Code Section I, Power Boilers   219

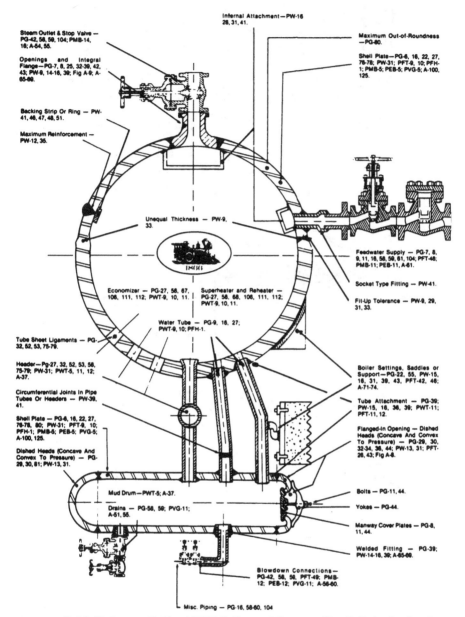

Figure 9.1 Quick Reference Guide: *ASME Boiler and Pressure Vessel Code*, Section I, Power Boilers. (*Courtesy of The Hartford Steam Boiler Inspection and Insurance Company.*)

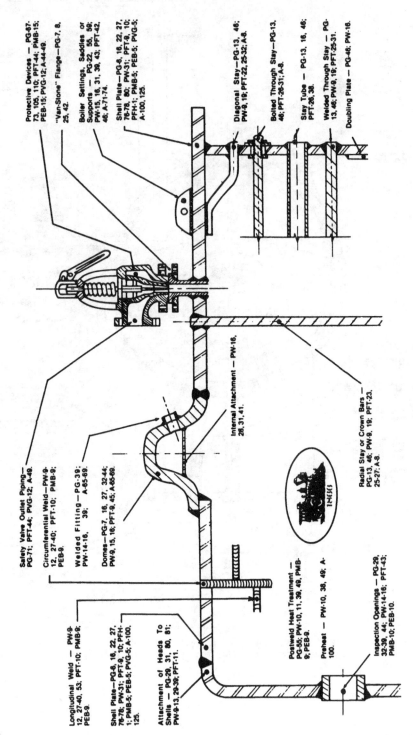

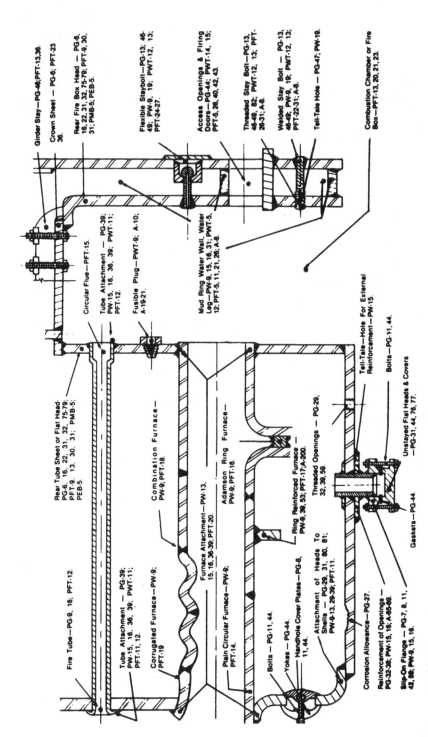

Figure 9.1 (*Continued*)

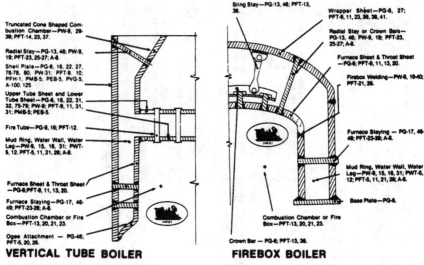

VERTICAL TUBE BOILER   **FIREBOX BOILER**

Figure 9.1 (*Continued*)

fied manufacturer or contractor. It must be inspected by the Authorized Inspector who conducts the final tests before the master data report is signed.

Once the responsible company has been identified, the Inspector must contact directly the person responsible for the erection to agree upon a schedule that will not interfere with the erection but will still permit a proper examination of the important fit-ups, chip-outs, and the like.

The first inspection will include a review of the company's quality control manual, welding procedures, and test results to make sure they cover all positions and materials used on the job. Performance qualification records of all welders working on parts of the boiler and piping under the jurisdiction of the Code must also be examined.

It is the manufacturer's, contractor's, or assembler's responsibility to submit drawings, calculations, and all pertinent design and test data so that the Inspector may make comments on the design or inspection procedures before the field work is started and thereby expedite the inspection schedule.

New material received in the field for field fabrication must be acceptable Code material and must carry proper identification markings for checking against the original manufacturer's certified mill test reports. Material may be examined at this time for thicknesses, tolerances, and possible defects.

The manufacturer who completes the boiler has the responsibility for the structural integrity of the whole or the part fabricated. Included in the responsibilities of the manufacturer are documentation of the quality control program to the extent specified in the Code.

General Notes

Quality Control System	—A-300
Material General:	—PG-5
a. Plate	—PG-6, 77
b. Forgings	—PG-7
c. Castings	—PG-8, 25
d. Pipe, Tubes & Pressure Containing Parts	—PG-9
e. Misc. Pressure Parts	—PG-11
f. Stays	—PG-13
Loadings	—PG-22; PW-43; A-71 - 74
Stress—Max. Allowable	—Tables PG-23.1, 23.2 and 23.3; A-150
Manufacturer's Responsibility	—PG-90, 104; PW-1; PEB-18
Inspector's Responsibility	—PG-90, 91
Pressure Tests	—PG-99, 100; PW-54; PMB-21; PEB-17; A-22, 63
Safety Devices	—PG-67 - 73; PMB-15; PEB-15; PVG-12, A-44-49
Trim:	
a. Gage Glasses	—PG-60; PFT-47; PMB-13; PEB-13; A-50 - 52
b. Pressure Gages	—PG-60; PEB-14; A-53
c. Gage Cocks	—PG-60; A-51
d. Water Columns	—PG-8, 60; PW-42; A-52
Service Restrictions	—PG-2; PMB-2; PEB-2
Welded Construction	—Part PW
Nameplates, Stamping, & Reports	—PG-104 - 113; PFH-1; PEB-19 ; A-350
NDE Requirements:	
a. MT & PT	—PG-25
b. RT	—PG-25; PW-51
c. UT	—PW-52
Porosity Charts	—A-250
Code Jurisdiction	—PG-58
Minimum Thickness	—PG-16; PFT-9
Feedwater Heater (Optional Requirements)	—PFH
Miniature Boilers	—PMB
Electric Boilers	—PEB
Organic Fluid Vaporizer Generators	—PVG

Special Reference to Particular Boiler Types

Traction, Stationary, or Portable Return Tube Firetube (Typically the Locomotive Type)	PG-60.1.5, 111.3; PFT-5.2. 23.2. 24.4, 29, 41, 42, 43.2, 43.3, 44, 45, 45.3. 45.7, 46.6
Horizontal Tube & Horizontal Return Tube	PG-59.3.7, 60.1.4. 111.1, 111.2; PFT-11.4.7, 42, 43.1, 43.2. 43.3, 44, 45.3, 46.3, 46.4, 46.5, 46.6, 47, 48.1
Vertical Tube	PG-111.4 PFT-5.2, 11.4.5, 20.5, 21.2, 21.3, 23.3, 23.4, 23.5, 42, 43.1, 43.5, 48.2
Fire Box	PFT-21, 42, 43.1
Scotch Marine	PG-111.6 PFT-20.2, 43.1, 43.4
Waste Heat	PG-106.4.1 PFT-43.1, 44
Forced Flow Steam Generator	PG-16.2, 21.1, 21.2, 58.3.3, 58.3.5, 58.3.6, 59.3.3, 60.1.2, 60.6.2, 61.5, 67.4, 106.3, 111.5.2, 112.1

Riveted Boilers
ASME Section I, Part PR, 1971 Edition

Figure 9.1 (*Continued*)

Chapter 10

Nuclear Vessels and Required Quality Assurance Systems

What is the difference between a fossil-fuel power plant and a nuclear power plant? In both types of plants, heat is used to produce steam in order to drive a turbine that turns a generator that, in turn, creates electricity. The basic difference between fossil-fuel power plants and nuclear power plants is the source of heat.

In a fossil-fuel power plant, coal, oil, or gas is burned in a furnace to heat water in a boiler in order to create steam. In nuclear plants no burning or combustion takes place. Nuclear fission is used, the fission reaction generates heat, and this heat is transferred, sometimes indirectly, to the water that provides the steam.

In a nuclear energy plant, the furnace is replaced by a reactor which contains a core of nuclear fuel consisting primarily of uranium. Energy is produced in the reactor by fission. A tiny particle, called a neutron, strikes the center or nucleus of a uranium atom, which splits into fragments releasing more neutrons. These fragments fly apart at great speed, generating heat as they collide with other atoms. The free neutrons collide with other uranium atoms causing a fission, which creates more neutrons that continue to split more atoms in a controlled chain reaction.

The chain fission reaction heats the water which moderates the reaction in the reactor creating steam, the force of which turns the turbine-generator. From this point on, nuclear energy plants operate essentially the same as fossil-fuel steam electric plants.

Power Plant Cycles

The purpose of this chapter is not to discuss nuclear technology but to let the reader know more about atomic power plant heat cycles to give a better understanding of the classifications and rules for construction and inspection of nuclear plant components. Code Classes 1, 2, 3, and MC allow a choice of rules that provide different levels of structural integrity and quality and explain why these components are subjected to a rigorous quality assurance inspection, bearing upon the relative importance assigned to the individual components of the nuclear power system to ensure their lasting performance.

There are several types of reactor coolant systems that are used. The two plant cycles most often found in utility service at present are the pressurized water reactor and the boiling water reactor.

Pressurized Water Reactor Plants

The pressurized water reactor (PWR), which is inside the containment vessel, is subjected to a high coolant water pressure. Here the pressurized water is heated. A pump circulates the heated water through a heat exchanger where steam for the turbine is generated. (See Fig. 10.1.)

The part of the pressurized water nuclear power plant which contains the reactor coolant is called the primary circuit. Included in the primary circuit is an important vessel called the pressurizer. The coolant volume varies when load changes require coolant temperature changes. When this occurs, the pressurizer serves as the expansion tank in the primary system, allowing the water to undergo thermal expansion and contraction and holding the primary circuit pressure nearly constant. If pressures were allowed to fluctuate too far, steam

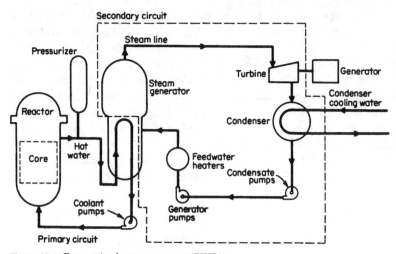

Figure 10.1 Pressurized water reactor (PWR).

bubbles might form at the reactor heating surfaces; these bubbles, or voids, if formed inside the reactor core, greatly alter reactor power output. The pressurizer has electric heating elements located low inside to provide the vapor needed to cushion the flowing liquid coolant. All of the above-mentioned items are located in the primary circuit.

The rest of the plant, called the secondary circuit, is essentially the same as a fossil-fuel steam electric plant. The heat exchangers make steam that passes through the turbine, condenser, condensate pump, feed pump, feedwater heater, and back to the heat exchanger.

Boiling Water Reactor Plants

In the boiling water reactor (BWR), the cooling water is inside the reactor core. The reactor makes steam that passes through the turbine, condenser, condensate pump, feedwater heater, and back to the reactor vessel.

In a boiling water reactor plant, steam generators and the pressurizer are not needed, as the plant has no secondary circuit, only a primary circuit. No primary to secondary heat exchange loss takes place. (See Fig. 10.2.)

The above explanation of the composition and operation of the boiling water plant (BWR) and the pressurized water plant (PWR) is meant to give the reader a basic understanding of nuclear power plant systems.

In addition to the two described, there are other types of plants, such as gas-cooled and liquid-cooled reactors, the number of which in operation is fewer.

Nuclear energy will only be viable if plants that make more nuclear fuel than they consume are built. These plants are called breeder reactors. If the breeder reactor is soon perfected and put in use, the energy

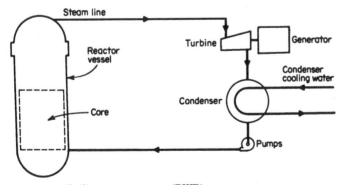

Figure 10.2 Boiling water reactor (BWR).

bank of United States uranium can be extended until a system is developed whereby energy is produced by fusion similar to the heat production method of the sun and other stars. If not, our uranium resources may be used up within a century.

Nuclear fusion is the joining together or fusion of atoms to form an atomic nucleus of higher atomic number and weight. Nuclear fission is the splitting apart of atoms. Both release enormous amounts of energy. All nuclear plants today operate by nuclear fission. Extensive research is being done to make electricity by nuclear fusion, which represents our best alternative for an abundant, accessible, and long-term energy source. When we know the secret of fusion energy, we will be able to duplicate the heat production methods of the sun.

All nuclear vessels must be manufactured under the appropriate part of the ASME Code Section III, Nuclear Vessels, Division 1, Nuclear Power Plant Components or Division 2, Concrete Reactor Vessels and Containments. Both divisions are broken down into subsections which are designated by capital letters preceded by the letter N for Division 1 and by the letter C for Division 2.

The ASME Code Section III, Division 1, Nuclear Power Plant Components, is divided into seven subsections covering all aspects of safety-related design and construction:

Subsection NCA: General Requirements for Division 1 and Division 2

Subsection NB: Class 1 Components

Subsection NC: Class 2 Components

Subsection ND: Class 3 Components

Subsection NE: Class MC Components (Metal Containment)

Subsection NF: Component Supports

Subsection NG: Core Support Structures

Section III, Division 1: Appendixes

Section III, Division 2: Code for Concrete Reactor Vessel and Containments; includes Subsection CB, Concrete Reactor Vessels, Subsection CC, Concrete Containments, and Division 2 Appendixes

These Code classes are intended to be applied to the classification of items of a nuclear power system and containment system. Within these systems the Code recognizes the different levels of importance associated with the functions of each item as related to the safe operation of the nuclear power plant.

The Code classes allow a choice of rules that provide assurance of structural integrity and quality commensurate with the relative importance assigned to the individual items of the nuclear power plant. Each subsection is intended to be an independent document with no cross-referencing of subsections necessary, with the exception of Subsection NCA, General Requirements, which is a mandatory requirement for those who intend to use one or more of the other subsections of Section III, Division 1 and Division 2. It includes Subsections NB through NG. Subsection NCA comprises all general requirements for manufacturers, installers, material manufacturers, suppliers, and owners of nuclear power plant components including the requirements for quality assurance, inspectors' duties, and stamping.

Section III, Division 2, Concrete Reactor Vessels and Containments, defines the requirements for the design and construction of concrete reactor vessels and concrete containments and Division 2 Appendixes. It covers requirements for materials, parts, and appurtenances of components that are designed to provide a pressure-retaining or -containing barrier.

The following explanation covers only Section III, Division 1, and does not include Division 2. *To avoid confusion it should be noted that all Code references in this chapter, unlike those in the rest of this book, refer to the ASME Code Section III, Division 1, Nuclear Vessels, rather than Section VIII, Division 1, Pressure Vessels.*

These seven subsections are intended to be applied to the classification of pressure-retaining or -containing components of a nuclear power system and containment system, component supports, and core support structures.

The *ASME Boiler and Pressure Vessel Code* Section III, Division 1, Nuclear Power Plant Components, differs from the other sections of the Code in that it requires a very detailed stress analysis of vessels classified as Class 1, Class 2, Class 3, and Class MC.

Class 1 includes reactor and primary circuit vessels, containing coolant for the core that cannot be separated from the core and its radioactivity.

Class 2 components are those which are not part of the reactor–coolant boundary but are used to remove the heat from the reactor–coolant pressure boundary.

Class 3 components support Class 2 without being part of them.

Class MC includes containment vessels.

All nuclear vessels and components require constant vigilance during fabrication and operation. The manufacturer is required to prepare, or have prepared, design specifications, drawings, and a complete stress analysis of pressure parts. The design report that will include all material has to be certified by a registered professional engineer. Copies of the certified stress report are required to be made available to the Authorized Nuclear Inspector at the manufacturing site as well as with the enforcement authority at the point of installation.

It is the duty of the Authorized Nuclear Inspector to verify that the equipment is fabricated according to the design specifications and drawings as submitted. Another duty is to make all inspections as specified by the Code rules.

Quality controls adequate in the past are unacceptable today. Pressure vessel fabricators view this period as the age of quality assurance. Today, the industry faces heavy demand for test reports and other supporting data, especially for the construction of nuclear vessels and parts. Fundamentally, there should be no difference in the level of quality between a high-pressure, high-temperature fossil-fuel boiler and a nuclear vessel. The basic differences are the amount of testing and documentation required in nuclear fabrication, whereas the power boiler code requires a quality control system (see Power Boiler Code, Appendix A-300).

The nuclear code requires that manufacturers of nuclear equipment have a more rigorous quality assurance program.* Excellence of each item produced must be the aim of each manufacturer.

The responsibility for the program lies with the manufacturer. A manufacturer who wants to fabricate nuclear vessels or parts covered by Section III obligates the company with respect to quality assurance and documentation. The manufacturer must have an understanding of the many Code details, a written quality assurance manual, and written implementing procedures describing the system of design, material, procurement, fabrication, testing, and documentation. The manufacturer must also have a clear understanding of nuclear quality assurance standards being used in every aspect of nuclear power plant construction from the design and construction phases to full operation.

The nuclear industry's good safety record is due to the rigorous standards of the quality assurance programs applied to all phases of information, creation, planning, and fabrication as well as to the design, construction, testing, inspection, and corrective action taken when nonconformities are discovered during construction and later when implementing the quality assurance programs in the operation and maintenance of a nuclear power plant.

---

*Note that portions of nuclear Code Article NCA-4000 of Section III, Divisions 1 and 2, General Requirements, 1980 edition, were deleted and replaced by reference to ANSI/ASME NQA-1, Quality Assurance Requirements for Nuclear Power Plants.

Quality Assurance

The quality assurance control program must be evaluated and approved for compliance by the ASME before issuance and upon each renewal of the Certificate of Authorization. This is done at the manufacturer's request and expense. The evaluation of the manufacturer's quality assurance capability will be made by a survey team designated by the ASME. The team will be made up of one or more consultants to the ASME as designated by the secretary of the Boiler and Pressure Vessel Committee, which includes the ASME team leader, a team member, a representative of the National Board of Boiler and Pressure Vessel Inspectors, a representative of the jurisdictional authority in which the manufacturing facility is located, the Authorized Nuclear Inspector Supervisor from the manufacturer's Authorized Inspection Agency, the Authorized Nuclear Inspector performing the Code inspection at the plant being surveyed, and a representative of a utility using nuclear energy.

The survey team's purpose will be to compile a report of the manufacturer's quality assurance capabilities at the plant or site. This report will be submitted to the ASME Subcommittee on Nuclear Accreditation and forwarded to the secretary of the Boiler and Pressure Vessel Committee, who will carry out the necessary procedures leading to the issuance of a Certificate of Authorization to the manufacturer for the use of an N Code symbol for the types of fabrication to be manufactured.

The intent of the ASME Code survey team will be to verify that the manufacturer has an acceptable quality assurance program describing the controls and manner in which nuclear items will be produced. In addition, the team will verify that an inspection contract or agreement with an Authorized Inspection Agency is in effect to review the manufacturer's quality assurance program and its compliance with the intent of Section II, Section III, NQA-1, Section V, and Section IX. As with Section VIII, Division 1, Section III also references Section V for the qualifications of nondestructive examination procedures and Section IX for welding qualification procedures. The team will examine the means and methods for the documentation of the manufacturer's quality assurance program, review the personnel involved in the manufacturer's quality assurance activities, and examine their status with respect to production and management functions; the team also will examine the manufacturer's relationship with the Authorized Nuclear Inspector with respect to the organization being audited and the means of keeping the Inspector informed and up-to-date on quality control matters; review all procedures as required by the Code with respect to qualifications of processes and individuals; and examine equipment and facilities employed for testing and carrying out quality assurance work.

It is the manufacturer's responsibility to aid the ASME survey team so that it can do its work. It would be advisable for the manufacturer to provide qualified personnel who are able to reply to the necessary inquiries generally made by the individuals of the survey team. The manufacturer should also provide a suitable workroom for the survey team. Facilities must be made accessible to the survey team so that an understanding of the manufacturer's operational capabilities is possible.

A management organization chart should be available as well as the organization chart of the quality assurance department. Also available to the survey team should be the manufacturer's qualification records of welding procedures and welders, nondestructive examination procedures and testing personnel, traveller or process sheet check-off lists, and other exhibits showing the methods used with respect to quality control of production. All nondestructive examining equipment and measuring and testing facilities must be available for examination. A written procedure of the training and indoctrination program for personnel performing work that affects quality control with names of persons involved and documentation records of examinations should be included.

Most importantly, the quality assurance manual, in which the company explains how its personnel control fabrication to meet Code requirements, will form the basis for understanding the manufacturer's system. It will be reviewed by the survey team. The manual will serve as the basis for continuous monitoring of the system by the Authorized Nuclear Inspector. In addition, the manual will serve as the document used by the Authorized Nuclear Inspector Supervisors during semiannual ANSI 626.0 audits and later survey reviews when the manufacturer's Certificate of Authorization is to be renewed. All manuals should be classified as either controlled or noncontrolled, as follows:

1. A controlled manual is one that is given a number, assigned to one individual, and kept up-to-date in accordance with the approved system described in the manual by the party responsible for manuals in the plant.

2. A noncontrolled manual is a copy of the manual distributed to customers and other interested parties, which will only be considered up-to-date at the time of distribution and should be marked as such by rubber stamping on the first page as well as several other pages throughout the book, indicating it will not be kept up-to-date. All pages of the manual should have the name of the manufacturer and a revision block including a space for revision levels and the dates of these revisions.

An outline of the procedures to be used by the manufacturer in describing a quality assurance system should include the following items.

1. *Organization*
   a. A statement of authority should be specified in this section. This statement should identify management's participation in the program—that the quality assurance manager shall have the authority and responsibility to establish and maintain a quality assurance program, and organizational freedom to identify quality problems in order to provide solutions to these problems. This statement of authority should be signed by the president of the company or someone else in top management.
   b. The manufacturer's organization chart must be included, giving titles and showing the relationship between management, quality assurance, purchasing, engineering, fabrication, testing, and inspection.
   c. Assurance that quality is achieved and maintained by those assigned the responsibility for completing the task should be given.
   d. A description of how quality achievement is verified should be listed.

2. *Quality assurance program*
   a. The manual will document a quality assurance program which includes procedures that will assure and provide that the plans and activities affecting quality are carried out.
   b. The quality assurance manual should give a program for indoctrination and training, which includes program descriptions, listing of personnel involved, and records of their performance. All personnel involved with quality control must be given training with excellence of performance in mind.

3. *Design control*
   a. The design shall be controlled, defined, and verified. Verification is required by the manufacturer that design specifications certified by a registered professional engineer are the owner's or the agent's certified design specifications, a copy of which is to be made available to the Authorized Nuclear Inspector before construction begins and to the enforcement authority at the location of installation before components are placed in service.
   b. Responsibility for structural integrity and the Design Reports for Class 1 components and component supports, Class 2 vessels designed under NC-3200, Class MC vessels and component supports, certain Class 2 and 3 components, and Class CS should be

certified by the registered professional engineer who makes the stress report and the owner or agent.

    c. Load capacity data sheets for Class 1 and Class MC component supports designed under NC-3200 must be certified by a registered professional engineer stating that the load capacity of the component support is rated in accordance with Code Subsection NF, with the data sheet specifying the organization responsible for retaining data that substantiates stated load capacity.

    d. When the class of a vessel or part does not require a design report, the design calculations must be made available to the Authorized Nuclear Inspector.

    e. The complete certified stress report by the manufacturer and owner shall be made available to the Authorized Nuclear Inspector.

    f. Design reviews and checking shall be made by individuals other than those who initiated the original design.

4. *Procurement document control*

    a. The manufacturer should have an order entry system which provides for competent review and remarks by applicable persons to assure correct translation of design specification requirements into specifications, drawings, procedures, and instructions.

    b. Document control of revisions is needed to assure that documents and changes are reviewed for adequacy, approved, documented, and released by authorized personnel and distributed and used wherever required.

5. *Instructions, procedures, and drawings*

    a. Activities affecting quality shall be described in accordance with written instructions, procedures, and drawings.

    b. The inclusion, generation, revision, and issuance of these documents within the quality assurance system should be explained.

6. *Document control*

    a. Documents must be controlled so that only the correct documents are issued. This system should include how documents are identified, controlled, prepared, reviewed, approved, issued, and corrected prior to issuance.

    b. Procedures identifying responsible personnel to assure that the required documents are reviewed and approved before release should be established.

    c. The revision system must provide for engineering review by qualified individuals, including the registered professional engineer responsible for stress report or design calculations.

    d. All revisions and replacements must be controlled and documented.

    e. Documents must be made available to the Authorized Nuclear Inspector.

7. *Control of purchased items and services*
   a. The manufacturer is responsible for surveying and qualifying vendor quality system programs, or the material manufacturer must hold a quality system certificate issued by the ASME.
   b. The system of material control used in purchasing to assure identification of all material used, and to check compliance with specifications, should be explained.
   c. Details of the vendor's evaluation should include a sample of the survey questions asked by the manufacturer, the vendor's responses to these questions, and how often the vendor is to be audited.
   d. The qualification and listing of a vendor should be described, as should revisions to and control of the accepted vendor list. The procedure of removing a vendor from the accepted vendor's list shall be explained.
   e. The control of material at the source—who issues and approves purchase documents—must be described.
   f. The way in which material manufacturer's and subcontractor's corrective action is verified and controlled should be given.

8. *Identification and control of items*
   a. The procedure of receiving inspection control—identification of material with receiving check list—must be explained.
   b. The control of nonconformity of purchased items should be discussed and the corrective action system should be stated.
   c. The manufacturer's certified materials test report, including welding and brazing materials, must be given. How the test report is examined and signed off to assure correlation with material used for traceability must be stated.

9. *Control of processes*
   a. A description of the system for assuring that all welding procedures and welders or welding operators are properly qualified (Code Par. NCA-5250) must be stated.
   b. Welding qualification records and identifying stamps, including the method of recording the continuance of valid performance qualification by production welding, should be presented.
   c. The method of welding electrode control, i.e., electrode ordering, inspection when received, storage, how electrodes are issued, and the in-process protection of the electrode, should be described.
   d. The method used on drawings to tell type and size of weld and electrode to use, i.e., whether the drawings show weld detail in full cross section or use welding symbols, should be given.
   e. The fit-up and weld inspection must be explained.
   f. The system for assuring that all nondestructive examination personnel are qualified should be described.

g. Written procedures to assure that nondestructive examination requirements are maintained should be listed.
  h. The required written training program should be explained.
  i. The examination methods in the following required areas should be detailed:
     (1) general
     (2) specific
     (3) practical
     (4) physical
  j. The personnel qualification record and Level III support documents for examinations, experience, education, and physical examinations should be presented.
  k. The equipment qualification and calibration should be described.
  l. For subcontracted NDE operations, similar controls as above and the contract terms with the NDE organization should be given.
  m. Procedures for preheating, postweld heat treatment, hot forming, etc., should be listed.
  n. How the furnace is loaded, the thermocouple placement, how and where thermocouples are placed, the recording system for time and how and where it is placed, the recording system for time and temperature, and how time and temperature are recorded for each thermocouple should be described.
  o. The methods of heating and cooling, and also temperature control, should be given.
  p. The time and temperature recordings should be described, including their review by the Authorized Nuclear Inspector.

10. *Inspection*
    a. A statement shall be included which states that personnel performing functions affecting the quality of the work shall not report directly to the persons responsible for the performance of that work.
    b. That each person who performs the work shall be properly qualified, indoctrinated, and trained to perform the work should be stated.
    c. This chapter could be included with the chapter entitled Inspections, Tests, and Operating Status.

11. *Test control*
    a. Tests which will require verification of conformance of an item must be planned and executed and must be defined in the quality assurance program.
    b. The results must be documented as outlined in the manual.

12. *Control of measurement and test equipment*
    a. The established and documented procedures used for assuring that gages and measuring and testing devices are calibrated and controlled should be described.
    b. Calibration should be to national standards where such standards exist; other standards, when used, should be designated.
    c. Procedures to assure that measuring and test equipment are calibrated and adjusted at specified periods must be established.
    d. The control of nonconforming equipment should be discussed. Corrective action taken when discrepancies in examination or testing equipment are found should be stated.
13. *Handling, storage, and shipping*
    a. The procedures used to control methods of handling, storage, cleaning, and preservation of materials, packaging, and shipping in order to prevent deterioration, damage, and loss should be described.
    b. Procedures for examination and inspection should be included on the traveller check-off list.
14. *Inspection, test, and operating status*
    a. Identification and control of material heat number transfer should be described.
    b. Use of a "traveller" or process control checklist indicating fabricating processes is needed.
    c. The preparation and use of a traveller or process control checklist with sign-off spaces for use by manufacturer should be described.
    d. Mandatory hold points (Code Par. NCA-5241) must be listed.
    c. The control of nonconformities, material, parts, and components should be explained.
    f. Methods used for corrective action and nonconformities should be presented.
    g. The manufacturer should be carefully acquainted with the provisions in the Code that establish the duties of the Authorized Nuclear Inspector. The third party in the manufacturer's plant, by virtue of authorization by the state jurisdiction to do Code inspections, is the legal representative. This party's presence permits the manufacturer to fabricate under state laws. Under the ASME Code rules, the Authorized Nuclear Inspector has certain duties as are specified in the Code. It is the obligation of the manufacturer to see that the Inspector is given the opportunity to perform as required by law. The details of Authorized Nuclear

Inspector visits at the plant or site are to be worked out between the Inspector and the manufacturer. The manufacturer's quality assurance manual should state the provisions for assisting the Authorized Nuclear Inspectors in the performance of their duties.

   h. The manual should state that the Authorized Nuclear Inspector, Authorized Nuclear Inspector Supervisor, and the Authorized Inspection Agency have free access to areas and documents as necessary to carry out the duties specified in the Code, such as monitoring the manufacturer's quality assurance system, reviewing qualification records, verifying materials, inspecting and verifying in-process fabrication, witnessing final tests, and certifying data reports.

   i. The Inspector should be given the opportunity to indicate "hold points" and to discuss these inspection points with the quality assurance manager as well as make arrangements for their desired inspections.

   j. The Authorized Nuclear Inspector must be provided with the current quality assurance manual with latest revisions which have been accepted by the Authorized Nuclear Inspector Supervisor. The final hydrostatic or pneumatic tests required shall be witnessed by the Authorized Nuclear Inspector, and any examinations that are necessary before and after these tests shall be made by the Inspector.

   k. The manufacturer's data reports shall be carefully reviewed by a responsible representative of the manufacturer or assembler. If the reports are properly completed, the representative shall sign them for the company. The Authorized Nuclear Inspector, after examining all documents and feeling assured that the requirements of the Code have been met and the manufacturer's data sheets properly completed, may sign the data sheets with his or her name and National Board commission number.

15. *Control of nonconforming items*
    * a. The definition of a nonconformity should be given.
    * b. Nonconforming materials, components, and parts that do not meet the required standards should be corrected, or they may not be used.
    * c. Identification and inventory of parts should be explained.
    * d. The methods of corrective action, return of material for use, and record keeping of traceability as well as material segregation to prevent inadvertent use should be included.
    * e. The names of appropriate levels of management involved with responsibilities and authority for control of nonconformities should be given.

f. When disposition of nonconformity involves design, design report, or dimensional changes, review should be provided by a responsible engineering person or group. (See Correction of Nonconformities, Chap. 4, item 6.)
16. *Corrective action.* The way in which nonconforming items will not continue to be a problem should be explained. An identification, cause, and corrective action plan shall be documented so that conditions adverse to quality do not reoccur.
17. *Quality assurance records*
    a. A written procedure should be established for the handling of records, permanent and nonpermanent, for the time specified in Code, showing how records will be kept. The procedure should be determined early and a list developed to serve as index to the document file.
    b. Permanent records: The owner has the responsibility for continued maintenance of all records and the traceability of records for the life of the plant at the power plant site, the manufacturer's plant, or other locations determined by mutual agreement.
    c. Nonpermanent records: State responsibility for who will maintain, store and distribute, where they will be kept, and how long they will be kept.
    d. The quality assurance system must verify records of materials, manufacturing, examination, and tests before and during construction. The responsibility for and location of all records should be stated.
    e. Storage and protection of records from damage should be explained.
18. *Audits*
    a. The system of planned, periodic audits to determine the effectiveness of the quality assurance program should be explained.
    b. Audits should be performed in accordance with written procedure using a checklist developed for each audit in order to assure that a consistent approach is used by audit personnel.
    c. A report of audit results should be given to top management, including the corrective action to be taken if necessary.
    d. The corrective action taken by the quality assurance manager to resolve deficiencies, if any, revealed by the audits should be monitored.
    e. The results of the audit must be made available to the Authorized Nuclear Inspector.
    f. The name and title of person responsible for the audit system should be listed.

When constructing nuclear pressure vessels according to the ASME Code, complete quality assurance is a must. In this chapter, an attempt has been made to acquaint the reader with the basic elements of a total quality assurance manual and system—from design specifications and control to auditing of the system, which includes document control, purchase control, vendor surveys and control, receiving inspection, identification and control of materials, control of manufacturing and/or installations, control of examination and tests, control of measuring and test equipment, handling, storage, shipping and preservation, examination of process status, nonconformity control of materials and items and corrective action to take, as well as quality assurance record retention and authorized inspection.

If a manufacturer uses this chapter as a guide, it will help in preparing for an ASME nuclear survey, or if one is contemplated, it will help in understanding the rigorous standards an organization and manufacturing system must meet to enable nuclear power plant components to be fabricated. Pursuit of excellence in the manufacture of nuclear vessel components must be the aim of each manufacturer seeking ASME certification to fabricate nuclear vessel components.

Chapter

11

Department of Transportation Requirements for Cargo Tanks

History and Overview of 49 CFR §107 to §180

In September 17, 1985, the Research and Special Programs Administration, known as RSPA, a sub-branch of the Department of Transportation (DOT), published a notice in the *Federal Register* which was listed as Docket (House Measure) HM-183 and HM-183A (50 FR 37766). This notice contained proposals to revise and clarify the existing Transportation of Hazardous Materials Regulation which specifically pertained to the manufacture, maintenance, requalification, and use of all over-the-road "specification" cargo tanks. These proposals were based on research findings, petitions for rule change, requests for interpretations of current regulations, and input from the Federal Highway Administration.

During 1985 and 1986, the Research and Special Programs Administration (Department of Transportation) and the Federal Highway Administration held numerous public briefings and public hearings to allow interested persons to play a role in the development of the new rules process. Additionally, interested persons were encouraged to submit written comments and recommendations which would enhance and/or improve the safe transportation of hazardous materials being transported in over-the-road cargo tanks.

Throughout 1987 and into February of 1988, additional public meetings were held with several special interest groups and other organizations that had submitted written comments to obtain clarification of the new proposed rules and to obtain additional supporting information on their alternative proposals.

On June 12, 1989, the RSPA issued a Final Rule in the June 12, 1989 *Federal Register* which amended the Hazardous Materials Regulation (HMR) pertaining to the manufacture of cargo tanks and the operation, maintenance, repair, and requalification of all specification cargo tanks. This Final Rule also established a new §180 section containing requirements governing the maintenance, use, inspection, repair, retest, and requalification of cargo tanks used for the transportation of hazardous materials. Even though this June 12, 1989, date was published as a Final Rule, the RSPA granted a 90-day limitation for the receipt of petitions for reconsideration of the June 12th Final Rule.

During the time period of June 12, 1989, through September 30, 1990, the RSPA responded to over 1,000 petitions for reconsideration of the rules published on June 12, 1989. The May 22, 1990 *Federal Register* published and announced a series of effective dates for various implementation elements of the Docket HM-183.

On September 7, 1990, the *Federal Register* published technical changes which will effect the design and stress calculations for cargo tanks, and in November of 1990 President Bush signed into law the Hazardous Material Transportation Act of 1990, thereby ending almost ten years of debate.

As a result of this new legislation, the manufacturer of specification cargo tanks must hold an ASME Code U symbol stamp and Certificate of Authorization. Repair, modification, stretching, and rebarreling of specification cargo tanks must be performed by the holder of a National Board R symbol stamp. (A new Code Sec. 11 is scheduled to be published in July 2004.)

For those cargo tanks which are not ASME Code stamped, the owner's Registered Inspector must witness and certify that the repairs were performed in accordance with the applicable specifications and their quality control program. Such repairs shall be documented on the forms established in the quality control manual (see Chap. 4).

In the event the cargo tank is an ASME Code–stamped tank, the repair concern with the R or U stamp must contact its Authorized Inspector to witness the repair. Any major modifications such as stretching or rebarreling of the tank "skin" and repairs to the tank skin design shall additionally be reviewed by a design certifying engineer. This will be discussed later in this chapter.

§178.345

These 15 paragraphs describe the general design and construction requirements applicable to the three current specifications DOT 406, DOT 407, and DOT 412.

The first paragraph, 345-1, contains definitions of terms used in this and other regulations. The prospective manufacturer is cautioned to read the definitions carefully. The paragraph also notes that the manufacturer of a cargo tank must hold a current ASME certificate and must be registered with the DOT in accordance with part 107, subpart F of 49 CFR (Code of Federal Regulations). In addition, construction must be certified by an authorized inspector or a registered inspector (defined later in this chapter). Paragraph 345-1 includes reference to part 393 of the Federal Motor Carrier Safety Regulations and the requirements in §173 of 49 CFR.

§178.345-2 allows certain ASTM materials in addition to all the materials in Section II, Parts A and B of the ASME Code. It additionally sets out certain do's and don'ts of material selection and thicknesses.

The third paragraph, 345-3, addresses structural integrity, in general defining stresses and their allowables. It notes in some detail that acceleration and deceleration forces must be taken into account, something that is rare at least in Code Section VIII.

The remaining paragraphs go on to describe in general differences and similarities among cargo tanks.

Characteristics of DOT 406, DOT 407, and DOT 412 cargo tanks, set forth in complete detail in §178.346, §178.347, and §178.348, respectively, are as follows:

1. The DOT 406 cargo tank has a maximum allowable working pressure (MAWF) no lower than 2.65 psig and no higher than 4 psig. This tank is obviously not an ASME pressure vessel, but it is still required to be "constructed in accordance with the ASME Code." However, §178.346 does list paragraphs of the Code that do not apply as well as allowed modifications.

2. The DOT 407 cargo tank must be of a circular cross-section and have an MAWP of at least 25 psig. Such a tank with an MAWP over 35 psig or having a vacuum in its design must follow the ASME Code; a vacuum tank must have an external design pressure of at least 15 psig. If the MAWP is equal to or less than 35 psig the tank must be "constructed in accordance with the ASME Code"; §178.347 also lists paragraphs of the Code that do not apply and allowed modifications.

3. The DOT 412 cargo tank must have an MAWP of at least 5 psig. This tank is an "in-betweener," because one with an MAWP of 5 psig would not fit the ASME Code and one with an MAWP of over 15 psig would. Additionally, if the tank has an MAWP of greater than 15 psig it must be of circular cross-section and be certified and constructed following the Code. The MAWP for a vacuum-loading tank must be at least 25 psig internal and 15 psig external. A tank with an MAWP greater than 5 and less than or equal to 15 psig must be "constructed in accordance with the ASME Code"; §178.348 lists, again, allowed modifications and paragraphs of the Code that do not apply.

One can tell from the above synopsis that these rules are complicated and that it would be unwise to venture into the field of the new DOT specifications without first studying them in detail. A manufacturer may not register with the DOT without months or years of experience. This area is not for the beginner.

Specification Cargo Tank MC-331 (§178.337)

Figure 11.1 depicts the general form of the MC-331 specification cargo tank. These cargo tanks are designed in accordance with the General and Specific Design Requirements of 49 CFR §178.337 and are generally fabricated from heat-treated carbon-steel materials. Additionally, the design of the tank itself shall be in accordance with ASME Code Section VIII, Division 1. The MC-330 specification was replaced by MC-331 in 1967. Note that these two tanks are almost identical except for the manway opening location on the head of the vessel. Construction of this type of specification cargo tank is authorized and will continue to be manufactured and will not be replaced as were the MC-306, MC-307, and MC-312 cargo tanks. This specification tank construction has not been affected by the revision of the new federal regulations and will continue to be manufactured under its existing specification number. However, new and additional test and inspection requirements do exist.

The typical MC-331 tank is primarily used to carry pressure gases such as propane, butane, chlorines, anhydrous ammonia, as well as other pressure gasses, all of which are under internal pressure.

The designer must design, construct, and certify this specification cargo tank in accordance with the ASME Code Section VIII, Division 1. The tank may be of a seamless or welded construction. The design pressure of this tank must not be less than the vapor pressure of the commodity being transported within the tank at 115°F. The maximum allowable working pressure of any tank shall never be less than 100 psig nor more than 500 psig.

These tanks shall be painted white, aluminum, or another reflective color when the tank is uninsulated on the upper two-thirds of area of the tank. This is for the purpose of reflecting the heat of the sun.

The MC-331 specification requires that the vessel be postweld heat-treated as prescribed in the ASME Code Section VIII, Division 1 in accordance with Code Part UHT. If used in chlorine service the tank must be 100 percent radiographed and postweld heat-treated. If the tank will transport anhydrous ammonia, postweld heat treatment must be performed. Welded attachments to pads may be made after postweld heat treatment. The postweld heat-treatment (PWHT) temperature shall be in accordance with ASME Code Section VIII, Division 1, but cannot be less than 1050°F at the skin temperature of the vessel during PWHT.

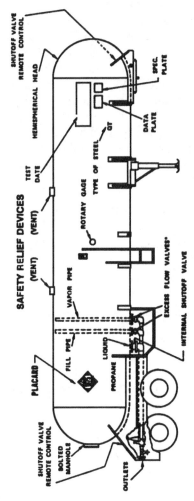

Figure 11.1 MC-331 specification cargo tank. (*Courtesy of National Tank Truck Carriers, Inc.*)

In addition to postweld heat treatment, materials are required to be impact-tested. Impact tests shall be made on a lot basis. A lot is defined as 100 tons or less of the same heat-treatment processing material lot having a thickness variation no greater than ±25 percent.

The minimum impact required for full specimens must be 20 ft · lb in the longitudinal direction at −30°F, Charpy V-notch test and 15 ft · lb in the transverse direction at −30°F, Charpy V-notch test. If the lot does not pass these tests, then the manufacturer may still use the material, provided this same test is applied to individual plates.

Welding procedures and welders shall be qualified in accordance with the ASME Code Section IX. Welded joints shall be made without undercutting in the shell and head. If undercutting is inadvertently done, such undercutting shall be repaired per Code Section VIII, Division 1.

In addition to the essential variables of Section IX, the following shall be considered essential and documented on the welding procedure specifications (WPS):

1. Number of passes (must not vary by more than 25 percent from the WPS)
2. Thickness of plate materials (must not vary by more than 25 percent from the WPS)
3. Heat input per weld pass (must not vary by more than 25 percent from the WPS)
4. Manufacturer's identification of welding rod and flux

Filler material must not contain more than 0.08 percent vanadium when fabrication is complete.

Welding documents and qualifications shall be retained for a minimum of five years by the tank manufacturer and repair concern and shall be available for review when so requested by a representative of the Department of Transportation and owner of the tank.

All longitudinal weld joints shall be placed so that they fall in the upper half of the tank. Mismatch and alignments shall conform to ASME Code Section VIII, Division 1.

If a tapered transition of adjoining base materials is not melted in the final welding process of a joint, the final 0.050 in of base material shall be removed by mechanical means. Substructures shall be properly fitted before attachment, and the welding sequence shall be designed to minimize stresses due to the shrinkage of welds.

Each cargo tank must conform in all respects to the specification for which it is certified. Under the new rules, all *manufacturers* and/or *assemblers* that perform welding on the barrel (tank wall) of a specification cargo tank or its supports must obtain the ASME Code U symbol stamp.

Those performing *repairs* to the tanks must hold and maintain the National Board R stamp or ASME U stamp. They will be required to document and demonstrate the implementation of an acceptable ASME Section VIII, Division 1, quality control system as outlined in Appendix 10 of the ASME Code (see Chap. 4).

It is very important to note at this time that not all specification cargo tanks will be required to be ASME Code–stamped; however, the manufacturer must possess a U stamp and repair concerns must possess R stamps. Additionally, ASME– and National Board–required quality control programs must be maintained and followed. This statement is the cause of the greatest confusion among manufacturers and assemblers of specification cargo tanks. It is hopeful that the above synopsis was helpful in determining these requirements.

Certification of Cargo Tanks (§178.340-10)

Certification is required for all specification cargo tanks which were designed, constructed, and tested in accordance with the applicable Department of Transportation specifications MC-306, MC-307, MC-312, and MC-331. The following marking requirements apply to these four popular specification tanks and give guidance for various configurations of specification tanks.

1. When cargo tanks are divided into compartments (*multipurpose tanks*), and each compartment meets and is constructed to a different MC specification, there shall be one metal specification plate mounted on the right-hand side, near the front of each compartment, detailing the information shown in Fig. 11.2.

2. If a cargo tank is constructed, for example, to the MC-307 specification, it may be physically *altered* to meet another MC specification or downgraded to a nonspecification tank. If this is the case, the manufacturer's Certificate of Compliance (Fig. 11.3) shall clearly indicate such alterations. A Certificate of Compliance is issued by the manufacturer of the cargo tank. This certification attests to the fact that the cargo tank has been designed, constructed, and tested in accordance with and complies with the applicable federal regulation MC specification. The manufacturer must submit the completed certified certificate to the owner who shall retain it for the duration of the ownership plus 1 year. In addition, the specification plate shall indicate the required specification number change. This includes downgrading to a nonspecification tank.

3. There are instances when the manufacturer of an MC specification tank produces a specification tank which is incomplete and

```
Vehicle manufacturer............................................
Manufacturer's serial number................................
Specification Identification¹...................................
DOT 406; or DOT 407; or DOT 412........................
Date of manufacture...........................................
Original test date..............................................
Certification date.............................................
Design pressure........................................... PSIG
Test pressure............................................. PSIG
Head material.................................................
Shell material.................................................
Weld material.................................................
Lining material...............................................
Nominal tank capacity by compartment (front to rear)....US Gal.
Maximum product load................................... lbs.
Loading limits......................... GPM and/or PSIG
Unloading limits....................... GPM and/or PSIG

¹ The following material designations (or combinations thereof) must be added:
Aluminum Alloy (AL); Mild Steel (MS); High Strength Low Alloy (HSLA);
Austenitic Stainless Steel (SS). For example "DOT MC 306-AL" for cargo tanks
made of aluminum. A multi-purpose cargo tank example would be "Combina-
tion MC 306 SS-307 SS."
```

Figure 11.2 Multiple tank specification plate markings.

requires *further manufacturing operations* (for example, the shell and head or barrel manufacturer). The manufacturer sends the completed barrel (shell and heads) to another vendor who attaches components, parts, appurtenances, or accessories. The manufacturer of the barrel may attach the specification plate to the shell or other supporting structure on the right-hand side prior to shipping to other facilities. The "Date of Certification_____" (see Fig. 11.2) line required on the stamping shall be left blank. The specification requirements which were not complied with shall be indicated on the Certificate of Compliance (Fig. 11.3). When the tank is complete and complies with all of the required MC specifications, the company completing the compliance shall stamp the certification plate with the date of certification and complete the Certificate of Compliance.

Maintenance of Specification Cargo Tanks— General Information (§180)

The new federal regulations made sweeping new changes regarding the maintenance and inspections of MC specification cargo tanks. A great part of a tank truck carrier's maintenance effort must be directed toward compliance with all of the relevant Department of Trans-

DEPARTMENT OF TRANSPORTATION
CERTIFICATE OF COMPLIANCE
Issued By

NAME OF SPECIFICATION TANK MANUFACTURER

TO _____(Name of purchaser)_____
              Purchaser

This certifies that the new ___(Name of tank/manufacturer)___ identified below was designed, constructed and tested in accordance with the Department of Transportation Motor Vehicle Cargo Tank Specification No. ___(DOT Specification No. MC-306/DOT-406, Etc.)___ for cargo tanks used in the transportation of classified liquids.
Vehicle Type: _____ Capacity: _____ Date shipped: _____
Year Fabricated: _____ Mfg. Serial No.: _____ Mfg. Registration No.: _____
Manufactured by: Name and address of tank manufacturer

--- CERTIFICATION ---

ITEMS NOT INSTALLED AT TIME OF SHIPMENT:
(List those items which were not installed at the time of manufacture of the tank.)

☐ Cargo Tank complies to Specification No. _____ as shipped.
☐ Cargo Tank complies to Specification No. _____ except for those items listed.

Above items installed _____
By _____ Date
   Authorized Signature

Name of Tank Manufacturer
Address of Tank Manufacturer and State of Manufacturer
By _____
   Authorized Signature

Figure 11.3 Manufacturer's Certificate of Compliance (typical).

portation regulations. Maintenance personnel must know enough about the current federal regulations to understand what tasks they must perform and when they must perform them. Additionally, carriers and repair concerns must have the latest copies of the Hazardous Materials Transportation Regulation and referencing documents available at their facility.

These federal regulations are available at the following address:

Approvals Branch

Office of Hazardous Materials Transportation

Department of Transportation

Washington, D.C. 20590-0001

Attention: DHM-32, Research and Special Programs Administration

Maintaining and reading these documents keep the persons responsible for implementing the new rules knowledgeable, will assist them when discussing maintenance requirements with their own companies, and will enable them to communicate accurately with outside repair concerns regarding compliance.

The new federal regulations are intended to give interested persons guidance and general information regarding such areas as hazardous materials program procedures, general registration information, direction in filing forms and reports with the DOT, placarding and marking requirements.

The impact of these new regulations on carrier maintenance is significant and can be classified into four areas.

1. The number of maintenance actions per tank have increased from one to possibly six. For example, a cargo tank which previously had a two-year external visual inspection cycle and was scheduled for shop work every two years may now require internal visual inspection, lining inspection, leakage tests, pressure tests, and thickness tests. This, of course, is subject to the service lading and particular tank specification. This same cargo tank could possibly be subject to as many as eight maintenance actions in a two-year period.
2. Depending on your present activities, a significant increase in technical capabilities will be required. The work force will be required to learn new and advanced technical nondestructive examination methods. In addition, greater expense will be incurred in the purchase of more sophisticated equipment. Examples which are pertinent to people and equipment are:
   a. Sonic leak detection
   b. Ultrasonic thickness measurement
   c. Pressure gaging of gages
   d. Certification of personnel and equipment
   e. Bench testing of valves and gages
   f. Magnetic particle testing
   g. Dye penetrant testing
   h. Quality control systems and procedures and safety specifications
   i. Record maintenance
   j. Safety equipment for personnel
   k. Verifying safety of tank linings and tank interiors
3. A new effort in the training and introduction of personnel will be required. Planning will be necessary to implement the new five-year upgrading program. This planning will essentially be the time required to complete the pressure tests of all specification cargo tanks and to perform the retrofit on selected manways. Planning, introduction, and documented training of personnel involves determining, scheduling, and managing a program both at management and shop levels to ensure that all procedures, both regulatory and safety, as well as the record-keeping requirements will be met for all maintenance services.
4. The most significant issue will be that all carriers must register their facilities with the Department of Transportation if they plan

to conduct any of the periodic tests, inspections, and examinations specified in §180.407. Registration of repair concerns, test and inspection facilities, and registered inspectors will be covered later.

Glossary of Terms

The new federal regulations now contain a vast number of new terms. To avoid possible misunderstandings of the vocabulary, the following terms are explained to help clarify the new requirements to the average user of 49 CFR:

1. *Authorized Inspector* means an Inspector who is currently commissioned by the National Board of Boiler and Pressure Vessel Inspectors and employed as an Inspector by an Authorized Inspection Agency.
2. *Authorized Inspection Agency* means:
    a. a jurisdiction which has adopted and administers one or more sections of the *ASME Boiler and Pressure Vessel Code* as a legal requirement and has a representative serving as a member of the ASME Conference Committee; or
    b. an insurance company which has been licensed or registered by the appropriate authority of a state of the United States or a province of Canada to underwrite boiler and pressure vessel insurance in such a state or province.
3. *Cargo tank* means a bulk packaging which:
    a. is a tank intended primarily for the carriage of liquids or gases and includes appurtenances, reinforcements, fittings, and closures (for tanks, see 49 CFR §173.345-1(c), §178.337-1, or §178.338-1, as applicable);
    b. is permanently attached to or forms a part of a motor vehicle, or is not permanently attached to a motor vehicle but which, by reason of its size, construction, or attachment to a motor vehicle is loaded or unloaded without being removed from the motor vehicle; and
    c. is not fabricated under a specification for cylinders, portable tanks, tank cars, or multiunit tank car tanks.
4. *Design certifying engineer* means a person registered with the Department of Transportation in accordance with §107.502(f) who has the knowledge and ability to perform stress analysis of pressure vessels, may otherwise determine if a cargo tank design and construction meet the applicable DOT specification, and has an engineering degree and one year of work experience in structural or mechanical design [see §107.502(f)]. Persons registered as profes-

sional engineers by the appropriate authority of a state of the United States or a province of Canada who have the requisite experience may be registered under this program.
5. *Manufacturer* means any person who certifies that packaging complies with a UN or DOT standard, including a person who applies or directs another to apply a DOT specification marking or a UN mark to a packaging.
6. *Registered Inspector* means a person registered with the Department of Transportation in accordance with §107.502(f) of the federal regulations who has the knowledge and ability to determine if a cargo tank conforms with the applicable DOT specification and has, at a minimum, any of the following combinations of education [see §107.502(f)] and work experience in cargo tank design, construction, inspection, or repair:
    a. an engineering degree and one year of work experience,
    b. an associate degree in engineering and two years of work experience, or
    c. a high school diploma (or General Equivalency Diploma) and three years of work experience.

Qualification and Maintenance of Specification Cargo Tanks (§180 to §180.417)

The new regulations have added an array of continuing proficiency qualification criteria, certification and training of personnel, qualification and certification of repair firms, and qualification and testing of cargo tanks to meet the requirements of the new specification.

The following is a breakdown by subpart and paragraph explaining what the new regulations mean.

Purpose, applicability, and general requirements (§180.1 to §180.3)

These first three paragraphs of the federal regulation are basic information statements detailing the requirements and the effect on the maintenance, reconditioning, repair, inspection, and testing of MC specification cargo tanks. They detail the fact that persons who perform functions which affect the maintenance, reconditioning, repair, inspection, and testing of MC specification cargo tanks must be qualified to do so in accordance with the regulations. These paragraphs state that it is in violation to imply in any way that a specification cargo tank is in compliance with the federal regulation when in fact the party making such claims is not registered with the Department of Transportation and unqualified to make the claim.

Definitions (§180.403)

In addition to the definitions contained elsewhere in this chapter, the following definitions apply:

1. *Modification* means any change to the original design and construction of a cargo tank or a cargo tank motor vehicle which affects its structural integrity or lading retention capability. Excluded from this category are the following:
   a. a change to motor vehicle equipment such as lights, truck or tractor power train components, steering and brake systems, and suspension parts, and changes to appurtenances, such as fender attachments, lighting brackets, ladder brackets; and
   b. replacement of components such as valves, vents, and fittings with a component of a similar design and of the same size.
2. *Owner* means the person who owns a cargo tank motor vehicle used for the transportation of hazardous materials, or that person's authorized agent.
3. *Rebarreling* means replacing more than 50 percent of the combined shell and head material of a cargo tank.
4. *Repair* means any welding on a cargo tank wall done to return a cargo tank or a cargo tank motor vehicle to its original design and construction specification, or to a condition prescribed for a later equivalent specification in effect at the time of repair. Excluded from this category are the following:
   a. a change to motor vehicle equipment such as lights, truck or tractor power train components, steering and brake systems, and suspension parts, and changes to appurtenances, such as fender attachments, lighting brackets, ladder brackets,
   b. replacement of components such as valves, vents, and fittings with a component of a similar design and of the same size, and
   c. replacement of an appurtenance by welding to a mounting pad.
5. *Stretching* means any change in length, width, or diameter of the cargo tank, or any change to a cargo tank motor vehicle's undercarriage that may affect the cargo tank's structural integrity.

Qualification of cargo tanks (§180.405)

This paragraph details how a cargo tank qualifies for use in the transportation of hazardous materials. Table 11.1 lists the various specification cargo tanks which were authorized for construction and for the transportation of hazardous materials. For a cargo tank made to a specification listed in the first column, such a tank may be used provided that it is marked and certified before the date in the second column.

TABLE 11.1 Effective Dates for Completing Construction

Specification	Date
MC-300	Sept. 2, 1967
MC-301	June 12, 1961
MC-302, MC-303, MC-304, MC-305, MC-310, MC-311	Sept. 2, 1967
MC-330	May 15, 1967
MC-306, MC-307, MC-312	Sept. 1, 1993

As seen in the table, after September 1, 1993, the MC-306, MC-307, and MC-312 specifications will no longer be authorized for construction; however, they will be permitted to be used in transporting hazardous materials after that date provided they were marked or certified on or before that date. After that date a new designation numbering system along with more stringent manufacturing and design requirements will be employed. These new specifications are called DOT-406, DOT-407, and DOT-412 and will replace MC-306, MC-307, and MC-312, as discussed earlier in this chapter.

Upgrading existing specification MC-306, MC-307, and MC-312 cargo tanks [§180.405(c) to (g)]

The upgrading of specification cargo tanks is referenced indirectly in §180.405(c) to (f). For example, an MC-306 tank may be upgraded to a DOT-406 tank. In order to upgrade from an MC-306 to a DOT-406 tank one must consider the new pressure relief systems, pressure relief valve settings, venting capacities, internal self-closing stop valves on loading–unloading outlets and rated flow capacity certification tests, a hydrostatic retesting, and numerous other items. There are specific requirements when performing the upgrade. As a note of caution, the Code of Federal Regulations should be consulted due to ongoing revisions in this area.

Persons certifying cargo tanks for upgrading to the new DOT-400 series, inspection to verify that existing specification cargo tanks meet the new requirements, and periodic tests and inspections must be designated by their employer as either Registered Inspectors or Design Certifying Engineers provided that they meet the minimum requirements for such assigned activities.

For example, manways on current MC-306, MC-307, and MC-312 specification cargo tanks for tanks certified after December 30, 1990, and the new DOT-406, DOT-407 and DOT-412 specification cargo tanks must now be equipped with manhole assemblies for tank capac-

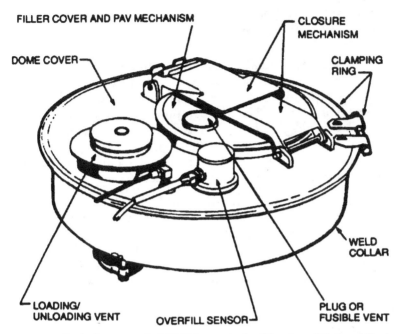

Figure 11.4 Typical manway for MC-306 cargo tanks. (*Courtesy of National Tank Truck Carriers, Inc.*)

ities greater than 400 gallons. These manholes must be at least 15 in in diameter and be capable of withstanding a pressure test of at least 36 psig. The manhole covers must be permanently marked by the manufacturer with stamping or other means such as a nameplate. The owner or user who purchases the manhole cover must ensure that a certification statement is received, indicating that the cover complies with the regulations (see §178.345-5).

These typical manways (see Fig. 11.4) must meet the above-noted criteria or they cannot be used in hazardous materials transportation. Each owner of five or more specification cargo tanks which were certified prior to December 30, 1990 or which are not equipped with these new manways must have the cargo tank inspected by a Registered Inspector to verify that manways meet the current specification. If it is determined that the cargo tank manways do not meet these specifications, they must be retrofitted and equipped with current allowable manways. Owners must retrofit 20 percent of their fleets each year beginning in 1991 until all affected manhole closures have been retrofitted and certified. If an owner has fewer than five cargo tanks requiring retrofitting, he or she has until August 31, 1995 to complete the retrofit for compliance.

Hydrostatic retesting is also required and is explained later in this chapter.

Requirements for tests and inspections of specification cargo tanks (§180.407)

Introduction. This is the "meat" of the new regulations. These regulations contain different and additional specific tests and inspection requirements which previously were left open to interpretation and local jurisdictional opinion. The old requirements listed certain visual inspections as well as special inspections on pressure gas tanks which should be performed.

Under the new rules all specification cargo tanks shall be inspected to some degree. Regulations specific to these inspections are listed below:

1. All specification cargo tanks shall not be subjected to pressures greater than their design pressure or maximum allowable working pressure established by the designer.

2. When tests and/or inspections required by §180 of the 49 CFR document have become due, the owner or carrier may not be allowed to fill the tank truck or ship product until the required tests and/or inspections have been completed.

3. Persons who perform, inspect, or witness these tests and/or inspections must meet the minimum qualification requirements for the tests and/or inspections they perform. Minimum requirements for Registered Inspectors are discussed later in this chapter.

4. If the specification cargo tank successfully passes the required tests and/or inspections, the tank truck shall be marked as described in §180.415 (see the section on test and inspection markings, §180.415 in this chapter).

5. If the specification fails to pass the required tests and/or inspections required, the tank truck must be repaired and retested. If a weld repair, modification, stretching, or rebarreling is required to the tank shell, it shall be performed by a Department of Transportation approved and registered facility. There are only two qualified and registered types of facilities where weld repair, modification, stretching, or rebarreling may be performed: (1) a National Board of Boiler and Pressure Vessel R stamp facility or (2) a facility which holds an ASME Code symbol U stamp. If the repairs are not completed, the specification tank shall be removed from service and the specification plate removed, obliterated, or covered in a secure manner.

The new requirements state that regardless of the time frame or frequency in which tests or inspections take place, additional retests and/or reinspections are mandatory prior to use if the specification tank shows signs of bad dents, corroded areas, leakage, or any other condition which might render it unsafe in transporting hazardous materials. Additionally, if a tank has been involved in an accident and damaged to such an extent that may render it unsafe to transport product, has been out of service for a period of more than one year, has been modified from the original design, or is determined to be unsafe by the Department of Transportation, such tanks shall be retested and/or reinspected for compliance to the requirements. It is important to remember that for tanks out of hazardous material transportation service for one year or more, a pressure test shall be applied.

Compliance dates for periodic tests and inspections have been upgraded and are clearly defined in the latest regulations. Table 11.2 indicates these compliance dates and defines which specification tanks require various inspections, tests, and retests.

External visual inspection and testing [§180.407(d)]. When an external visual inspection is a requirement and there is an external covering such as insulation on the external surfaces of the tank, an internal visual inspection shall be performed. This does not include pressure gas specification cargo tanks of specification MC-330 or MC-331 variety. Upon completion of the internal inspection, a hydrostatic or pneumatic test must also be performed when the following conditions occur: (a) when visual inspection is precluded by both internal coating and external insulation, and (b) when inspection is precluded by external insulation and the cargo tank is not equipped with a manhole or inspection opening.

External visual inspection and testing requirements are much more detailed, documented, and planned than in earlier versions of the federal regulations. These new regulations mandate strict external visual inspection and testing requirements and are listed here as follows:

1. External surfaces of cargo tank heads and shells, piping, valves, and gaskets are required to be inspected for violations of safety concerned with corroded or abraded areas, dents, distortions, defects in welds, and leakage.

2. Emergency devices and valves including self-closing stop valves, excess flow valves, and remote closure devices all must be free from corrosion, distortion, erosion, and any external damage that will prevent safe operation of such safety devices.

3. Missing nuts, bolts, and fusible links or elements must be replaced as well as loose bolts tightened.

4. The markings on the specification plates attached to the cargo tank shall be inspected to ensure that they are legible.

5. The major appurtenances and exterior structural attachments of the specification cargo tank such as suspension system attachments, connecting structures, and those elements of the "fifth wheel" (upper coupler) assembly must be inspected for signs of corrosion which might

TABLE 11.2 Compliance Dates—Inspections and Retests under §180.407(c)

Test or inspection (cargo tank specification, configuration, and service)	Date by which first test must be completed*	Interval period after first test
External visual inspection:		
All cargo tanks designed to be loaded by vacuum with full opening rear heads	September 1, 1991	6 months
All other cargo tanks	September 1, 1991	1 year
Internal visual inspection:		
All insulated cargo tanks, except MC-330, MC-331, MC-338	September 1, 1991	1 year
All cargo tanks transporting lading corrosive to the tank	September 1, 1991	1 year
All other cargo tanks, except MC-338	September 1, 1995	5 years
Lining inspection:		
All lined cargo tanks transporting lading corrosive to the tank	September 1, 1991	1 year
Leakage test:		
All cargo tanks except MC-338	September 1, 1991	1 year
Pressure retest:		
Hydrostatic or pneumatic[†]		
All cargo tanks which are insulated with no manhole or insulated and lined, except MC-338	September 1, 1991	1 year
All cargo tanks designed to be loaded by vacuum with full opening rear heads	September 1, 1992	2 years
MC-330 and MC-331 cargo tanks in chlorine service	September 1, 1992	2 years
All other cargo tanks	September 1, 1995	5 years
Thickness test:		
All unlined cargo tanks in corrosive service, except MC-338	September 1, 1992	2 years

*If a cargo tank is subject to an applicable inspection or test requirement under the regulations in effect on December 30, 1990, and the due date (as specified by a requirement in effect on December 30, 1990) for completing the required inspection or test occurs before the compliance date listed in Table 11.1, the earlier date applies.

[†]Pressure testing is not required for MC-330 and MC-331 cargo tanks in dedicated sodium metal service. Also, pressure testing is not required for uninsulated lined cargo tanks, with a design pressure or MAWP 15 psig or less, which receive an external visual inspection and lining inspection at least once each year.

prevent safe operation of the tanker. The upper coupler "fifth wheel" assembly need not be removed for this inspection. For specification tankers transporting corrosive materials which are corrosive to the tank, the upper coupler must be inspected at least once in a two-year period. The upper coupler shall be removed for this external inspection. This inspection shall be a detailed inspection for corrosion, abraded areas, dents, distortions, and defects in welds.

6. Reclosing pressure relief valves shall be externally inspected for corrosion. If the valve is attached to a cargo tank in corrosive service, the valve shall be removed and inspected. When the valve has been removed it shall be placed in a test jig and tested to ensure that it opens at the required pressure and reseats to a leaktight condition at 90 percent of the set-to-discharge pressure or the pressure prescribed for the applicable specification cargo tank.

7. Thickness testing using an ultrasonic thickness measuring device of the external tank wall and heads is required when corroded or abraded areas are noted during external visual inspection. These thickness tests shall be performed using written procedures which establish acceptance requirements for the inspection.

8. For specification cargo tanks where the rear head fully opens, gaskets must be visually inspected for cracks, splits, or cuts which exceed a depth of $\frac{1}{2}$ in or which cause leakage. These gaskets shall be replaced if required.

9. External inspections shall be recorded and documented on forms from the inspecting company and shall include the name and registration number of the company and/or Registered Inspector who completed the inspections.

Internal visual inspections [§180.407(e)]. When a specification cargo tank is not equipped with a manhole opening or other means with which to perform the internal tests and inspections required by the regulation, a hydrostatic or pneumatic test shall be applied. The visual internal inspection shall include as a minimum all of the following.

1. Internal areas of cargo tank heads and shells, piping, valves, and gaskets are required to be inspected for violations of safety concerned with corroded or abraded areas, dents, distortions, defects in welds, and leakage. Additionally, internal tank liners shall be inspected when the tank is lined.

2. Thickness testing (using an ultrasonic thickness measuring device) of the internal tank wall and heads is required when corroded or abraded areas are noted during internal visual inspection. These thickness tests shall be performed using written procedures which

establish acceptance requirements for the inspection. Degraded or defective areas of the tank liner must be removed or the tank repaired prior to placing it back into service.

3. Internal visual inspections shall be recorded and documented on forms from the inspecting company and shall include the name and registration number of the company and/or Registered Inspector who completed the inspections.

Lining inspection [§180.407(f)]. For specification cargo tanks which are lined to prevent corrosion or deterioration of the tank shell, heads, and appurtenances of the tank materials, such linings shall be inspected and verified at least once each year. This inspection shall include the following.

1. For specification tanks lined with rubber, a high-frequency spark tester capable of producing sufficient voltage shall be used to test the liner for holes.
2. For linings made from materials other than rubber, the required tests to determine tightness and to find leaks shall be as defined in the lining manufacturer's or lining installer's materials.
3. Degraded or defective lining material shall be removed and corroded areas below the tank wall or head shall be ultrasonically tested for thickness. After acceptable cleaning and repairs, the lining shall be replaced with new lining.
4. Lining inspections shall be recorded and documented on forms of the inspecting company and shall include the name and registration number of the company and/or registered inspector who completed the inspections.

Pressure tests [§180.407(g)]. All components of a specification cargo tank shall be pressure-tested in accordance with 49 CFR §180.417(b). During the pressure test the Registered Inspector must perform an external inspection to verify the tightness and leaktight integrity of the cargo tank. As part of the test an internal inspection shall be made prior to the filling of the test medium and application of the required pressure. Internal visual inspection is not required for specification tanks not equipped with manways or inspection openings.

All reclosing relief valves must be removed from the cargo tank for inspection and testing. After testing, if found acceptable, they shall be replaced on the tank. If found defective, they shall be replaced with new valves.

The hydrostatic or pneumatic test requirements for each specification cargo differs with the applicable specification to which it conforms.

Each owner of specification cargo tanks for specification designation numbers MC-300, MC-301, MC-302, MC-303, MC-304, MC-305, MC-306, MC-307, MC-310, MC-311, or MC-312, must pressure-test at least 20 percent of their fleet each year beginning September 1, 1991 when the owner has more than five tankers in his fleet. The owner with less than five cargo tanks has until August 31, 1995 to complete the pressure retest. When the cargo tank is of a multitank design, each chamber must be tested with the adjacent cargo tank empty and at atmospheric pressure. Pressure relief devices must be attached during the test.

Hydrostatic test method [§180.407(g)(viii)]. The entire cargo tank, including domes, must be completely filled with water. There must be a method of venting air during filling located at the highest point of the tank. Water or other liquids having similar viscosity shall be utilized for the test medium. The temperature of the test medium shall not exceed 100°F.

The tank shall have test gages attached directly to the tank. The dial indication range of these gages shall neither be less than $1\frac{1}{2}$ nor more than 4 times the test pressure. The tank and all closures shall hold pressure for at least 10 minutes during which time it shall be inspected for leakage, bulging, or other defects. Tests and test pressures shall be inspected by a Registered Inspector or Authorized Inspector as applicable.

Pneumatic test method [§180.407(g)(ix)]. Pneumatic testing should be performed with caution and involves a greater risk to life and property than does hydrostatic testing. Pneumatic testing is not recommended; however if necessary, testing shall be performed as follows.

1. The test medium must be either compressed air or an inert gas.

2. After the tank is filled with air or gas the pressure shall be applied, gradually increasing the pressure to one-half the test pressure. Upon reaching this predetermined level the pressure shall then be increased in steps of approximately one-tenth the test pressure until the required test pressure is reached.

3. The test pressure must be held for a minimum of five minutes after which the test pressure shall be reduced to the maximum allowable working pressure stamped on the specification plate or ASME Code data plate.

4. It is at this pressure that the entire shell and tank surface shall be inspected using a method such as a soap and water solution which would show evidence of leakage. The entire surface of all welded joints must be thoroughly inspected with the solution.

In the event that a cargo tank is insulated and a pneumatic test is performed, the insulation need not be removed. This is permissible only if the test pressure cannot be reached or the vacuum integrity could not be maintained in the insulated space.

If specification cargo tanks used in transporting flammable gases, oxygen, or refrigerated liquids are opened for any reason, internal cleanliness must be verified prior to closure. The inspector shall verify that foreign debris is removed from the tank and the tank shall be wiped clean.

Generally, specification cargo tanks which transport pressure gases and which have manholes such as those that conform to the MC-330 and MC-331 specifications are required to be nondestructively examined prior to the pneumatic test. This test shall include the wet fluorescent magnetic particle method immediately prior to the pneumatic test. The procedures for performing such tests shall be in accordance with the ASME Code Section V, Nondestructive Examination, and the Compressed Gas Association's *Technical Bulletin TB-2*.

Upon completion of all pneumatic and magnetic particle tests, a report shall be filled out and certified by the Registered Inspector. This report shall include the results of the magnetic particle test, a statement as to the nature and severity of the defects found, location of such defects, such as in the weld, heat–affected zone, and head to shell seam, and so forth.

Acceptance criteria [§180.407(g)(ix)(6)]. Acceptance criteria is not clearly defined in the new regulations and is left primarily to the reasonable discretion of the testing and inspection concern certifying the required inspection or test. Such acceptance is based on the following criteria.

1. Cargo tanks that leak during the pressure tests shall be repaired by welding, utilizing a repair concern which holds the National Board R or ASME U stamp and Certificate of Authorization. Any leakage in any amount is unacceptable.
2. Any cargo tanks which fail to retain test pressure or pneumatic test pressure shall be investigated to ensure that no leaks are present. In the event that leakage is discovered, repairs to the cargo tank shall be performed by a National Board R or ASME U stamp and Certificate of Authorization holder.
3. If during visual internal or external inspection deterioration such as corrosion, pitting, and erosion of the shell, heads, and other appurtenances carrying lading is found, methods for determining the minimum design metal thickness shall be employed. Such methods most frequently used are ultrasonic thickness testing devices. The ultrasonic thickness test procedure and acceptance criteria for such thickness readings are as follows.

a. The thickness of shells and heads of all unlined cargo tanks which transport materials that are corrosive to the tank must be measured at least once every two years. If the tank measures less than the sum of the minimum design metal thickness plus one-fifth of the original thickness, such cargo tanks shall be tested at least once a year.
b. Ultrasonic measuring equipment shall be sensitive and capable of accurately measuring thicknesses to within ±0.002 of an inch.
c. Personnel performing ultrasonic thickness testing shall be trained in the use of such equipment per the manufacturer's instruction using written procedures for the use, calibration, and handling of ultrasonic test equipment.
d. Thickness testing shall be performed as a minimum, on areas of the tank head and shells in addition to those areas around any piping used or those parts containing lading.
e. Ultrasonic thickness readings shall be taken in areas of high stress concentrations which are caused by bending moments of the tank and those areas near openings such as nozzle openings, inspection openings and manways, shell reinforcements, and areas around appurtenance attachments.

Additionally the upper coupler (fifth wheel) assembly attachments, areas near suspension systems attachments, connecting structures, and areas which by the very nature of the design of the specification tank are known to be thin areas in the shell and liquid level lines must be tested.

If the thickness tests indicate that the minimum design metal thickness is below that established by the specification as stamped on the specification plate and/or ASME Code Data plate for the maximum lading density, the cargo tank may still be used provided the cargo tank transports lading of a lower density. The specification plate shall be restamped or changed to indicate the new service limits and a Design Certifying Engineer (see the section titled Glossary of Terms in this chapter) certifies that the design and thickness are acceptable for a lower density lading. The engineer shall issue a new manufacturers Certificate of Compliance. If this "recertification" is not performed as outlined, the cargo tank may not be returned to hazardous materials service. The tank's specification plate shall be removed, obliterated, or covered in a secure manner.

Upon completion of all thickness testing, the Registered Inspector shall complete a report and must record the results of the test and certify the report with the required DOT registration number.

4. For cargo tanks which appear to have bulged due to evidence of weakness or the application of excessive pressure which may have

been applied to the shell of the tank during filling, pressure testing, or other tests, the tank dimensions shall be taken and verified against the original cargo tank drawing dimensions to verify that the tank continues to meet the requirements of the original design.

Leak testing [§180.407(h)]. As of September 1, 1991, federal regulation required separate pressure and leakage tests to be performed on all specification cargo tanks except MC-338 cargo tanks. *Pressure tests do not satisfy the leak test requirement.* The Department of Transportation states that the pressure test evaluates the tank's structural integrity, and the leakage test evaluates the tightness of the vehicle.

All specification cargo tanks are required to be leak-tested except as noted above. The leak test shall include the shell, heads, piping, and all valves and accessories which contain lading. The leak test pressure shall not be less than 80 percent of the tank design pressure or maximum allowable working pressure (MAWP), whichever is marked on the certification or specification plates. There are further exceptions which apply to pressure gas cargo tanks. The Environmental Protection Agency's Method 27, 40 CFR Part 60, Appendix A is an acceptable alternative test.

Minimum Qualifications for Inspectors and Testers (§180.409)

The new regulations state that any persons who perform or witness the inspections and tests mandated in the federal regulations must be registered with the Department of Transportation. Any organization that seeks to certify, inspect, and/or witness tests or inspections to specification cargo tanks which meet the requirements of the regulations shall apply in writing to the DOT. This application shall be in the form of a letter such as in Fig. 11.5. The letter shall contain enough information to indicate that the inspector is familiar with the DOT specification cargo tanks and is trained in the methods of testing and inspection.

Upon acceptance of the application for certification to DOT, a letter will be sent to the cargo tank owner, repair concern, or carrier stating the functions which may be performed based on the original registration statement. This letter of acceptance from DOT is valid for three years from the date of issue of the original registration statement. Renewal is accomplished by submission of an updated registration statement. In the event that there is a change made in the company or change in the scope of work which is different than that of the original

To: Office of Hazardous Materials Transportation
Attn.: DHM-32
Research and Special Programs Administration
Department of Transportation
400 Seventh Street SW
Washington, D.C. 20590

1. Name.

2. Street address, mailing address, and telephone number for each facility or business.

3. Certification by the person responsible for compliance with 49 CFR, Part 180, using the following language which is pertinent only to inspection and testing and not to manufacture, assembly, certification, or repair:

"I certify that all Registered Inspectors used when required in performance of all company inspection and testing functions meet the minimum qualification requirements set forth in 49 CFR, ¶171.8."

"I am the person responsible for ensuring compliance with requirements for inspection and testing set forth in 49 CFR, Part 180."

"I have knowledge of the requirements applicable to the inspections and tests to be performed.

4. These inspections and tests will be performed: (Then list the specific services to be performed. Quotes of the titles of the inspections and tests covered in ¶180.407 are suggested, e.g., external visual inspection, pressure retest, lining inspection, etc.)

5. The above maintenance services will be performed on the following cargo tanks or cargo tank motor vehicles: (List tanks by MC/DOT specification and/or DOT exemption number.)

6. (Name of carrier) does/does not employ Registered Inspectors to conduct the inspections and tests identified in Item 4 above, or

(Name of carrier) utilizes the services of the following person (non-employee) to conduct the inspections and tests identified in Item 4 above:

Name _____

Address _____

_____

DOT Registration No. _____

7. Signature of responsible person making certifications in Item 3 above.

8. If a repair concern, state the nature of repairs in Item 4 above and submit a copy of A.S.M.E. Certificate of Authorization or National Board "R" Certificate with this letter.

Figure 11.5 Sample registration statement [§107.503 and §180.407(b)].

registration statement, a revised statement must be submitted to the DOT in writing indicating the new information.

Personnel performing ultrasonic thickness testing, pressure tests which include hydrostatic and pneumatic tests, or leak testing must be trained in conducting the test or inspection method. Training shall be documented. Personnel designated as Registered Inspectors by the company must have letters designating them as Registered Inspectors on file along with documentation indicating that they meet the minimum requirements set forth in the standard.

If the owner elects not to perform the periodic tests and inspections him or herself or has no Registered Inspectors, the owner may use an employee of another company to perform or witness the tests and inspections of the regulation, provided the owner verifies that the inspector is trained and experienced in the use of the equipment to be used, retains a copy of the designation for the inspector as well as other qualifications, and submits a registration statement to the Department of Transportation stating and certifying that the inspector meets the minimum requirements of the Hazardous Materials Regulation. This special case also applies to the owner who wishes to use an employee to perform pressure tests without designation as a Registered Inspector.

Acceptable Results of Tests and Inspections (§180.411)

As stated earlier in this chapter, acceptance criteria for tests, inspections, and damage to tanks is ambiguous and not straightforward. This section, §180.411, contains the additional acceptance criteria stipulated by the new regulations. The following additional acceptance criteria resulting from the tests and inspections is stated:

Corroded or abraded areas

1. Corroded or abraded areas which encroach on the minimum thickness of the tank or other lading carrying items are cause for rejection and must be repaired.

2. The maximum allowable depth of dents that are in a weld or are dented at the weld is $\frac{1}{2}$ in. Dents greater than this are required to be brought into compliance.

3. For dents in all other locations the maximum allowable depth of dents is $\frac{1}{10}$ the greatest dimension of the dent. In no case shall the depth be greater than 1 in.

4. For cuts, digs, or gouges, the thickness remaining below such defects shall not be less than the minimum design material thickness of the cargo tank.

Weld or structural defects. Weld defects such as pinholes, cracks, incomplete penetration, or lack of fusion and structural defects are unacceptable regardless of size and must be repaired by a qualified repair concern, either the holder of a National Board R stamp or an ASME Code U stamp and Certificate of Authorization, and which is registered with the Department of Transportation. The tank shall be taken out of hazardous materials transportation until repairs can be made.

Leakage. Any leakage during either testing or operation shall be grounds for removing the cargo tank from service.

Relief valves. Defective pressure relief valves shall be removed from service and repaired or replaced.

Pressure tests. Specification cargo tanks which fail to meet the acceptance criteria found in the individual specification shall be repaired as stated above.

Repair, Modification, Stretching, or Rebarreling of a Cargo Tank (§180.413)

These terms are mentioned throughout the regulations. It is important to clarify several side issues which are a cause of some confusion. The terms listed here are commonly misunderstood due to the ambiguous nature of such repair conditions on the cargo tank.

Stretching, as concerns the text of §180.413, is not considered a modification. *Rebarreling* is not considered a repair. The federal rules state that any repair, modification, stretching, or rebarreling of specification cargo tanks shall be performed by a holder of an ASME Code U or National Board R stamp who has registered with the Department of Transportation and has been certified to perform such repairs. This requirement became mandatory beginning July 1, 1992.

Repair and modifications

1. When DOT-406, DOT-407, and DOT-412 specification cargo tanks are repaired or modified, the repair or modification shall be performed in accordance with the specification requirements in effect at the time of initial manufacture or at the time of repair.

2. MC-300 series cargo tanks must be repaired or modified in accordance with the specification of manufacture or may be upgraded to the DOT-400 series in effect at the time of repair or modification.

3. The exception to item 2 above is the MC-304 and MC-307 specification cargo tanks, which shall be repaired or modified in accordance with

the specification of manufacture or may be upgraded to the DOT-407 series in effect at the time of repair or modification. Additionally the MC-310 and MC-312 specification cargo tanks must be repaired or modified in accordance with the specification of manufacture or may be upgraded to the DOT-412 series in effect at the time of repair or modification.

4. The MC-338 cargo tank must be repaired or modified in accordance with the original specification of manufacture or in accordance with the specification in effect at the time of repair or modification.

5. Pressure gas tanks such as the MC-330 or MC-331 must be repaired in accordance with the National Board Inspection Code and the referenced parts and subsections of the ASME Code Section VIII, Division 1.

6. Records shall be maintained by the owner at the principal place of business. Such records shall include the repair or modification reports and shall be maintained during the time the tank is in service and for one year after that date. The motor carrier shall receive copies of the repair reports in the event the carrier is not the owner, in case the motor carrier is requested by an enforcement authority for evidence of the qualification of such repairs.

Stretching and rebarreling if needed shall be performed using all new materials and equipment affected by the stretching or rebarreling. The structural integrity of the tank shall be certified upon completion of such work by a Design Certifying Engineer registered with the Department of Transportation.

Test and Inspection Markings (§180.415)

All specification cargo tanks which have undergone the tests and inspections stated in this manual must be marked as shown in Fig. 11.6. The markings shall be in English with the test date (month first followed by the year), followed by the test or type of inspection performed. The letters and numbers shall be a minimum height of 1.250 in. The marking shall be placed on the tank shell near the specification plate or anywhere on the front head of the tank. In the event that multiple tests are performed at the same time on a cargo tank constructed under multiple specifications, one set of test markings may be placed on the tank. If the tests are performed at different times, separate markings shall be placed near the front of each compartment of the cargo tank. Figure 11.7 demonstrates unacceptable markings.

Reporting and Record Retention Requirements (§180.417)

The new regulations have mandated new and additional requirements for the retention of records and the reporting of certain functions. The

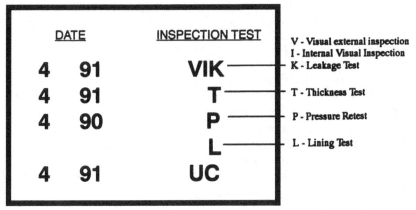

Figure 11.6 Acceptable correct markings of cargo tanks upon completion of all required tests and inspections [§180.415(b)].

INSPECTION TEST	DATE
VIK	5-91
UC	5-91
T	5-91
P	5-91

Figure 11.7 Unacceptable markings.

owner shall keep copies of all ASME Code Manufacturer's Data Reports. Additionally, the Certificate of Compliance from the original tank manufacturer shall be retained. These documents shall be retained for as long as the owner maintains ownership of the cargo tank plus one year after transfer of the tank to a new party.

If the owner does not have these certification documents or they have been stolen or lost or become illegible, the owner must perform or have performed appropriate tests and inspections under the direction and supervision of a Registered Inspector to determine that the cargo tank conforms to the applicable specification. Upon satisfactory completion of the required tests and inspections, the owner along with the Registered Inspector must certify that the cargo tank meets the specification and complete a Certificate of Compliance. Retention of this certification is as listed in previous sections of this manual.

For DOT specification cargo tanks manufactured before September 1, 1993 either of the following applies:

1. If the specification cargo tank is ASME Code–stamped, and the owner does not have the manufacturer's certificate and data report as required by the specification, the owner may contact the National Board for a copy of the manufacturer's data report if the cargo tank was registered with the National Board.

2. The information contained on the cargo tank's identification and ASME Code plates can be copied. Additionally, both the owner and the Registered Inspector must certify that the cargo tank fully conforms with the specification. The owner must retain such documents, as specified in this section.

How to comply. Compliance is twofold. First, the owner may subcontract all cargo tanks to qualified concerns registered with the Department of Transportation to perform the periodic tests and inspections, and second, just simply comply.

The first is simple. You must find a concern that has registered its facility to perform the periodic tests, inspections, and examinations specified in the Code of Federal Regulations §180.407. Registration with the Department of Transportation is a simple procedure. §107.501 and §107.502 of the regulations states that "any person who manufactures, assembles, inspects and tests, certifies, or repairs Department of Transportation specification cargo tanks . . . must be registered with the Research and Special Projects Administration (RSPA) of the United States Department of Transportation." A person may only perform those periodic tests, inspections, repairs, and examinations that are filed with the DOT in the registration statement. Figure 11.5 shows a typical registration statement.

In addition to the registration of the company with the Department of Transportation, those persons employed by the company who will perform the periodic tests, inspections, and examinations must be qualified and certified as to their experience and qualifications specifically in regard to procedures, equipment, knowledge of cargo tanks, and inspection procedures. The employer must submit certification to the DOT that the employee or tester is qualified in accordance with §180.409(b). The carrier must designate certain personnel to perform the periodic tests, inspections, and examinations and maintain and keep these designations on file.

Chapter

12

ISO 9000 Quality Assurance Requirements

As we move into the next century, the terms *quality assurance, quality management,* and *total quality control* are becoming the new buzz words. Actually they are more than buzz words; they are reality for many American companies as well as foreign ones. The concept of total quality control (TQC) has been widely practiced in Japan for over a decade and is a way of life for Korean companies as well.

With the collapse of the Berlin Wall, the unification of Germany, the European Common Market, the dismantling of the Communist party, and Asian communities beginning to band together, one international quality assurance standard for all nations seems to be most practical. This chapter will discuss the latest requirements for these international standards.

Introduction to ISO 9000 Standards

The International Organization for Standardization (ISO) is a worldwide federation of national standards bodies (ISO member bodies) that is headquartered in Geneva, Switzerland. The American National Standards Institute (ANSI) is the representative organization for the United States within the International Organization for Standardization federation. The purpose of the International Organization for Standardization is to develop internationally recognized standards to facilitate commerce worldwide and to enhance product safety.

The work of preparing international standards is normally carried out through ISO technical committees. Each member body or country interested in a subject for which a technical committee has been established has the right to be represented on the committee.

In 1979, Technical Committee 176, Quality Assurance, was formed and in 1985 the ISO 9000 Series of Standards was circulated in draft form to the member bodies for approval. In 1987, following approval by a majority of the member bodies, the ISO 9000 Series of Standards became an international standard recognized worldwide.

Since the United States is a member of the International Organization for Standardization, the ISO 9000 series was concurrently adopted as an ANSI standard, the Q90 Series of Standards. ISO 9000 series standards and the ANSI Q90 series are identical in content. The ISO 9000–Q90 series consists of five standards:

ISO Series	ANSI–ASQC*
ISO 9000	Q90
ISO 9001	Q91
ISO 9002	Q92
ISO 9003	Q93
ISO 9004	Q94

*ASQC = The American Society for Quality Control.

The first document is ISO 9000, which is essentially an overview and guide for the series. This standard is entitled "Quality Management and Quality Assurance Standards—Guidelines for Selection and Use." This document lists the reference standards applicable to the other standards, definitions useful for the establishment of an ISO quality assurance system, and characteristics of quality systems and gives generic quality assurance and control requirements regarding quality management, quality assurance, and quality control.

The next three documents (9001, 9002, 9003) provide three levels of generic quality system requirements that must be addressed within an ISO 9000 quality assurance program. The most stringent, ISO 9001, entitled "Quality Systems—Model for Quality Assurance in Design/Development; Production, Installation, and Servicing," establishes a model or guide for the manufacturer of pressure vessels to use in establishing a quality assurance program for design and/or development of their product. This ISO 9001 program includes 20 different quality points. Table 12.1 lists the 20 attributes and compares the requirements of ISO 9001 quality assurance requirements with those of ASME Code Section III, Nuclear Quality Assurance, requirements. There are close similarities. Note that the nuclear quality assurance program does not list statistical process control (SPC) nor does it address service (after service). Generally speaking, a holder of an ASME Code Section III Code symbol stamp already has many facets of an ISO 9001 program in place.

TABLE 12.1 Comparison of Quality Assurance Requirements between ISO 9001 and ASME NQA-1

ISO 9001 Quality System Requirements	NQA-1 (Nuclear Quality Assurance) II basic requirements
4.1 Management responsibility	1.0 Organization (1S-1)
4.2 Quality system	2.0 Quality assurance program (2S-1, 2S-2, 2S-3)
4.3 Contract review	Not addressed as a unique item
4.4 Design control	3.0, 5.0 Design control
4.5 Document control	5.0, 6.0 Document control (Instructions, procedures, and drawings, 3S-1, 6S-1)
4.6 Purchasing	4.0, 7.0 Procurement document control,
4.7 Purchaser-supplied product	Control of purchased items and services (4S-1, 7S-1)
4.8 Product identification and traceability	8.0 Identification and control of items
4.9 Process control	9.0 Control of processes (9S-1)
4.10 Inspection and testing	10.0, 11.0 Inspection, Test control (11S-1)
4.11 Inspection, measuring, and test equipment	12.0 Control of measuring and test equipment
4.12 Inspection and test status	14.0 Inspection, test, and operating status
4.13 Control of nonconforming product	15.0 Control of nonconforming items (15S-1)
4.14 Corrective action	16.0 Corrective action
4.15 Handling, storage, packaging, and delivery	13.0 Handling, storage, and shipping (13S-1)
4.16 Quality records	17.0 Quality assurance records (17S-1)
4.17 Internal quality audits	18.0 Audits (18S-1)
4.18 Training	2.0, 17.0 Suppl. 2S-1, 2S-2
4.19 Servicing	Not addressed as a unique item
4.20 Statistical techniques	Not addressed as a unique item

Many pressure vessel manufacturers in the United States are under the impression that if they perform any design functions for pressure vessels, then they must obtain ISO 9001 certification. This is not the case. If a pressure vessel manufacturer is manufacturing vessels under known, existing nationally recognized standards such as TEMA or ASME, where the design formula and function are known and accepted, that manufacturer may certify under the ISO 9002 model. If the owner is designing vessels or entire systems where no known standards exist, ISO 9001 is appropriate.

The ISO 9002 model is for an organization in which design is not a significant function in its operation; however, some design does take place. The ISO 9002 is entitled "Quality Systems—Model for Quality Assurance in Production and Installation." Note that design, development, and servicing have been deleted. This model has only 18 attributes which must be described in the quality assurance program.

The ISO 9003 is a model for an organization primarily concerned with final inspection or testing of a product. This standard is entitled "Quality Systems—Model for Quality Assurance in Final Inspection and Test." This is the lowest level of an ISO 9000 quality assurance system with only 12 attributes. Very few companies hold or will hold an ISO 9003 certification.

It is important that a company must select the model, ISO 9001, ISO 9002, or ISO 9003, that is most appropriate for the product or service they are providing. It would not make sense for a company only engaged in final testing of products such as a weld testing laboratory to seek certification for ISO 9001 if very few of the quality functions required by an ISO 9001 quality system are performed by the company.

The last and final ISO 9000 standard, ISO 9004, is a guide to quality management planning and implementation for a producer of products or services. It attempts to specify what types of controls and information should be provided to meet the various requirements of the particular model the company has selected. Some manufacturers may find ISO 9004 an excellent guide for the preparation of checklists for auditing ISO 9001, ISO 9002, and ISO 9003 quality assurance programs.

Assessment of the Quality Assurance Program

As a part of establishing an ISO 9000 type quality assurance (QA) program, the company must pass a conformity assessment by a recognized third-party agency. This is where things get complicated. This is the area which causes the most confusion within the industry.

In the United States, compliance with ISO 9000 is nonmandatory. Consequently, there is currently no governmental or local jurisdictional requirements for conformance. Current market forces, primarily foreign, are dictating the need for establishing an ISO 9000 type quality assurance program. The American Society for Quality Control has formed a subsidiary corporation called the Registrar Accreditation Board (RAB) to accredit companies that will then register suppliers under the appropriate Q90–ISO 9000 series standard. An accredited company could then perform the conformity assessment and, if acceptable, issue a Certification of Compliance.

Worldwide, the situation is decidedly different. Increasingly, nations around the globe are adopting the ISO 9000 Series of Standards as the acceptable model for a QA program and some have made conformity to ISO 9000 a condition for doing business. Most prominent among these foreign markets is the European Community. As a member of the International Organization for Standardization, the European Community also adopted the ISO 9000 series as a Community standard in 1987. In the European Community (EC) it is known as the EN 29000 Series of Standards. Once again, EN 29000, ISO 9000, and Q90 are identical.

In 1957 the European Economic Community was established by the Treaty of Rome. Its objective was to create a union of European nations with common economic ideas and goals. Although noble in thought, in practice it met with only limited success due to the insistence by the member countries of retaining national standards and economic laws. In 1979, however, the European Court of Justice ruled that one member state could not block the import of a product sold in another member state except for health or safety reasons. This established the principal of mutual recognition of national regulations and industrial standards. As a result, the necessity for developing standards and directives to this end became apparent. Thus, the requirement in the European community revolves around product conformance. The European Economic Community has technical committees working to develop industry-specific directives. These industry specific directives describe the requirements that must be met for a particular type of product as well as how the company may prove conformance to such industry specific requirements. As part of the framework for this, the company must have a quality assurance program that meets one of the models of the ISO 9000 series discussed above.

Conformance is accomplished in accordance with options provided for the type of product. A *Large Pressure Vessel Directive* is in the final stages of approval and draft copies have been distributed to interested parties. In Annex II (see Table 12.2) of the *Large Pressure Vessel Directive* is a list of the classifications and the required ISO 9000 quality assurance modules to be used by manufacturers of different classes of pressure vessels. For example, in the case of large pressure vessels, the specific directive offers several choices for manufacturers to demonstrate conformance and gives information to assist manufacturers in the proper selection of the ISO 9000 model for establishing the quality assurance program (see Table 12.2). For class A vessels, manufacturers may wish to be assessed by a combination of modules B and F or modules B and D or module H plus design examination. Table 12.3 details the global chart used to assist manufacturers in selecting the correct quality assurance program for their products and companies. It is important to note that pressure vessel manufacturers must know their clients and products when establishing their quality systems. By choosing modules B and F, manufacturers submit prototypes of their finished products to a Notified Body for type examination and statistical verification. By choosing modules B and D, they undergo a quality

TABLE 12.2 Classification Table for Vessels for Gases, Liquefied Gases, and Liquids Above Atmospheric Pressure Boiling Point
P = pressure; V = volume

		Module	
		Series or regularly produced	Single unit
Class A For substances as specified	$P > 0.5$ bar and $P \cdot V > 20$ bar L	B + F (4) or B + D or H + design examination	G
Class B For substances as specified	$P \cdot V > 3000$ bar L and $P > 2$ bar	B + F (4) or B + D or H + design examination	G or H + design examination
	$P \cdot V < 3000$ bar L or $P > 2$ bar $P > 0.3$ bar $P \cdot V > 200$ bar L	B + C or B + E	Aa (1) + design examination
Class C For substances other than Classes A and B	$P \cdot V > 3000$ bar L and $P > 2$ bar	B + F or B + D	G or H + design examination
	$200 \leq P \cdot V \leq 3000$ bar L or $P \leq 2$ and $P \cdot V > 200$ bar L	B + C or B + E	Aa (1)
	$50 \leq P \cdot V \leq 200$ bar L and $P \geq 0.5$ bar $P < -0.3$ bar and $V > 200$ L	A	A

NOTES: For module D audits are to be carried out every six months.

For modules C, D, and G the notified body shall check in particular the requirements of clauses 4, 2, and 5 of Annex I.

For Aa(1) specific tests are to be defined.

assurance program assessment by a Notified Body with respect to the requirements of ISO 9002 and also type testing by the Notified Body (see Table 12.3). By choosing module H plus design examination, manufacturers are assessed by a Notified Body with respect to the requirements of ISO 9001, plus the design documents for the vessel are checked for adequacy.

TABLE 12.3 Modules for Certification and Testing—Global Approach

	Module A	Module B				Module G	Module H
DESIGN	Internal manufacturing check	Type examination (finished product testing) by Notified Body				Unit verification by Notified Body	EN 29001, ISO 9001, Q91 Full quality assurance
		+	+	+	+		
		Module C	Module D	Module E	Module F		
PRODUCTION	Technical documentation about products and tests	Random checks by Notified Body	Quality assurance production ISO 9002, EN 29002, Q92 Audit by Notified Body	Product quality assurance Final inspection ISO 9003, EN 29003, Q93 Audit by Notified Body	Statistical verification by Notified Body		• Design • Production • Final inspection Audited by Notified Body
			CE Mark				

Notified Bodies and the CE Mark

Only companies that are certified as third-party accreditation bodies and that are registered firms within their respective countries and within the EC may audit, certify, and register firms for application of the CE mark. On the national level, each European country has its own standardization organization. Regardless of which option, either product verification or complete ISO 9000 certification, manufacturers choose, successful assessment by a Notified Body will result in manufacturers being registered with that Notified Body and will allow them to place the CE mark on their products. This mark is the European Economic Community symbol for compliance to the directive of that particular product. Each member state establishes recognized Notified Bodies. Mutual recognition of each members' Notified Body is mandatory. It is important to realize that Notified Bodies can only exist within a member state of the European Community. It is also important to verify when selecting a Notified Body that it is a recognized agency for the type of product being manufactured. Not all Notified Bodies are qualified to certify pressure vessels. There is currently no provision for mutual recognition of Notified Bodies outside of the EC. Consequently, assessment by an agency outside the European Community will not guarantee that a Certificate of Compliance will be recognized by a member state. Once again, this is still market-driven. If a customer within the European Community will accept a pressure vessel built to the ASME Code, there is no requirement that manufacturers meet the requirements of the European directive. Compliance to the directive would, however, guarantee universal acceptance of the product throughout the EC.

As a direct result of the EC 1992 initiative, and the increasing acceptance of ISO 9000 as an internationally recognized model for a quality assurance system, the ASME has indicated a need to evaluate ISO 9000 criteria with the possibility of bringing ASME's quality control requirements closer to those of ISO 9000.

Appendix A

Cylinder Volume Tables and Diagrams

The most economical method of constructing cylindrical pressure vessels is to keep the length-to-diameter ratio between 2 and 5 while using a diameter that will provide the lowest number of courses. Occasionally a customer will order a vessel of a given capacity but allow the fabricator some leeway in its diameter and length. The fabricator may then select the size that will be easiest and least expensive to construct. If a 1000-gal tank has been ordered, for example, a course 8 ft long with two hemispherical heads 48 in in diameter might be used. The cylinder would then have a capacity of 752 gal and the two heads a capacity of 125 gal each, for a total of 1002 gal. By fabricating the vessel in one course rather than two, moreover, the manufacturer will save time in laying out the shell and in handling as well as the cost of joint preparation and welding on the girth seam. It is important to choose a length and diameter that will give the required capacity with as few seams as possible. The volume tables for cylinders and heads in this appendix, Tables A.1 and A.2, are based on the diameters and lengths most used in the pressure vessel industry. To use the cylinder tables, locate the inside diameter of the vessel in the first column on the left-hand side and move across horizontally to the column headed by the length of the cylinder. Read the capacity in gallons at this point. If the vessel has dished heads, the length of the cylinder is measured between the tangent lines of each head. The accompanying volume diagram allows for larger diameters and longer lengths and also for rapid estimation of tank lengths and diameters for a known capacity. To use the diagram, simply rotate a straight edge around the required capacity in the middle column and read the desired length in the right-hand column and the corresponding diameter for this length and capacity in the left-hand column. To illustrate: A tank is wanted that will hold 8000 gal. If the shell is made 20 ft long from tangent line to tangent line of two torispherical heads and 96 in in inside diameter, the capacity will be 8016 gal (7520 gal for the cylinder and 298 gal for each head). The volume diagram, Fig. A.1, can also be used to determine the length of a tank of which the capacity and the diameter are both fixed. In fact, as long as any two factors are known, the third can easily be found. For this reason, many prefer to use the diagram rather than the tables.

TABLE A.1 Cylinder Volume
Volume = $D^2 \times L \times 0.0034$, in which D = internal diameter, in inches, and L = length, in inches

Inside diameter, in.	Cubic feet per foot of cylinder	Gallons per inch of cylinder	Gallons per foot of cylinder	Length of shell, ft										
				3	4	5	6	7	8	9	10	11	12	13
				Volume, gal										
12	0.79	0.49	5.88	17.64	23.52	29.40	35.28	41.16	47.04	52.92	58.80	64.68	70.56	76.44
18	1.77	1.10	13.22	39.66	52.88	66.10	79.32	92.54	105.7	118.9	132.2	145.4	158.6	171.8
24	3.14	1.96	23.58	70.74	94.32	117.9	141.4	165.0	188.6	212.2	235.8	259.3	282.9	306.5
30	4.9	3.06	38.72	116.1	154.8	193.6	232.3	271.0	309.7	348.4	387.2	425.9	464.6	503.3
36	7.1	4.41	52.88	158.6	211.5	264.4	317.2	370.1	423.0	475.9	528.8	581.6	634.5	687.4
42	9.6	6.00	71.97	215.9	287.8	359.8	431.8	503.7	575.7	647.3	719.7	791.6	863.6	935.6
48	12.6	7.83	94.00	282.0	376.0	470.0	564.0	658.0	752.0	846.0	940.0	1,034	1,128	1,222
54	15.9	9.91	118.9	356.9	475.8	594.8	713.8	832.7	951.7	1,070	1,189	1,308	1,427	1,546
60	19.6	12.24	146.8	440.6	587.5	734.4	881.2	1,028	1,175	1,321	1,468	1,615	1,762	1,909
66	23.8	14.81	177.7	533.1	710.8	888.6	1,066	1,244	1,421	1,599	1,777	1,954	2,132	2,310
72	28.3	17.63	211.5	634.5	846.0	1,057	1,269	1,480	1,692	1,903	2,115	2,326	2,538	2,749
78	33.2	20.69	248.2	744.6	992.8	1,241	1,489	1,737	1,985	2,233	2,482	2,730	2,978	3,226
84	38.5	23.99	287.8	863.6	1,151	1,439	1,727	2,015	2,303	2,591	2,878	3,166	3,454	3,742
90	44.2	27.54	330.4	991.4	1,321	1,652	1,982	2,313	2,643	2,974	3,304	3,635	3,965	4,296
96	50.3	31.33	376.0	1,128	1,504	1,880	2,256	2,632	3,008	3,384	3,760	4,136	4,512	4,888
102	56.8	35.37	424.4	1,273	1,697	2,122	2,546	2,971	3,395	3,820	4,244	4,669	5,093	5,518
108	63.6	39.66	475.8	1,427	1,903	2,379	2,855	3,331	3,807	4,283	4,758	5,234	5,710	6,186
114	70.9	44.19	530.2	1,590	2,120	2,651	3,181	3,711	4,241	4,772	5,302	5,832	6,362	6,893
120	78.5	48.96	587.5	1,762	2,350	2,937	3,525	4,112	4,700	5,287	5,875	6,462	7,050	7,637
126	86.3	53.98	647.7	1,943	2,590	3,238	3,886	4,534	5,181	5,829	6,477	7,125	7,772	8,420
132	95.	59.24	710.9	2,132	2,843	3,554	4,265	4,976	5,687	6,398	7,109	7,819	8,530	9,241
138	103.6	64.75	777.0	2,331	3,108	3,885	4,662	5,439	6,216	6,993	7,770	8,547	9,324	10,101
144	113.1	70.50	846.0	2,538	3,384	4,230	5,076	5,922	6,768	7,614	8,460	9,306	10,152	10,998

TABLE A.1 Cylinder Volume (*Continued*)
Volume = $D^2 \times L \times 0.0034$, in which D = internal diameter, in inches, and L = length, in inches

Inside diameter, in.	Cubic feet per foot of cylinder	Gallons per inch of cylinder	Gallons per foot of cylinder	Length of shell, ft											
				14	15	16	17	18	19	20	25	30	35	40
				Volume, gal										
12	0.79	0.49	5.88	82.32	88.20	94.08	99.96	105.8	111.7	117.6	147.0	176.4	205.8	235.2
18	1.77	1.10	13.22	185.0	198.3	211.5	224.7	237.9	251.1	264.4	330.5	396.6	462.7	528.8
24	3.14	1.96	23.58	330.1	353.7	377.2	400.8	424.4	448.0	471.6	589.5	707.4	825.3	943.2
30	4.9	3.06	38.72	542.0	580.8	619.5	658.2	696.9	735.6	774.4	968.0	1,161	1,355	1,548
36	7.1	4.41	52.88	740.3	793.2	846.0	898.9	951.8	1,004	1,057	1,322	1,586	1,850	2,115
42	9.6	6.00	71.97	1,007	1,079	1,151	1,223	1,295	1,367	1,439	1,799	2,159	2,518	2,878
48	12.6	7.83	94.00	1,316	1,410	1,504	1,598	1,692	1,786	1,880	2,350	2,820	3,290	3,760
54	15.9	9.91	118.9	1,665	1,784	1,903	2,022	2,141	2,260	2,379	2,974	3,569	4,163	4,758
60	19.6	12.24	146.8	2,056	2,203	2,350	2,496	2,643	2,790	2,937	3,672	4,406	5,140	5,875
66	23.8	14.81	177.7	2,488	2,665	2,843	3,021	3,198	3,376	3,554	4,443	5,331	6,220	7,108
72	28.3	17.63	211.5	2,961	3,172	3,384	3,595	3,807	4,018	4,230	5,287	6,345	7,402	8,460
78	33.2	20.69	248.2	3,475	3,723	3,971	4,219	4,467	4,716	4,964	6,205	7,446	8,687	9,928
84	38.5	23.99	287.8	4,030	4,318	4,606	4,894	5,182	5,469	5,757	7,197	8,636	10,076	11,515
90	44.2	27.54	330.4	4,626	4,957	5,287	5,618	5,948	6,279	6,609	8,262	9,914	11,566	13,219
96	50.3	31.33	376.0	5,264	5,640	6,016	6,392	6,768	7,144	7,520	9,400	11,280	13,160	15,040
102	56.8	35.37	424.4	5,942	6,367	6,791	7,216	7,640	8,489	8,489	10,612	12,734	14,856	16,979
108	63.6	39.66	475.8	6,662	7,138	7,614	8,090	8,566	9,041	9,517	11,897	14,276	16,656	19,035
114	70.9	44.19	530.2	7,432	7,953	8,483	9,014	9,544	10,074	10,604	13,256	15,907	18,558	21,209
120	78.5	48.96	587.5	8,225	8,812	9,400	9,987	10,575	11,162	11,750	14,688	17,625	20,563	23,500
126	86.3	53.98	647.7	9,068	9,716	10,363	11,011	11,659	12,307	12,954	16,193	19,432	22,670	25,909
132	95.	59.24	710.9	9,952	10,663	11,374	12,085	12,796	13,507	14,218	17,772	21,327	24,881	28,436
138	103.6	64.75	777.0	10,878	11,655	12,432	13,209	13,986	14,763	15,540	19,425	23,310	27,195	31,080
144	113.1	70.50	846.0	11,844	12,690	13,536	14,382	15,228	16,074	16,920	21,150	25,380	29,611	33,841

TABLE A.2 Head Volume
Volumes listed are approximate volumes of dished portion of head; straight flanges are not included

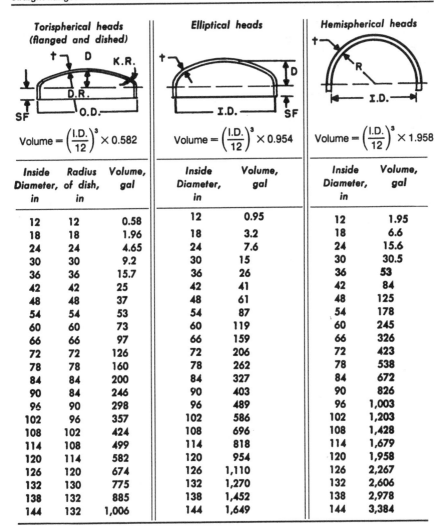

Torispherical heads (flanged and dished)			Elliptical heads		Hemispherical heads	
Volume = $\left(\frac{I.D.}{12}\right)^3 \times 0.582$			Volume = $\left(\frac{I.D.}{12}\right)^3 \times 0.954$		Volume = $\left(\frac{I.D.}{12}\right)^3 \times 1.958$	
Inside Diameter, in	Radius of dish, in	Volume, gal	Inside Diameter, in	Volume, gal	Inside Diameter, in	Volume, gal
12	12	0.58	12	0.95	12	1.95
18	18	1.96	18	3.2	18	6.6
24	24	4.65	24	7.6	24	15.6
30	30	9.2	30	15	30	30.5
36	36	15.7	36	26	36	53
42	42	25	42	41	42	84
48	48	37	48	61	48	125
54	54	53	54	87	54	178
60	60	73	60	119	60	245
66	66	97	66	159	66	326
72	72	126	72	206	72	423
78	78	160	78	262	78	538
84	84	200	84	327	84	672
90	84	246	90	403	90	826
96	90	298	96	489	96	1,003
102	96	357	102	586	102	1,203
108	102	424	108	696	108	1,428
114	108	499	114	818	114	1,679
120	114	582	120	954	120	1,958
126	120	674	126	1,110	126	2,267
132	130	775	132	1,270	132	2,606
138	132	885	138	1,452	138	2,978
144	132	1,006	144	1,649	144	3,384

284 Appendix A

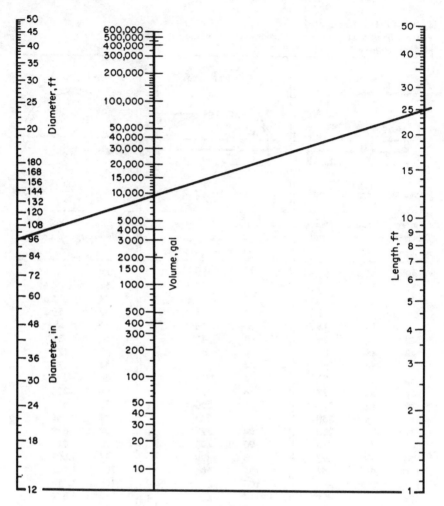

Figure A.1 Cylinder volume diagram. A cylinder 25 ft long and 96 in in diameter will hold 9400 gal.

Appendix B

Circumferences of Cylinders

The length of a shell plate to be cut for a cylindrical vessel is measured by the mean or bend-line circumference of the shell so that the plate will equal a required inside diameter after being rolled. The circumferences given in Table B.1 are for the inside diameter.

When plates are to be cut to provide a given inside diameter, an amount equal to the plate thickness multiplied by 3.1416 must be added to the circumference shown in the table for that diameter. For example, suppose a vessel is to be 60 in in inside diameter and is to have a shell thickness of 1 in. The 60-in inside diameter generates a circumference of 188.496 in. To this figure must be added the plate thickness multiplied by 3.1416, or 3.1416 in. The length of plate to be cut to make a cylinder 60 in in inside diameter is therefore 188.496 plus 3.1416, or 191.6376 in.

A simpler method is to add the thickness of plate to the inside diameter and obtain the circumference directly from the table. For example, for a vessel requiring a 60-in inside diameter and a 1-in shell plate, add 1 in to 60 in for a total of 61 in. The circumference given in the table for a 61-in diameter equals 191.637 in.

The plate for cylinders with dished heads should never be cut until the heads have been taped if a good fit is to be secured at the head joint. To illustrate: If the inside circumference of a head has been taped at 188.5 in and the thickness of the shell and head is 1 in, an amount equal to the plate thickness multiplied by 3.1416 will have to be added to the taped measurement for the shell to meet the inside diameter of the head when the plate is cut. Therefore, the plate will have to be cut to 191.64 in.

If the head has been taped on the outside circumference, subtract an amount equal to the plate thickness multiplied by 3.1416 from the total measurement before cutting the shell plate.

(Text continues on page 296.)

TABLE B.1 Circumferences and Areas of Circles

Areas are approximate but sufficiently accurate for estimating

Dia.	Circum.	Area	Dia.	Circum.	Area	Dia.	Circum.	Area	Dia.	Circum.	Area
3.	9.4248	7.0686	11.	34.558	95.033	19.	59.690	283.53	27.	84.823	572.56
1/8	9.8175	7.6699	1/8	34.950	97.205	1/8	60.083	287.27	1/8	85.216	577.87
1/4	10.210	8.2958	1/4	35.343	99.402	1/4	60.476	291.04	1/4	85.608	583.21
3/8	10.603	8.9462	3/8	35.736	101.62	3/8	60.868	294.83	3/8	86.001	588.57
1/2	10.996	9.6211	1/2	36.128	103.87	1/2	61.261	298.65	1/2	86.394	593.96
5/8	11.388	10.321	5/8	36.521	106.14	5/8	61.654	302.49	5/8	86.786	599.37
3/4	11.781	11.045	3/4	36.914	108.43	3/4	62.046	306.35	3/4	87.179	604.81
7/8	12.174	11.793	7/8	37.306	110.75	7/8	62.439	310.24	7/8	87.572	610.27
4.	12.566	12.566	12.	37.699	113.10	20.	62.832	314.16	28.	87.965	615.75
1/8	12.959	13.364	1/8	38.092	115.47	1/8	63.225	318.10	1/8	88.357	621.26
1/4	13.352	14.186	1/4	38.485	117.86	1/4	63.617	322.06	1/4	88.750	626.80
3/8	13.744	15.033	3/8	38.877	120.28	3/8	64.010	326.05	3/8	89.143	632.36
1/2	14.137	15.904	1/2	39.270	122.72	1/2	64.403	330.06	1/2	89.535	637.94
5/8	14.530	16.800	5/8	39.663	125.19	5/8	64.795	334.10	5/8	89.928	643.55
3/4	14.923	17.728	3/4	40.055	127.68	3/4	65.188	338.16	3/4	90.321	649.18
7/8	15.315	18.665	7/8	40.448	130.19	7/8	65.581	342.25	7/8	90.713	654.84
5.	15.708	19.635	13.	40.841	132.73	21.	65.973	346.36	29.	91.106	660.52
1/8	16.101	20.629	1/8	41.233	135.30	1/8	66.366	350.50	1/8	91.499	666.23
1/4	16.493	21.648	1/4	41.626	137.89	1/4	66.759	354.66	1/4	91.892	671.96
3/8	16.886	22.691	3/8	42.019	140.50	3/8	67.152	358.84	3/8	92.284	677.71
1/2	17.279	23.758	1/2	42.412	143.14	1/2	67.544	363.05	1/2	92.677	683.49
5/8	17.671	24.850	5/8	42.804	145.80	5/8	67.937	367.28	5/8	93.070	689.30
3/4	18.064	25.967	3/4	43.197	148.49	3/4	68.330	371.54	3/4	93.462	695.13
7/8	18.457	27.109	7/8	43.590	151.20	7/8	68.722	375.83	7/8	93.855	700.98
6.	18.850	28.274	14.	43.982	153.94	22.	69.115	380.13	30.	94.248	706.86
1/8	19.242	29.465	1/8	44.375	156.70	1/8	69.508	384.46	1/8	94.640	712.76
1/4	19.635	30.680	1/4	44.768	159.48	1/4	69.900	388.82	1/4	95.033	718.69
3/8	20.028	31.919	3/8	45.160	162.30	3/8	70.293	393.20	3/8	95.426	724.64
1/2	20.420	33.183	1/2	45.553	165.13	1/2	70.686	397.61	1/2	95.819	730.62
5/8	20.813	34.472	5/8	45.946	167.99	5/8	71.079	402.04	5/8	96.211	736.62
3/4	21.206	35.785	3/4	46.338	170.87	3/4	71.471	406.49	3/4	96.604	742.64
7/8	21.598	37.122	7/8	46.731	173.78	7/8	71.864	410.97	7/8	96.997	748.69

7.	21.991	38.485	15.	47.124	176.71	23.	72.257	415.48	31.	97.389	754.77
1/8	22.384	39.871	1/8	47.517	179.67	1/8	72.649	420.00	1/8	97.782	760.87
1/4	22.776	41.282	1/4	47.909	182.65	1/4	73.042	424.56	1/4	98.175	766.99
3/8	23.169	42.718	3/8	48.302	185.66	3/8	73.435	429.13	3/8	98.567	773.14
1/2	23.562	44.179	1/2	48.695	188.69	1/2	73.827	433.74	1/2	98.960	779.31
5/8	23.955	45.664	5/8	49.087	191.75	5/8	74.220	438.36	5/8	99.353	785.51
3/4	24.347	47.173	3/4	49.480	194.83	3/4	74.613	443.01	3/4	99.746	791.73
7/8	24.740	48.707	7/8	49.873	197.93	7/8	75.006	447.69	7/8	100.138	797.98
8.	25.133	50.265	16.	50.265	201.06	24.	75.398	452.39	32.	100.531	804.25
1/8	25.525	51.849	1/8	50.658	204.22	1/8	75.791	457.11	1/8	100.924	810.54
1/4	25.918	53.456	1/4	51.051	207.39	1/4	76.184	461.86	1/4	101.316	816.86
3/8	26.311	55.088	3/8	51.444	210.60	3/8	76.576	466.64	3/8	101.709	823.21
1/2	26.704	56.745	1/2	51.836	213.82	1/2	76.969	471.44	1/2	102.102	829.58
5/8	27.096	58.426	5/8	52.229	217.08	5/8	77.362	476.26	5/8	102.494	835.97
3/4	27.489	60.132	3/4	52.622	220.35	3/4	77.754	481.11	3/4	102.887	842.39
7/8	27.882	61.862	7/8	53.014	223.65	7/8	78.147	485.98	7/8	103.280	848.83
9.	28.274	63.617	17.	53.407	226.98	25.	78.540	490.87	33.	103.673	855.30
1/8	28.667	65.397	1/8	53.800	230.33	1/8	78.933	495.79	1/8	104.065	861.79
1/4	29.060	67.201	1/4	54.192	233.71	1/4	79.325	500.74	1/4	104.458	868.31
3/8	29.452	69.029	3/8	54.585	237.10	3/8	79.718	505.71	3/8	104.851	874.85
1/2	29.845	70.882	1/2	54.978	240.53	1/2	80.111	510.71	1/2	105.243	881.41
5/8	30.238	72.760	5/8	55.371	243.98	5/8	80.503	515.72	5/8	105.636	888.00
3/4	30.631	74.662	3/4	55.763	247.45	3/4	80.896	520.77	3/4	106.029	894.62
7/8	31.023	76.589	7/8	56.156	250.95	7/8	81.289	525.84	7/8	106.421	901.26
10.	31.416	78.540	18.	56.549	254.47	26.	81.681	530.93	34.	106.814	907.92
1/8	31.809	80.516	1/8	56.941	258.02	1/8	82.074	536.05	1/8	107.207	914.61
1/4	32.201	82.516	1/4	57.334	261.59	1/4	82.467	541.19	1/4	107.600	921.32
3/8	32.594	84.541	3/8	57.727	265.18	3/8	82.860	546.35	3/8	107.992	928.06
1/2	32.987	86.590	1/2	58.119	268.80	1/2	83.252	551.55	1/2	108.385	934.82
5/8	33.379	88.664	5/8	58.512	272.45	5/8	83.645	556.76	5/8	108.778	941.61
3/4	33.772	90.763	3/4	58.905	276.12	3/4	84.038	562.00	3/4	109.170	948.42
7/8	34.165	92.886	7/8	59.298	279.81	7/8	84.430	567.27	7/8	109.563	955.25

TABLE B.1 Circumferences and Areas of Circles (Continued)

Dia.	Circum.	Area	Dia.	Circum.	Area	Dia.	Circum.	Area	Dia.	Circum.	Area
35.	109.956	962.11	43.	135.088	1452.2	51.	160.221	2042.8	59.	185.354	2734.0
1/8	110.348	969.00	1/8	135.481	1460.7	1/8	160.614	2052.8	1/8	185.747	2745.6
1/4	110.741	975.91	1/4	135.874	1469.1	1/4	161.007	2062.9	1/4	186.139	2757.2
3/8	111.134	982.84	3/8	136.267	1477.6	3/8	161.399	2073.0	3/8	186.532	2768.8
1/2	111.527	989.80	1/2	136.659	1486.2	1/2	161.792	2083.1	1/2	186.925	2780.5
5/8	111.919	996.78	5/8	137.052	1494.7	5/8	162.185	2093.2	5/8	187.317	2792.2
3/4	112.312	1003.8	3/4	137.445	1503.3	3/4	162.577	2103.3	3/4	187.710	2803.9
7/8	112.705	1010.8	7/8	137.837	1511.9	7/8	162.970	2113.5	7/8	188.103	2815.7
36.	113.097	1017.9	44.	138.230	1520.5	52.	163.363	2123.7	60.	188.496	2827.4
1/8	113.490	1025.0	1/8	138.623	1529.2	1/8	163.756	2133.9	1/8	188.888	2839.2
1/4	113.883	1032.1	1/4	139.015	1537.9	1/4	164.148	2144.2	1/4	189.281	2851.0
3/8	114.275	1039.2	3/8	139.408	1546.6	3/8	164.541	2154.5	3/8	189.674	2862.9
1/2	114.668	1046.3	1/2	139.801	1555.3	1/2	164.934	2164.8	1/2	190.066	2874.8
5/8	115.061	1053.5	5/8	140.194	1564.0	5/8	165.326	2175.1	5/8	190.459	2886.6
3/4	115.454	1060.7	3/4	140.586	1572.8	3/4	165.719	2185.4	3/4	190.852	2898.6
7/8	115.846	1068.0	7/8	140.979	1581.6	7/8	166.112	2195.8	7/8	191.244	2910.5
37.	116.239	1075.2	45.	141.372	1590.4	53.	166.504	2206.2	61.	191.637	2922.5
1/8	116.632	1082.5	1/8	141.764	1599.3	1/8	166.897	2216.6	1/8	192.030	2934.5
1/4	117.024	1089.8	1/4	142.157	1608.2	1/4	167.290	2227.0	1/4	192.423	2946.5
3/8	117.417	1097.1	3/8	142.550	1617.0	3/8	167.683	2237.5	3/8	192.815	2958.5
1/2	117.810	1104.5	1/2	142.942	1626.0	1/2	168.075	2248.0	1/2	193.208	2970.6
5/8	118.202	1111.8	5/8	143.335	1634.9	5/8	168.468	2258.5	5/8	193.601	2982.7
3/4	118.596	1119.2	3/4	143.728	1643.9	3/4	168.861	2269.1	3/4	193.993	2994.8
7/8	118.988	1126.7	7/8	144.121	1652.9	7/8	169.253	2279.6	7/8	194.386	3006.9
38.	119.381	1134.1	46.	144.513	1661.9	54.	169.646	2290.2	62.	194.779	3019.1
1/8	119.773	1141.6	1/8	144.906	1670.9	1/8	170.039	2300.8	1/8	195.171	3031.3
1/4	120.166	1149.1	1/4	145.299	1680.0	1/4	170.431	2311.5	1/4	195.564	3043.5
3/8	120.559	1156.6	3/8	145.691	1689.1	3/8	170.824	2322.1	3/8	195.957	3055.7
1/2	120.951	1164.2	1/2	146.084	1698.2	1/2	171.217	2332.8	1/2	196.350	3068.0
5/8	121.344	1171.7	5/8	146.477	1707.4	5/8	171.609	2343.5	5/8	196.742	3080.3
3/4	121.737	1179.3	3/4	146.869	1716.5	3/4	172.002	2354.3	3/4	197.135	3092.6
7/8	122.129	1186.9	7/8	147.262	1725.7	7/8	172.395	2365.0	7/8	197.528	3104.9

39.	122.522	1194.6	47.	147.655	1734.9	55.	172.788	2375.8	63.	197.920	3117.2
1/8	122.915	1202.3	1/8	148.048	1744.2	1/8	173.180	2386.6	1/8	198.313	3129.6
1/4	123.308	1210.6	1/4	148.440	1753.5	1/4	173.573	2397.5	1/4	198.706	3142.0
3/8	123.700	1217.7	3/8	148.833	1762.7	3/8	173.966	2408.3	3/8	199.098	3154.5
1/2	124.093	1225.4	1/2	149.226	1772.1	1/2	174.358	2419.2	1/2	199.491	3166.9
5/8	124.486	1233.2	5/8	149.618	1781.4	5/8	174.751	2430.1	5/8	199.884	3179.4
3/4	124.878	1241.0	3/4	150.011	1790.8	3/4	175.144	2441.1	3/4	200.277	3191.9
7/8	125.271	1248.8	7/8	150.404	1800.1	7/8	175.536	2452.0	7/8	200.669	3204.4
40.	125.664	1256.6	48.	150.796	1809.6	56.	175.929	2463.0	64.	201.062	3217.0
1/8	126.056	1264.5	1/8	151.189	1819.0	1/8	176.322	2474.0	1/8	201.455	3229.6
1/4	126.449	1272.4	1/4	151.582	1828.5	1/4	176.715	2485.0	1/4	201.847	3242.2
3/8	126.842	1280.3	3/8	151.975	1837.9	3/8	177.107	2496.1	3/8	202.240	3254.8
1/2	127.235	1288.2	1/2	152.367	1847.5	1/2	177.500	2507.2	1/2	202.633	3267.5
5/8	127.627	1296.2	5/8	152.760	1857.0	5/8	177.893	2518.3	5/8	203.025	3280.1
3/4	128.020	1304.2	3/4	153.153	1866.5	3/4	178.285	2529.4	3/4	203.418	3292.8
7/8	128.413	1312.2	7/8	153.545	1876.1	7/8	178.678	2540.6	7/8	203.811	3305.6
41.	128.805	1320.3	49.	153.938	1885.7	57.	179.071	2551.8	65.	204.204	3318.3
1/8	129.198	1328.3	1/8	154.331	1895.4	1/8	179.463	2563.0	1/8	204.596	3331.1
1/4	129.591	1336.4	1/4	154.723	1905.0	1/4	179.856	2574.2	1/4	204.989	3343.9
3/8	129.983	1344.5	3/8	155.116	1914.7	3/8	180.249	2585.4	3/8	205.382	3356.7
1/2	130.376	1352.7	1/2	155.509	1924.4	1/2	180.642	2596.7	1/2	205.774	3369.6
5/8	130.769	1360.8	5/8	155.902	1934.2	5/8	181.034	2608.0	5/8	206.167	3382.4
3/4	131.161	1369.0	3/4	156.294	1943.9	3/4	181.427	2619.4	3/4	206.560	3395.3
7/8	131.554	1377.2	7/8	156.687	1953.7	7/8	181.820	2630.7	7/8	206.952	3408.2
42.	131.947	1385.4	50.	157.080	1963.5	58.	182.212	2642.1	66.	207.345	3421.2
1/8	132.340	1393.7	1/8	157.472	1973.3	1/8	182.605	2653.5	1/8	207.738	3434.2
1/4	132.732	1402.0	1/4	157.865	1983.2	1/4	182.998	2664.9	1/4	208.131	3447.2
3/8	133.125	1410.3	3/8	158.258	1993.1	3/8	183.390	2676.4	3/8	208.523	3460.2
1/2	133.518	1418.6	1/2	158.650	2003.0	1/2	183.783	2687.8	1/2	208.916	3473.2
5/8	133.910	1427.0	5/8	159.043	2012.9	5/8	184.176	2699.3	5/8	209.309	3486.3
3/4	134.303	1435.4	3/4	159.436	2022.8	3/4	184.569	2710.9	3/4	209.701	3499.4
7/8	134.696	1443.8	7/8	159.829	2032.8	7/8	184.961	2722.4	7/8	210.094	3512.5

TABLE B.1 Circumferences and Areas of Circles (*Continued*)

Dia.	Circum.	Area	Dia.	Circum.	Area	Dia.	Circum.	Area	Dia.	Circum.	Area
67.	210.487	3525.7	75.	235.619	4117.9	83.	260.752	5410.6	91.	285.885	6503.9
1/8	210.879	3538.8	1/8	236.012	4132.6	1/8	261.145	5426.9	1/8	286.278	6521.8
1/4	211.272	3552.0	1/4	236.405	4147.4	1/4	261.538	5443.3	1/4	286.670	6539.7
3/8	211.665	3565.2	3/8	236.798	4162.2	3/8	261.930	5459.6	3/8	287.063	6557.6
1/2	212.058	3578.5	1/2	237.190	4177.0	1/2	262.323	5476.0	1/2	287.456	6575.5
5/8	212.450	3591.7	5/8	237.583	4191.8	5/8	262.716	5492.4	5/8	287.848	6593.5
3/4	212.843	3605.0	3/4	237.976	4206.7	3/4	263.108	5508.8	3/4	288.241	6611.6
7/8	213.236	3618.3	7/8	238.368	4221.5	7/8	263.501	5525.3	7/8	288.634	6629.6
68.	213.628	3631.7	76.	238.761	4236.5	84.	263.894	5541.8	92.	289.027	6647.6
1/8	214.021	3645.0	1/8	239.154	4251.4	1/8	264.286	5558.3	1/8	289.419	6665.7
1/4	214.414	3658.4	1/4	239.546	4266.4	1/4	264.679	5574.8	1/4	289.812	6683.8
3/8	214.806	3671.8	3/8	239.939	4281.3	3/8	265.072	5591.4	3/8	290.205	6701.9
1/2	215.199	3685.3	1/2	240.332	4296.3	1/2	265.465	5607.9	1/2	290.597	6720.1
5/8	215.592	3698.7	5/8	240.725	4311.4	5/8	265.857	5624.5	5/8	290.990	6738.2
3/4	215.984	3712.2	3/4	241.117	4326.4	3/4	266.250	5641.2	3/4	291.383	6756.4
7/8	216.377	3725.7	7/8	241.510	4341.5	7/8	266.643	5657.8	7/8	291.775	6774.7
69.	216.770	3739.3	77.	241.903	4356.6	85.	267.035	5674.5	93.	292.168	6792.9
1/8	217.163	3752.8	1/8	242.295	4371.8	1/8	267.428	5691.2	1/8	292.561	6811.2
1/4	217.555	3766.4	1/4	242.688	4386.9	1/4	267.821	5707.9	1/4	292.954	6829.5
3/8	217.948	3780.0	3/8	243.081	4402.1	3/8	268.213	5724.7	3/8	293.346	6847.8
1/2	218.341	3793.7	1/2	243.473	4417.3	1/2	268.606	5741.5	1/2	293.739	6866.1
5/8	218.733	3807.3	5/8	243.866	4432.5	5/8	268.999	5758.3	5/8	294.132	6884.5
3/4	219.126	3821.0	3/4	244.259	4447.8	3/4	269.392	5775.1	3/4	294.524	6902.9
7/8	219.519	3834.7	7/8	244.652	4463.1	7/8	269.784	5791.9	7/8	294.917	6921.3
70.	219.911	3848.5	78.	245.044	4478.4	86.	270.177	5808.8	94.	295.310	6939.8
1/8	220.304	3862.2	1/8	245.437	4493.7	1/8	270.570	5825.7	1/8	295.702	6958.2
1/4	220.697	3876.0	1/4	245.830	4509.0	1/4	270.962	5842.6	1/4	296.095	6976.7
3/8	221.090	3889.8	3/8	246.222	4524.4	3/8	271.355	5859.6	3/8	296.488	6995.3
1/2	221.482	3903.6	1/2	246.615	4539.8	1/2	271.748	5876.5	1/2	296.881	7013.8
5/8	221.875	3917.5	5/8	247.008	4555.2	5/8	272.140	5893.5	5/8	297.273	7032.4
3/4	222.268	3931.4	3/4	247.400	4570.7	3/4	272.533	5910.6	3/4	297.666	7051.0
7/8	222.660	3945.3	7/8	247.793	4586.2	7/8	272.926	5927.6	7/8	298.059	7069.6

71.		223.053	3959.2	79.		248.186	4901.7	87.		273.319	5944.7	95.		298.451	7088.2
	1/8	223.446	3973.1		1/8	248.579	4917.2		1/8	273.711	5961.8		1/8	298.844	7106.9
	1/4	223.838	3987.1		1/4	248.971	4932.7		1/4	274.104	5978.9		1/4	299.237	7125.6
	3/8	224.231	4001.1		3/8	249.364	4948.3		3/8	274.497	5996.0		3/8	299.629	7144.3
	1/2	224.624	4015.2		1/2	249.757	4963.9		1/2	274.889	6013.2		1/2	300.022	7163.0
	5/8	225.017	4029.2		5/8	250.149	4979.5		5/8	275.282	6030.4		5/8	300.415	7181.8
	3/4	225.409	4043.3		3/4	250.542	4995.2		3/4	275.675	6047.6		3/4	300.807	7200.6
	7/8	225.802	4057.4		7/8	250.935	5010.9		7/8	276.067	6064.9		7/8	301.200	7219.4
72.		226.195	4071.5	80.		251.327	5026.5	88.		276.460	6082.1	96.		301.593	7238.2
	1/8	226.587	4085.7		1/8	251.720	5042.3		1/8	276.853	6099.4		1/8	301.986	7257.1
	1/4	226.980	4099.8		1/4	252.113	5058.0		1/4	277.246	6116.7		1/4	302.378	7276.0
	3/8	227.373	4114.0		3/8	252.506	5073.8		3/8	277.638	6134.1		3/8	302.771	7294.9
	1/2	227.765	4128.2		1/2	252.898	5089.6		1/2	278.031	6151.4		1/2	303.164	7313.8
	5/8	228.158	4142.5		5/8	253.291	5105.4		5/8	278.424	6168.8		5/8	303.556	7332.8
	3/4	228.551	4156.8		3/4	253.684	5121.2		3/4	278.816	6186.2		3/4	303.949	7351.8
	7/8	228.944	4171.1		7/8	254.076	5137.1		7/8	279.209	6203.7		7/8	304.342	7370.8
73.		229.336	4185.4	81.		254.469	5153.0	89.		279.602	6221.1	97.		304.734	7389.8
	1/8	229.729	4199.7		1/8	254.862	5168.9		1/8	279.994	6238.6		1/8	305.127	7408.9
	1/4	230.122	4214.1		1/4	255.254	5184.9		1/4	280.387	6256.1		1/4	305.520	7428.0
	3/8	230.514	4228.5		3/8	255.647	5200.8		3/8	280.780	6273.7		3/8	305.913	7447.1
	1/2	230.907	4242.9		1/2	256.040	5216.8		1/2	281.173	6291.2		1/2	306.305	7466.2
	5/8	231.300	4257.4		5/8	256.433	5232.8		5/8	281.565	6308.8		5/8	306.698	7485.3
	3/4	231.692	4271.8		3/4	256.825	5248.8		3/4	281.958	6326.4		3/4	307.091	7504.5
	7/8	232.085	4286.3		7/8	257.218	5264.9		7/8	282.351	6344.1		7/8	307.483	7523.7
74.		232.478	4300.8	82.		257.611	5281.0	90.		282.743	6361.7	98.		307.876	7543.0
	1/8	232.871	4315.4		1/8	258.003	5297.1		1/8	283.136	6379.4		1/8	308.269	7562.2
	1/4	233.263	4329.9		1/4	258.396	5313.3		1/4	283.529	6397.1		1/4	308.661	7581.5
	3/8	233.656	4344.5		3/8	258.789	5329.4		3/8	283.921	6414.9		3/8	309.054	7600.8
	1/2	234.049	4359.2		1/2	259.181	5345.6		1/2	284.314	6432.6		1/2	309.447	7620.1
	5/8	234.441	4373.8		5/8	259.574	5361.8		5/8	284.707	6450.4		5/8	309.840	7639.5
	3/4	234.834	4388.5		3/4	259.967	5378.1		3/4	285.100	6468.2		3/4	310.232	7658.9
	7/8	235.227	4403.1		7/8	260.359	5394.3		7/8	285.492	6486.0		7/8	310.625	7678.3

TABLE B.1 Circumferences and Areas of Circles (Continued)

Dia.	Circum.	Area	Dia.	Circum.	Area	Dia.	Circum.	Area	Dia.	Circum.	Area
99.	311.018	7697.7	107.	336.15	8992	115.	361.28	10387	123.	386.42	11882
1/8	311.410	7717.1	1/8	336.54	9014	1/8	361.68	10410	1/8	386.81	11907
1/4	311.803	7736.6	1/4	336.94	9035	1/4	362.07	10432	1/4	387.20	11931
3/8	312.196	7756.1	3/8	337.33	9056	3/8	362.46	10455	3/8	387.60	11956
1/2	312.588	7775.6	1/2	337.72	9077	1/2	362.86	10477	1/2	387.99	11980
5/8	312.981	7795.2	5/8	338.12	9098	5/8	363.25	10500	5/8	388.38	12004
3/4	313.374	7814.8	3/4	338.51	9119	3/4	363.64	10522	3/4	388.77	12028
7/8	313.767	7834.4	7/8	338.90	9140	7/8	364.03	10545	7/8	389.17	12052
100.	314.16	7854	108.	339.29	9161	116.	364.43	10568	124.	389.56	12076
1/8	314.55	7873	1/8	339.69	9183	1/8	364.82	10590	1/8	389.95	12101
1/4	314.95	7893	1/4	340.08	9204	1/4	365.21	10613	1/4	390.34	12125
3/8	315.34	7913	3/8	340.47	9225	3/8	365.60	10636	3/8	390.74	12150
1/2	315.73	7933	1/2	340.86	9246	1/2	366.00	10659	1/2	391.13	12174
5/8	316.12	7952	5/8	341.26	9268	5/8	366.39	10682	5/8	391.52	12199
3/4	316.52	7972	3/4	341.65	9289	3/4	366.78	10705	3/4	391.92	12223
7/8	316.91	7992	7/8	342.04	9310	7/8	367.18	10728	7/8	392.31	12248
101.	317.30	8012	109.	342.43	9331	117.	367.57	10751	125.	392.70	12272
1/8	317.69	8032	1/8	342.83	9353	1/8	367.96	10774	1/8	393.09	12297
1/4	318.09	8052	1/4	343.22	9374	1/4	368.35	10798	1/4	393.49	12321
3/8	318.48	8071	3/8	343.61	9396	3/8	368.75	10821	3/8	393.88	12346
1/2	318.87	8091	1/2	344.01	9417	1/2	369.14	10844	1/2	394.27	12370
5/8	319.27	8111	5/8	344.40	9439	5/8	369.53	10867	5/8	394.66	12395
3/4	319.66	8131	3/4	344.79	9460	3/4	369.92	10890	3/4	395.06	12419
7/8	320.05	8151	7/8	345.18	9481	7/8	370.32	10913	7/8	395.45	12444
102.	320.44	8171	110.	345.58	9503	118.	370.71	10936	126.	395.84	12469
1/8	320.84	8191	1/8	345.97	9525	1/8	371.11	10960	1/8	396.23	12494
1/4	321.23	8211	1/4	346.36	9546	1/4	371.49	10983	1/4	396.63	12518
3/8	321.62	8231	3/8	346.75	9568	3/8	371.89	11007	3/8	397.02	12543
1/2	322.01	8252	1/2	347.15	9589	1/2	372.28	11030	1/2	397.41	12568
5/8	322.41	8272	5/8	347.54	9611	5/8	372.67	11053	5/8	397.81	12593
3/4	322.80	8292	3/4	347.93	9633	3/4	373.07	11076	3/4	398.20	12618
7/8	323.19	8312	7/8	348.33	9655	7/8	373.46	11099	7/8	398.59	12643

103.		323.59	8332	111.		348.72	9677	119.		373.85	11122	127.		398.98
	⅛	323.98	8352		⅛	349.11	9698		⅛	374.24	11146		⅛	399.38
	¼	324.37	8372		¼	349.50	9720		¼	374.64	11169		¼	399.77
	⅜	324.76	8393		⅜	349.90	9742		⅜	375.03	11193		⅜	400.16
	½	325.16	8413		½	350.29	9764		½	375.42	11216		½	400.55
	⅝	325.55	8434		⅝	350.68	9786		⅝	375.81	11240		⅝	400.95
	¾	325.94	8454		¾	351.07	9808		¾	376.21	11263		¾	401.34
	⅞	326.33	8474		⅞	351.47	9830		⅞	376.60	11287		⅞	401.73
104.		326.73	8495	112.		351.86	9852	120.		376.99	11310	128.		402.13
	⅛	327.12	8515		⅛	352.25	9874		⅛	377.39	11334		⅛	402.52
	¼	327.51	8536		¼	352.65	9897		¼	377.78	11357		¼	402.91
	⅜	327.91	8556		⅜	353.04	9919		⅜	378.17	11381		⅜	403.30
	½	328.30	8577		½	353.43	9941		½	378.56	11404		½	403.70
	⅝	328.69	8597		⅝	353.82	9963		⅝	378.96	11428		⅝	404.09
	¾	329.08	8618		¾	354.22	9985		¾	379.35	11451		¾	404.48
	⅞	329.48	8638		⅞	354.61	10007		⅞	379.74	11475		⅞	404.87
105.		329.87	8659	113.		355.00	10029	121.		380.13	11499	129.		405.27
	⅛	330.26	8679		⅛	355.39	10052		⅛	380.53	11522		⅛	405.66
	¼	330.65	8700		¼	355.79	10074		¼	380.92	11546		¼	406.05
	⅜	331.05	8721		⅜	356.18	10097		⅜	381.31	11570		⅜	406.44
	½	331.44	8741		½	356.57	10119		½	381.70	11594		½	406.84
	⅝	331.83	8762		⅝	356.96	10141		⅝	382.10	11618		⅝	407.23
	¾	332.22	8783		¾	357.36	10163		¾	382.49	11642		¾	407.62
	⅞	332.62	8804		⅞	357.75	10185		⅞	382.88	11666		⅞	408.02
106.		333.01	8825	114.		358.14	10207	122.		383.28	11690	130.		408.41
	⅛	333.40	8845		⅛	358.54	10230		⅛	383.67	11714		⅛	408.80
	¼	333.80	8866		¼	358.93	10252		¼	384.06	11738		¼	409.19
	⅜	334.19	8887		⅜	359.32	10275		⅜	384.45	11762		⅜	409.59
	½	334.58	8908		½	359.71	10297		½	384.85	11786		½	409.98
	⅝	334.97	8929		⅝	360.11	10320		⅝	385.24	11810		⅝	410.37
	¾	335.37	8950		¾	360.50	10342		¾	385.63	11834		¾	410.76
	⅞	335.76	8971		⅞	360.89	10365		⅞	386.02	11858		⅞	411.16

	12668
	12693
	12718
	12743
	12768
	12793
	12818
	12843
	12868
	12893
	12919
	12944
	12970
	12995
	13020
	13045
	13070
	13096
	13121
	13147
	13172
	13198
	13223
	13248
	13273
	13299
	13324
	13350
	13375
	13401
	13426
	13452

TABLE B.1 Circumferences and Areas of Circles (Continued)

Dia.	Circum.	Area	Dia.	Circum.	Area	Dia.	Circum.	Area	Dia.	Circum.	Area
131.	411.55	13478	139.	436.68	15175	147.	461.82	16972	155.	486.95	18869
1/8	411.94	13504	1/8	437.08	15203	1/8	462.21	17000	1/8	487.34	18900
1/4	412.34	13529	1/4	437.47	15230	1/4	462.60	17029	1/4	487.73	18930
3/8	412.73	13555	3/8	437.86	15258	3/8	462.99	17058	3/8	488.13	18961
1/2	413.12	13581	1/2	438.25	15285	1/2	463.39	17087	1/2	488.52	18991
5/8	413.51	13607	5/8	438.65	15313	5/8	463.78	17116	5/8	488.91	19022
3/4	413.91	13633	3/4	439.04	15340	3/4	464.17	17145	3/4	489.30	19052
7/8	414.30	13659	7/8	439.43	15367	7/8	464.56	17174	7/8	489.70	19083
132.	414.69	13685	140.	439.82	15394	148.	464.96	17203	156.	490.09	19113
1/8	415.08	13711	1/8	440.22	15422	1/8	465.35	17232	1/8	490.48	19144
1/4	415.48	13737	1/4	440.61	15449	1/4	465.74	17262	1/4	490.88	19174
3/8	415.87	13763	3/8	441.00	15477	3/8	466.14	17291	3/8	491.27	19205
1/2	416.26	13789	1/2	441.40	15504	1/2	466.53	17321	1/2	491.66	19235
5/8	416.66	13815	5/8	441.79	15532	5/8	466.92	17350	5/8	492.05	19266
3/4	417.05	13841	3/4	442.18	15559	3/4	467.31	17379	3/4	492.45	19297
7/8	417.44	13867	7/8	442.57	15587	7/8	467.71	17408	7/8	492.84	19328
133.	417.83	13893	141.	442.97	15615	149.	468.10	17437	157.	493.23	19359
1/8	418.23	13919	1/8	443.36	15642	1/8	468.49	17466	1/8	493.62	19390
1/4	418.62	13946	1/4	443.75	15670	1/4	468.88	17496	1/4	494.02	19421
3/8	419.01	13972	3/8	444.14	15697	3/8	469.28	17525	3/8	494.41	19452
1/2	419.40	13999	1/2	444.54	15725	1/2	469.67	17555	1/2	494.80	19483
5/8	419.80	14025	5/8	444.93	15753	5/8	470.06	17584	5/8	495.20	19514
3/4	420.19	14051	3/4	445.32	15781	3/4	470.46	17614	3/4	495.59	19545
7/8	420.58	14077	7/8	445.72	15809	7/8	470.85	17643	7/8	495.98	19576
134.	420.97	14103	142.	446.11	15837	150.	471.24	17672	158.	496.37	19607
1/8	421.37	14130	1/8	446.50	15865	1/8	471.63	17702	1/8	496.77	19638
1/4	421.76	14156	1/4	446.89	15893	1/4	472.03	17731	1/4	497.16	19669
3/8	422.15	14183	3/8	447.29	15921	3/8	472.42	17761	3/8	497.55	19701
1/2	422.55	14209	1/2	447.68	15949	1/2	472.81	17790	1/2	497.94	19732
5/8	422.94	14236	5/8	448.07	15977	5/8	473.20	17820	5/8	498.34	19763
3/4	423.33	14262	3/4	448.46	16005	3/4	473.60	17849	3/4	498.73	19794
7/8	423.72	14288	7/8	448.86	16033	7/8	473.99	17879	7/8	499.12	19825

135.	424.12	14314	143.	449.25	16061	151.	474.38	17908	159.	499.51	19856
⅛	424.51	14341	⅛	449.64	16089	⅛	474.77	17938	⅛	499.91	19887
¼	424.90	14367	¼	450.03	16117	¼	475.17	17967	¼	500.30	19919
⅜	425.29	14394	⅜	450.43	16145	⅜	475.56	17997	⅜	500.69	19950
½	425.69	14420	½	450.82	16173	½	475.95	18026	½	501.09	19982
⅝	426.08	14447	⅝	451.21	16201	⅝	476.35	18056	⅝	501.48	20013
¾	426.47	14473	¾	451.61	16229	¾	476.74	18086	¾	501.87	20044
⅞	426.87	14500	⅞	452.00	16258	⅞	477.13	18116	⅞	502.26	20075
136.	427.26	14527	144.	452.39	16286	152.	477.52	18146	160.	502.66	20106
⅛	427.65	14553	⅛	452.78	16314	⅛	477.92	18175	⅛	503.05	20138
¼	428.04	14580	¼	453.18	16342	¼	478.31	18205	¼	503.44	20169
⅜	428.44	14607	⅜	453.57	16371	⅜	478.70	18235	⅜	503.83	20201
½	428.83	14633	½	453.96	16399	½	479.09	18265	½	504.23	20232
⅝	429.22	14660	⅝	454.35	16428	⅝	479.49	18295	⅝	504.62	20264
¾	429.61	14687	¾	454.75	16456	¾	479.88	18325	¾	505.01	20295
⅞	430.01	14714	⅞	455.14	16485	⅞	480.27	18355	⅞	505.41	20327
137.	430.40	14741	145.	455.53	16513	153.	480.67	18385	161.	505.80	20358
⅛	430.79	14768	⅛	455.93	16542	⅛	481.06	18415	⅛	506.19	20390
¼	431.19	14795	¼	456.32	16570	¼	481.45	18446	¼	506.58	20421
⅜	431.58	14822	⅜	456.71	16599	⅜	481.84	18476	⅜	506.98	20453
½	431.97	14849	½	457.10	16627	½	482.24	18507	½	507.37	20484
⅝	432.36	14876	⅝	457.50	16656	⅝	482.63	18537	⅝	507.76	20516
¾	432.76	14903	¾	457.89	16684	¾	483.02	18567	¾	508.15	20548
⅞	433.15	14930	⅞	458.28	16713	⅞	483.41	18597	⅞	508.55	20580
138	433.54	14957	146.	458.67	16742	154.	483.81	18627	162.	508.94	20612
⅛	433.93	14984	⅛	459.07	16770	⅛	484.20	18658	⅛	509.33	20644
¼	434.33	15012	¼	459.46	16799	¼	484.59	18688	¼	509.73	20675
⅜	434.72	15039	⅜	459.85	16827	⅜	484.99	18719	⅜	510.12	20707
½	435.11	15067	½	460.24	16856	½	485.38	18749	½	510.51	20739
⅝	435.50	15094	⅝	460.64	16885	⅝	485.77	18779	⅝	510.90	20771
¾	435.90	15121	¾	461.03	16914	¾	486.16	18809	¾	511.30	20803
⅞	436.29	15148	⅞	461.42	16943	⅞	486.56	18839	⅞	511.69	20835

Some may find it easier to use Table B.1 to arrive at a measurement. For example, if the outside circumference of a head has been taped at 194.78 in, the diameter shown for this measurement is 62 in. If the shell plate is 1 in thick, subtract 1 in from 62 in, for a diameter of 61 in. The circumference given in the table for this diameter equals 191.637 in. This figure represents the mean or bend-line circumference and therefore the length of plate required to make a cylinder 60 in in inside diameter.

Appendix C

Decimal Equivalents and Theoretical Weights of Steel Plates

TABLE C.1

Plate thickness		Weight, psf	Plate thickness		Weight, psf
Fractions of an inch	Decimal equivalent		Fractions of an inch	Decimal equivalent	
1/64	0.01562	0.637	33/64	0.51562	21.037
1/32	0.03125	1.274	17/32	0.53125	21.675
3/64	0.04687	1.911	35/64	0.54687	22.312
1/16	0.0625	2.55	9/16	0.5625	22.950
5/64	0.07812	3.1875	37/64	0.57812	23.587
3/32	0.09375	3.825	19/32	0.59375	24.225
7/64	0.10937	4.4625	39/64	0.60937	24.862
1/8	0.125	5.10	5/8	0.625	25.50
9/64	0.14062	5.7375	41/64	0.64062	26.137
5/32	0.15625	6.375	21/32	0.65625	26.775
11/64	0.17187	7.0125	43/64	0.67187	27.412
3/16	0.1875	7.65	11/16	0.6875	28.050
13/64	0.20312	8.2875	45/64	0.70312	28.687
7/32	0.21875	8.925	23/32	0.71875	29.325
15/64	0.23437	9.5625	47/64	0.73437	29.962
1/4	0.25	10.20	3/4	0.75	30.60
17/64	0.26562	10.837	49/64	0.76562	31.237
9/32	0.28125	11.475	25/32	0.78125	31.875
19/64	0.29687	12.112	51/64	0.79687	32.512
5/16	0.3125	12.75	13/16	0.8125	33.150
21/64	0.32812	13.387	53/64	0.82812	33.787
11/32	0.34375	14.025	27/32	0.84375	34.425
23/64	0.35937	14.662	55/64	0.85937	35.062
3/8	0.375	15.3	7/8	0.875	35.70
25/64	0.39062	15.937	57/64	0.89062	36.337
13/32	0.40625	16.575	29/32	0.90625	36.975
27/64	0.42187	17.212	59/64	0.92187	37.612
7/16	0.4375	17.85	15/16	0.9375	38.250
29/64	0.45312	18.487	61/64	0.95312	38.887
15/32	0.46875	19.125	31/32	0.96875	39.525
31/64	0.48437	19.762	63/64	0.98437	40.162
1/2	0.50	20.40	1	1.00	40.80

Appendix D

Pipe Wall Thicknesses

TABLE D.1
All wall thicknesses conform to ANSI B36.10 excepting 5S and 10S for stainless steel pipe, which conforms to ANSI B36.19. The thicknesses shown represent the nominal or average pipe wall dimensions and include an allowance for mill tolerance of 12.5 percent. Where minimum thickness is required, $t_n \times 0.875 = t_m$

Nominal pipe size	Outside diameter, in.	5S	10S	20	30	40	STD.	60	80	Extra heavy	100	120	140	160	Double extra heavy
1/2	0.840	0.065	0.083			0.109	0.109		0.147	0.147				0.187	0.294
3/4	1.050	0.065	0.083			0.113	0.113		0.154	0.154				0.218	0.308
1	1.315	0.065	0.109			0.133	0.133		0.179	0.179				0.250	0.358
1 1/4	1.66	0.065	0.109			0.140	0.140		0.191	0.191				0.250	0.382
1 1/2	1.90	0.065	0.109			0.145	0.145		0.200	0.200				0.281	0.400
2	2.375	0.065	0.109			0.154	0.154		0.218	0.218				0.343	0.436
2 1/2	2.875	0.083	0.120			0.203	0.203		0.276	0.276				0.375	0.552
3	3.5	0.083	0.120			0.216	0.216		0.300	0.300				0.437	0.600
3 1/2	4.0	0.083	0.120			0.226	0.226		0.318	0.318					0.636
4	4.5	0.083	0.120			0.237	0.237		0.337	0.337				0.531	0.674
5	5.0						0.247			0.355					0.710
5	5.563	0.109	0.134			0.258	0.258		0.375	0.375				0.625	0.750
6	6.625	0.109	0.134			0.280	0.280		0.432	0.432				0.718	0.864
7	7.675						0.301			0.500					0.875
8	8.625	0.109	0.148	0.250	0.277	0.322	0.322	0.406	0.500	0.500	0.593	0.718	0.812	0.906	0.875
9	9.625						0.342			0.500					
10	10.75	0.134	0.165	0.250	0.307	0.365	0.365	0.500	0.593	0.500	0.718	0.843	1.000	1.125	
11	11.75						0.375			0.500					
12	12.75	0.165	0.180	0.250	0.330	0.406	0.375	0.562	0.687	0.500	0.843	1.000	1.125	1.312	
14 O.D.	14.0			0.312	0.375	0.437	0.375	0.593	0.750	0.500	0.937	1.093	1.250	1.406	
16 O.D.	16.0			0.312	0.375	0.500	0.375	0.656	0.843	0.500	1.031	1.218	1.437	1.593	
18 O.D.	18.0			0.312	0.437	0.562	0.375	0.750	0.937	0.500	1.156	1.375	1.562	1.781	
20 O.D.	20.0			0.375	0.500	0.593	0.375	0.812	1.031	0.500	1.280	1.500	1.750	1.968	
24 O.D.	24.0			0.375	0.562	0.687	0.375	0.968	1.218	0.500	1.531	1.812	2.062	2.343	
30 O.D.	30.0			0.500	0.625										

Appendix E

Dry Saturated Steam Temperatures

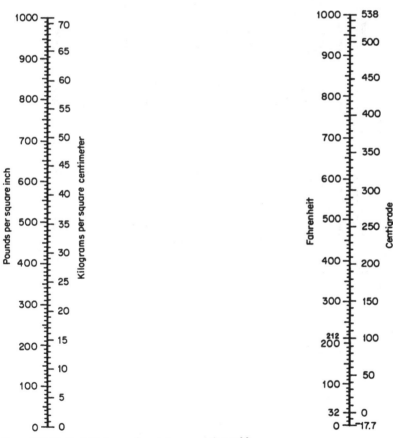

Figure E.1 United States and metric conversion table.

TABLE E.1 Dry Saturated Steam Temperatures

Temperature, deg F	Absolute pressure, psi	Specific volume			Enthalpy			Entropy		
		Saturated liquid	Evaporation	Saturated vapor	Saturated liquid	Evaporation	Saturated vapor	Saturated liquid	Evaporation	Saturated vapor
32	0.08854	0.01602	3306	3306	0.00	1075.8	1075.8	0.0000	2.1877	2.1877
35	0.09995	0.01602	2947	2947	3.02	1074.1	1077.1	0.0061	2.1709	2.1770
40	0.12170	0.01602	2444	2444	8.05	1071.3	1079.3	0.0162	2.1435	2.1597
45	0.14752	0.01602	2036.4	2036.4	13.06	1068.4	1081.5	0.0262	2.1167	2.1429
50	0.17811	0.01603	1703.2	1703.2	18.07	1065.6	1083.7	0.0361	2.0903	2.1264
60	0.2563	0.01604	1206.6	1206.7	28.06	1059.9	1088.0	0.0555	2.0393	2.0948
70	0.3631	0.01606	867.8	867.9	38.04	1054.3	1092.3	0.0745	1.9902	2.0647
80	0.5069	0.01608	633.1	633.1	48.02	1048.6	1096.6	0.0932	1.9428	2.0360
90	0.6982	0.01610	468.0	468.0	57.99	1042.9	1100.9	0.1115	1.8972	2.0087
100	0.9492	0.01613	350.3	350.4	67.97	1037.2	1105.2	0.1295	1.8531	1.9826
110	1.2748	0.01617	265.3	265.4	77.94	1031.6	1109.5	0.1471	1.8106	1.9577
120	1.6924	0.01620	203.25	203.27	87.92	1025.8	1113.7	0.1645	1.7694	1.9339
130	2.2225	0.01625	157.32	157.34	97.90	1020.0	1117.9	0.1816	1.7296	1.9112
140	2.8886	0.01629	122.99	123.01	107.89	1014.1	1122.0	0.1984	1.6910	1.8894
150	3.718	0.01634	97.06	97.07	117.89	1008.2	1126.1	0.2149	1.6537	1.8685
160	4.741	0.01639	77.27	77.29	127.89	1002.3	1130.2	0.2311	1.6174	1.8485
170	5.992	0.01645	62.04	62.06	137.90	996.3	1134.2	0.2472	1.5822	1.8293
180	7.510	0.01651	50.21	50.23	147.92	990.2	1138.1	0.2630	1.5480	1.8109
190	9.339	0.01657	40.94	40.96	157.95	984.1	1142.0	0.2785	1.5147	1.7932
200	11.526	0.01663	33.62	33.64	167.99	977.9	1145.9	0.2938	1.4824	1.7762
210	14.123	0.01670	27.80	27.82	178.05	971.6	1149.7	0.3090	1.4508	1.7598
212	14.696	0.01672	26.78	26.80	180.07	970.3	1150.4	0.3120	1.4446	1.7566
220	17.186	0.01677	23.13	23.15	188.13	965.2	1153.4	0.3239	1.4201	1.7440
230	20.780	0.01684	19.365	19.382	198.23	958.8	1157.0	0.3387	1.3901	1.7288
240	24.969	0.01692	16.306	16.323	208.34	952.2	1160.5	0.3531	1.3609	1.7140
250	29.825	0.01700	13.804	13.821	218.48	945.5	1164.0	0.3675	1.3323	1.6998
260	35.429	0.01709	11.746	11.763	228.64	938.7	1167.3	0.3817	1.3043	1.6860
270	41.858	0.01717	10.044	10.061	238.84	931.8	1170.6	0.3958	1.2769	1.6727
280	49.203	0.01726	8.628	8.645	249.06	924.7	1173.8	0.4096	1.2501	1.6597
290	57.556	0.01735	7.444	7.461	259.31	917.5	1176.8	0.4234	1.2238	1.6472

Temperature, deg F	Absolute pressure, psi	Specific volume			Enthalpy			Entropy		
		Saturated liquid	Evaporation	Saturated vapor	Saturated liquid	Evaporation	Saturated vapor	Saturated liquid	Evaporation	Saturated vapor
300	67.013	0.01745	6.449	6.466	269.59	910.1	1179.7	0.4369	1.1980	1.6350
310	77.68	0.01755	5.609	5.626	279.92	902.6	1182.5	0.4504	1.1727	1.6231
320	89.66	0.01765	4.896	4.914	290.28	894.9	1185.2	0.4637	1.1478	1.6115
330	103.06	0.01776	4.289	4.307	300.68	887.0	1187.7	0.4769	1.1233	1.6002
340	118.01	0.01787	3.770	3.788	311.13	879.1	1190.1	0.4900	1.0992	1.5891
350	134.63	0.01799	3.324	3.342	321.63	870.7	1192.3	0.5029	1.0754	1.5783
360	153.04	0.01811	2.939	2.957	332.18	862.2	1194.4	0.5158	1.0519	1.5677
370	173.37	0.01823	2.606	2.625	342.79	853.5	1196.3	0.5286	1.0287	1.5573
380	195.77	0.01836	2.317	2.335	353.45	844.6	1198.1	0.5413	1.0059	1.5471
390	220.37	0.01850	2.0651	2.0836	364.17	835.4	1199.6	0.5539	0.9832	1.5371
400	247.31	0.01864	1.8447	1.8633	374.97	826.0	1201.0	0.5664	0.9608	1.5272
410	276.75	0.01878	1.6512	1.6700	385.83	816.3	1202.1	0.5788	0.9386	1.5174
420	308.83	0.01894	1.4811	1.5000	396.77	806.3	1203.1	0.5912	0.9166	1.5078
430	343.72	0.01910	1.3308	1.3499	407.79	796.0	1203.8	0.6035	0.8947	1.4982
440	381.59	0.01926	1.1979	1.2171	418.90	785.4	1204.3	0.6158	0.8730	1.4887
450	422.6	0.0194	1.0799	1.0993	430.1	774.5	1204.6	0.6280	0.8513	1.4793
460	466.9	0.0196	0.9748	0.9944	441.4	763.2	1204.6	0.6402	0.8298	1.4700
470	514.7	0.0198	0.8811	0.9009	452.8	751.5	1204.3	0.6523	0.8083	1.4606
480	566.1	0.0200	0.7972	0.8172	464.4	739.4	1203.7	0.6645	0.7868	1.4513
490	621.4	0.0202	0.7221	0.7423	476.0	726.8	1202.8	0.6766	0.7653	1.4419
500	680.8	0.0204	0.6545	0.6749	487.8	713.9	1201.7	0.6887	0.7438	1.4325
520	812.4	0.0209	0.5385	0.5594	511.9	686.4	1198.2	0.7130	0.7006	1.4136
540	962.5	0.0216	0.4434	0.4649	536.6	656.6	1193.2	0.7374	0.6568	1.3942
560	1133.1	0.0221	0.3647	0.3868	562.2	624.2	1186.4	0.7621	0.6121	1.3742
580	1325.8	0.0228	0.2989	0.3217	588.9	588.4	1177.3	0.7872	0.5659	1.3532
600	1542.9	0.0236	0.2432	0.2668	617.0	548.5	1165.5	0.8131	0.5176	1.3307
620	1786.6	0.0247	0.1955	0.2201	646.7	503.6	1150.3	0.8398	0.4664	1.3062
640	2059.7	0.0260	0.1538	0.1798	678.6	452.0	1130.5	0.8679	0.4110	1.2789
660	2365.4	0.0278	0.1165	0.1442	714.2	390.2	1104.4	0.8987	0.3485	1.2472
680	2708.1	0.0305	0.0810	0.1115	757.3	309.9	1067.2	0.9351	0.2719	1.2071
700	3093.7	0.0369	0.0392	0.0761	823.3	172.1	995.4	0.9905	0.1484	1.1389
705.4	3206.2	0.0503	0	0.0503	902.7	0	902.7	1.0580	0	1.0580

Abridged from "Thermodynamic Properties of Steam" by Joseph H. Keenan and Frederick G. Keyes. John Wiley & Sons, New York. 1937

Appendix F

Metric (SI) Units; English and Metric Conversions

Metric notation and terminology employs a base term, preceded by a prefix, which indicates a multiple of that base term. For example, mass is measured in grams. To indicate 1000 grams, the prefix *kilo* is placed in front of gram, and thus, we obtain kilogram. The multiples of the base are always expressed as powers of ten, as follows:

micro	1/1,000,000
milli	1/1,000
centi	1/100
deci	1/10
deca	10
hecta	100
kilo	1,000
mega	1,000,000

An example of pressure vessel thickness calculations is demonstrated for both English and SI metric units as follows:

	English units	Metric units
Pressure P	250 psi	1723.7 kPa
Diameter D	24 in	609.6 mm
Radius R	12 in	304.8 mm
Allowable stress S	13,800 psi	95,148.2 kPa
Joint efficiency E	1.0	1.0
Crown radius L	24 in	609.6 mm

Appendix F

TABLE F.1 Sample Calculations

English	Metric
Shell Thickness	
$T = \dfrac{P(R)}{SE - 0.6P}$	$T = \dfrac{P(R)}{SE - 0.6P}$
$= \dfrac{250(12)}{13,800(1) - 0.6(250)}$	$= \dfrac{1723.7(304.8)}{95,148.2(1) - 0.6(1723.7)}$
$= 0.220$ in	$= 5.582$ mm
Ellipsoidal Head Thickness	
$T = \dfrac{P(D)}{2SE - 0.2P}$	$T = \dfrac{P(D)}{2SE - 0.2P}$
$= \dfrac{250(24)}{2(13,800)(1) - 0.2(250)}$	$= \dfrac{1723.7(609.6)}{2(95,148.2)(1) - 0.2(1723.7)}$
$= 0.218$ in	$= 5.532$ mm
Torispherical Head Thickness	
$T = \dfrac{0.885\, PL}{SE - 0.1P}$	$T = \dfrac{0.885\, PL}{SE - 0.1P}$
$= \dfrac{0.885(250)(24)}{13,800(1) - 0.1(250)}$	$= \dfrac{0.885(1723.7)(609.6)}{95,148(1) - 0.1(1723.7)}$
$= 0.385$ in	$= 9.791$ mm

TABLE F.2 English and Metric Conversions

Multiply	By	To obtain
	Length	
Inches	25.4	Millimeters
Inches	0.0254	Meters
Inches	2.540	Centimeters
Feet	304.8	Millimeters
Feet	30.48	Centimeters
Feet	0.3048	Meters
Millimeters	0.03937	Inches
Centimeters	0.3937	Inches
Meters	39.37	Inches
Meters	3.2808	Feet
	Area	
Square inches	645.2	Square millimeters
Square inches	0.0006452	Square meters
Square feet	0.09290	Square meters
Square inches	6.4516	Square centimeters
Square millimeters	0.00155	Square inches
Square centimeters	0.1550	Square inches
Square meters	10.764	Square feet
	Volume and volumetric flow rates	
Cubic inches	0.01639	Liters
Cubic inches	16390.0	Cubic millimeters
Gallons	3.785	Liters
Liters	0.2642	Gallons
Gallons/minute	0.06308	Liters/second
Gallons	0.003785	Cubic meters
Gallons	3785000.0	Cubic millimeters
Cubic feet	0.02832	Cubic meters
Cubic feet	28.32	Liters
Liters	0.03531	Cubic feet
Liters	61.02	Cubic inches
Cubic meters	264.2	Gallons
Cubic millimeters	0.00006102	Cubic inches
Cubic meters	35.31	Cubic feet
Liters/second	15.8529	Gallons/minute
	Weight or mass	
Pounds	0.4536	Kilograms
Pounds	453.6	Grams
Kilograms	2.2046	Pounds
Grams	0.0022046	Pounds

TABLE F.2 English and Metric Conversions (*Continued*)

Multiply	By	To obtain
Force		
Pounds force	4.448	Newtons
Newtons	0.2248	Pound force
Bending (torque)		
Pound-foot	1.3558	Newton-meters
Pound-inch	0.11298	Newton-meters
Newton-meters	0.73757	Pound-foot
Newton-meters	8.8511	Pound-foot
Pressure, stress		
Pound-square inch (psi)	6894.757	Pascals
Kips/square inch	6894757	Pascals
Pound/square inch	6.8948	Kilopascals
Kips/square inch	6.8948	Megapascals
Kilopascals	0.14504	Pounds/square inch
Kilogram/square centimeter	14.22	Pounds/square inch
Pounds/square inch	0.0703	Kilograms/square inch
Pound/square foot	47.8803	Pascals
Kilogram/square meter	9.8067	Pascals
Fracture toughness		
Thousand pounds-$\sqrt{\text{inch}}$	1.0988×10^6	Pascal-$\sqrt{\text{meter}}$

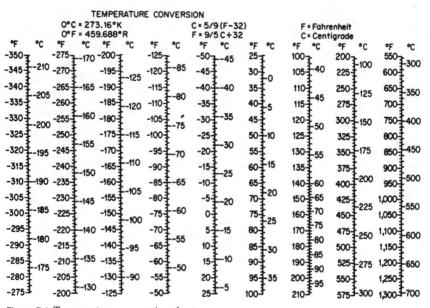

Figure F.1 Temperature conversion chart.

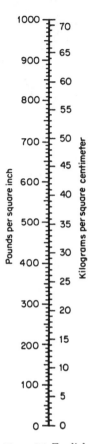

Figure F.2 English and metric conversion table.

Bibliography

"American Society of Mechanical Engineers Boiler Code," *The Locomotive*, Hartford Steam Boiler Inspection and Insurance Co., Hartford, CT. April 1915.

ASME Boiler and Pressure Vessel Code, The American Society of Mechanical Engineers, New York, 1980.

ASME Boiler and Pressure Vessel Code, The American Society of Mechanical Engineers, New York, 1992.

Cargo Tank Maintenance Manual, National Tank Truck Carriers, Inc., Alexandria, VA, 1991.

Code of Federal Regulations, 49 CFR Part 100 through 180 as amended under *Docket No. HM-183*, Research and Special Projects Administration, United States Department of Transportation, Washington, D.C., September 1990, April 1992, August 1992.

Eisenberg, G. M., "Elements of Joint Design," Transactions of the ASME—Series J, *Journal of Pressure Vessel Technology*, February 1975 and February 1977.

Federal Register, September 7, 1990, October 1991.

Garvin, W. L., "Testing, Maintenance, and Installation of Safety Valve and Safety Relief Valves," *National Board Bulletin*, Columbus, OH, July 1971.

Gillissie, John G., *The Authorized Inspector and the Manufacturer's Quality Control System*, The National Board of Boiler and Pressure Vessel Inspectors, Columbus, OH, 1975.

Greene, Arthur M., Jr., *History of the ASME Boiler Code*, The American Society of Mechanical Engineers, New York, NY, 1955.

Harrison, S. F., "Inspector's Responsibility," *National Board Bulletin*, Columbus, OH, October 1969.

Industrial Radiography, Gevaert, Teterboro, NJ, 1965.

Introductory Welding Metallurgy, The American Welding Society, Miami, FL, 1968.

Mooney, John L., "Application of the New ASME Section VII, Division 1 Toughness Requirements to a Typical Pressure Vessel," *Mechanical Engineering*, March 1989, (originally ASME Paper 88–PVP–9, presented at the Pressure Vessels and Piping Conference, June 19–23, 1988 Pittsburgh, PA).

The National Board Inspection Code, The National Board of Boiler and Pressure Vessel Inspectors, Columbus, OH, 1992.

"The National Board of Boiler and Pressure Vessel Inspectors Organizes," *Power*, New York, NY, February 15, 1921.

Panel discussion by Canadian Inspectors, *Proceedings of the Twenty-Third General Meeting of the National Board of Boiler and Pressure Vessel Inspectors,* Columbus, OH, May 1954.

"Revised Rules for Stress Multipliers," *National Board Bulletin,* The National Board of Boiler and Pressure Vessel Inspectors, Columbus, OH, April 1987.

Recommended Practice No. SNT-TC-1A, American Society for Nondestructive Testing, Columbus, OH, 1980.

Recommended Practice No. SNT-TC-1A, American Society for Nondestructive Testing, Columbus, OH, 1986.

"Set Screw Failure," *The Locomotive,* Hartford Steam Boiler Inspection and Insurance Co., Hartford, CT, April 1959.

"Tabulation of the Boiler and Pressure Vessel Laws of the United States and Canada," *Data Sheet,* Uniform Boiler and Pressure Vessel Laws Society, Louisville, KY, October 1991.

Welding Handbook, 7th ed., American Welding Society, Miami, FL, 1982.

Index

Abrasion, 110
Access openings, 106, 107, 109–110, 257, 260, 263
(*See also* Inspection openings)
Accidents:
  basic causes of, 105, 106
  explosions, 1–3
  during hydrostatic tests, 161–163
  prevention of, during welding, 188–189, 200
  with quick-opening doors, 111–113
Addenda, 5, 10, 38, 40, 90
Agitators, design of vessels with, 108
Alteration procedure:
  cargo tanks, 8, 16–27, 247
  pressure vessels, 8, 16–27
  report of, 17, 19–20
American National Standard Institute (ANSI), 4, 230, 232, 271, 272
American Society for Nondestructive Testing (ASNT), 201, 204, 214
(*See also* Nondestructive examinations)
American Society of Mechanical Engineers (ASME), 1
  Boiler and Pressure Vessel Committee, 1, 4–5
  Code developed by (*see* ASME Boiler and Pressure Vessel Code)
  contact information for, 9
American Welding Society, 183, 184
American Welding Society *Standard Welding Symbols,* 183–189
Annealing, 190, 193, 196
Areas of circles, 285–296
ASME Boiler and Pressure Vessel Code:
  alternative rules, 211–215

ASME Boiler and Pressure Vessel Code (*Cont.*):
  Certificate of Authorization, 115–116, 139, 231, 242, 262, 267
  Code Cases, 9–10, 13
  descriptive guide to, 33
  Divisions of, 211
  first committees of, 1–2
  history of, 1–3
  Inspector's duties, 138
  interpretation of, 9–10, 13, 33
  joint review, 6, 8, 102, 116–117
  obtaining a certificate, 5–6
  renewal of a certificate, 6, 116, 139, 231
  Sections, 3–4, 33, 34, 139, 167, 199–200, 211–212, 217, 228
  states/cities/counties/territories requiring construction by, 10–14
  survey, 139–140, 232
  tolerances, 144
Assemblers:
  cargo tank, 16, 246
  pressure vessels and boilers, 16, 218
  qualified, 16
Assessment, ISO 9000 quality, 274–277
Atomic power plants (*see* Nuclear vessels)
Audits, 138–139, 232, 239–240
Austenitic materials:
  effects of heat on, 147, 190–191
  testing of, 97
  in weld deposits, 151, 175, 179
Authorized Inspection Agency, 6, 17, 102, 231, 238, 251
Authorized Inspectors, 7, 9, 16–17, 29–31, 102–103, 212–213, 237, 238, 251

313

Backing strip:
  joint efficiency with use of, 42, 98
  use of, 133–134, 144, 154, 157, 161, 170, 179–180, 200
Bench tests:
  gages, 250
  valves, 250
Bend tests, 41, 167–169, 173–174, 177–178
Bolt stress table, 77
Bolted flat cover:
  plates, 72–73, 77
  thickness chart for, 74–75
Bolted heads and flange connections, calculations for, 76
Braced and stay surfaces:
  calculations for, 81, 82, 84
  thickness chart for, 82, 84
Brazing, 167, 169
Brittle fracture, 89, 90, 97, 126, 213
Buttons, corrosion resistant, 105

Calculations for Code vessels, 118
  (*See also under specific topics*)
Canadian pressure vessel requirements, 10, 12–15
Cargo tanks, 241–270
  acceptable test/inspection results, 266–267
  acceptance criteria, 262–264
  assembly upper coupler, 257, 263
  certification of, 247–249
  characteristics of, 243–244
  compliance date chart, 258
  DOT definitions of terms, 242–243
  glossary of terms, 251–252
  history, 241–242
  hydrostatic testing, 261
  inspector/tester qualifications, 264, 266
  leak testing, 264
  lining inspection, 260
  maintenance of, 248–250, 252–253
  materials for, 243
  pneumatic tests, 261–262
  pressure tests, 260–261, 267
  qualification, 253–255
  registration for repairs/tests of, 265
  repair/modifications to, 267–268
  stamping of, 268–269
  stretching/rebarreling, 268
  structural integrity of, 243
  test and inspections, 256–259

Cargo tanks (*Cont.*):
  ultrasonic thickness testing, 262–264
  upgrading of, 254–255
Castings, radiography of, 135, 153, 178, 200
Cast-iron vessels, 153
Categories of joints, 33–46, 97–99
CE mark, 277
Certificate of Compliance, 247–249
Certification modules, 275–277
  classification table, 276
  global approach, 275, 277
Checklist, inspection, 124–125
Circumferences of cylinders, 285, 296
Circumferential joints:
  radiographic examination of, 33, 38, 151, 153
  single-welded butt joints, 179
  tolerances for, 184, 212
Cities requiring Code vessels, 12
Clad material, 109, 151–152, 170, 197
Closures, quick-opening, 111, 113
Code Cases, ASME, 9–10, 13
Code inspector (*see* Authorized Inspectors)
Code symbol, 218
  DOT requirements for, 15
  Hazardous Material Transportation Act of 1990 requirement for, 242
  and ISO 9001 program requirements, 272
  procedure for obtaining, 5–6
  as quality assurance symbol, 231
  for repairs, 247, 256
Compliance date table, 258
Conical heads, calculations for, 78
Corrosion:
  allowances for, 42–43, 58, 73, 83, 117, 141, 164
  cargo tanks, 259
  correcting for, 87–88
  design precautions against, 107–109
  failures caused by, 105
Counties (U.S.) requiring Code vessels, 12
Cover plates, 110
Critical range of steels, 190, 193
Cryogenic vessels, 89–90, 98
Cyclic service, 212
Cylindrical shells, calculations for, 54–58

Data reports, 123, 136, 148, 238
  data sheets, 123, 128, 137, 148

Data reports (*Cont.*):
  for Division 2, 214
  field assembly, 149–150
  filed with National Board, 148
  impact test exemptions in, 96–97
  joint efficiencies in, 42
  manufacturer's, 238, 270
  pressure protection statements in, 113, 149
  service restrictions in, 99
Decimal equivalents for plate thicknesses, 297
Department of Transportation (DOT) cargo tank requirements, 241–270
Design Certifying Engineer, 251–252
Design control, 148, 239
  report, 148, 212, 215, 229, 233–234
  review of, 148, 213–215
Design of pressure vessels:
  choice of materials, 35–37
  precautions, 89
  specifications for, 215, 229–230, 233
  (*See also specific topics*)
Dimpled jackets, 101
Division 2 vessels, 211–215
Division 3 vessels, 215
Document control, 234–235, 240
DOT 406 cargo tanks, 243, 254–255
DOT 407 cargo tanks, 243, 254–255
DOT 412 cargo tanks, 243, 254–255

Eccentricity of shells (*see* Tolerances)
Eddy current testing, 201, 202
Effective dates, 242, 258
Efficiency of joints, 3, 38–39, 43–46, 59–70, 151, 159
Ellipsoidal heads, 46–53, 58, 65–70
  calculations for, 71–73
  external pressure, 73
  internal pressure, 71–72
Engineering:
  control of design, 233–234
  control of material, 119–120, 225, 237
  quality assurance, 231–232
Erosion (*see* Corrosion)
Etching of sectioned specimen, 103, 159–160
External inspection:
  hydrostatic test, 261
  pneumatic test, 261–262
  visual, 200, 257, 259

External pressure:
  cylindrical shells, 285–296
  heads, 73, 80

Fatigue, 106
Fatigue analysis, 212
Ferritic material, 95, 170, 190, 192, 196
Field assembly inspection, 149–150
Fillet welds, size of, 134, 144
Flame cutting, 184
Flat heads:
  calculations for, 72–75
  thickness chart for, 76
Flued openings (*see* Openings and reinforced openings)
Forgings:
  inspection of, 120, 200, 213
  nondestructive examination of, 134, 153
  postweld treatment of, 197
  welded repairs of, 153

Gages:
  calibration of, 135–136
  requirements of, 162, 250, 261
Gamma-ray radiography, 153–155
Girth joints (*see* Circumferential joints)
Grain size, effects of heat on, 189–193

Handling:
  shipping, 237, 240
  storage, 237, 240
Heads:
  conical, 78
  elements of joint design, 46–53, 58, 72
  ellipsoidal, 46–53, 58, 65–72
  flat, 72–75
  hemispherical, 80
  knuckle radius, 72
  skirt length, 71
  toriconical, 78–79
  torispherical, 46–53, 58–64, 71–72
  volumes of, 283
Heat treatment, 120, 126, 135, 147, 164, 172, 189–190, 193–198, 213
  (*See also* Postweld heat treatment)
Hemispherical heads, 80
HM-183 (House Measure 183), 241
Hold points, 123, 238
Hold tags, 129
Hydrostatic testing:
  under alternative rules, 214
  cargo tanks, 260–261

Hydrostatic testing (*Cont.*):
  cast iron, 142
  pressures for, 126, 136, 142, 161, 214
  water temperature, 126

Identification:
  material and plate, 120–121, 126, 128
  trepanning specimens, 160
  of welds, 132
Impact test, 89–97
  to avoid brittle fracture, 90
  of cargo tanks, 244, 246
  changes to rules, 90
  exemptions from, 86–88, 90–97
Inspection:
  checklist, 124–125
  in Division 2, 213
  field assembly, 149–150
  hydrostatic, 137–138, 142, 260–261
  internal, 109–110, 259–260
  of materials, 120–123
  seven simplified steps to, 163–165
  shop, 116–117, 141, 143–146
  vessels exempt from, 102–103
  of welded repairs, 16–20, 195
Inspection agencies:
  Authorized, 251
  contract with, 6
Inspection openings, 109–110, 257, 260, 263
  access to, 106, 107
  minimum sizes for, 110
Inspectors (*see* Authorized Inspectors)
Integrally clad plate, 109, 151–152, 170, 197
Interlocking device for quick-opening doors, 111, 113
Internal inspection, visual, 109–110, 259–260
Internal pressure (heads), 71–72, 80
International Organization for Standardization (ISO), 271–272
  (*See also* ISO 9000 standards)
ISO 9000 standards, 271–278

Jacketed vessels, 101
Job shops, 22, 28–31
Joints:
  bolted, 76, 77
  categories of, 33–46, 97–99
  design of heads, 76, 77
  efficiency of (*see* Efficiency of joints)

Joints (*Cont.*):
  offset, 167
  welded (*see* Welded joints)

Knuckle radius, 72

Laminations, 106, 121, 143, 159
Layered construction, 213
Leak testing, 264
Lethal substances, vessels containing, 98–99, 151, 196
Lining inspection (cargo tanks), 260
Liquid penetrant inspection, 143, 213
Loadings:
  machinery, 107–108
  piping systems, 107
  snow, 108
  thermal expansion, 108
  wind, 107
Longitudinal joints, tolerance for, 145–146
Low-temperature materials, 43–44
Low-temperature vessels, 89–91

Magnetic particle test, 149, 200, 262
Manholes (manways):
  for cargo tanks, 244, 250, 254–255, 262
  for pressure vessels, 111–113
  retrofit, 257, 260, 263
  testing, 106, 107, 110
Manufacturer:
  of cargo tanks, DOT requirements for, 242–243
  data report of (*see* Data reports)
  design report of, 211–215
  for field assembly, 149–150
  organization chart of, 116
  responsibility of, 5–6, 8–9, 13–15
  statement of authority from, 116
  welding qualification of, 130–131
Markings:
  of plates, 121, 123, 126, 128–129
  of tank trucks, 247, 248, 254, 268
  (*See also* Stamping)
Materials:
  allowable stresses for, 43–44
  austenitic (*see* Austenitic materials)
  for cargo tanks, 243
  certification, 31
  choice of, 211, 213
  clad plate, 109, 151–152, 170, 197
  classes of, 35–37

Materials (*Cont.*):
  control, 119–120, 213, 235
  corrosion of (*see* Corrosion)
  defective, 106
  ductility of, 89, 106, 193, 195
  fatigue, 106, 212
  forgings, 120, 134, 153, 197, 200, 213
  identification of, 120–121, 126, 128
  inspection of, 120–123
  low-temperature, 89–91
  notch toughness, 86–88, 90–97, 246
  procurement control of, 116, 119–128
  stamping of (*see* Stamping)
  stress values of, 35–37
Maximum allowable working pressure (MAWF), 243
MC-331 cargo tank, 244–247
Metric conversion table, 305–309
Mill orders, 29–30
Mill test reports, 121, 128, 143, 150, 164
Minimum design metal temperature (MDMT), 90, 92–95
Minimum thickness:
  of heads, 46–53, 58–70, 72
  of pipe, 299–300
  of plate, 109, 266
  of stayed surfaces, 82, 84
Modification, 242, 253, 267–268

Nameplates, 17, 20, 27, 103–104, 270
National Board of Boiler and Pressure Vessel Inspectors, 7–9
National Board R symbol, 8, 15–27
  alterations, 17, 20, 27
  DOT requirements for cargo tanks, 242
  weld repairs, 17, 27
Noncircular cross section, 99–100
Nonconformance, 238–239
  correction of, 128–130, 238–239
  report of, 130
Nondestructive examinations, 134–136, 199–209
  acoustic emissions, 208–209
  under alternative rules, 213–214
  certification of personnel, 134
  dye penetrant, 146, 203
  leak test, 205, 208, 264
  magnetic particle, 143, 146, 203, 207, 262
  qualification of procedures, 134–135
  radiography, 33, 38–42, 144, 151–159, 202, 206–207, 214

Nondestructive examinations (*Cont.*):
  spark testing, 260
  ultrasonic, 205, 208, 259, 262–264
Normalizing, 193
Notified bodies, 275–278
Nozzle openings, 83, 85–89, 106–107, 110, 263
  (*See also* Openings and reinforced openings)
Nuclear vessels:
  boiling water reactor, 227
  chain reaction, 225
  coolant system, 226–228
  fission, 225–226
  fusion, 225–226
  metal containments, 229
  neutrons, role of, 225–227
  nonpermanent records, 239
  permanent records, 239
  plant components, 229
  power plant cycles, 225
  pressurized water reactor, 226–227
  primary circuit, 225–227
  quality assurance manual, 232–240
  secondary circuit, 226–227
  steel liner, 229

Offset, 145–146
  (*See also* Tolerances)
Openings and reinforced openings:
  availability of, 260
  calculation sheet for, 83, 87–88
  in or adjacent to welds, 184
  requirements/Code paragraphs for, 88, 89
  ultrasonic thickness readings near, 263
  (*See also* Inspection openings)
Organization chart, 116, 232
Out-of-roundness in cylindrical shells, 145–146

Peening, 172
Penetrameters, 154–156
Pipe:
  boiler, 218, 222
  code jurisdiction over, 218, 222
  wall thickness of, 299–300
  welded, 213
Plates:
  alignment of, 146
  cutting of, 16–17, 286

Plates (*Cont.*):
  identification of, 120–121, 126, 128
  impact test of, 86–97
  inspection of, 120–121, 123
  integrally clad, 109, 151–152, 170, 197
  marking of, 121, 123, 126, 128–129
  mill orders, 29
  warehouse orders, 30
  weights of, 297
Pneumatic tests, 261–264
Porosity, excessive, 106, 134, 144
Postweld heat treatment, 194–198
  condition requiring, 196–198
  furnace examination, 195–196
  stamping for, 104
Power Boiler Code, 217–223
Preheating, 193–194
Prequalified standard welding procedures specifications, 169–170
Pressure relief devices, 8, 106, 113–115
Pressure tests:
  cargo tanks, 260–261, 267
  tank trucks, 261
  (*See also* Hydrostatic testing; Pneumatic tests)
Process control, 123, 235–236
Production schedule, 22, 28

Q90 series standards, 272
Quality control manuals:
  Code requirements for, 116
  under Division 2, 215
  for nuclear vessels, 232–240
  writing/implementing, 115–140
Quality control/assurance:
  under Division 2, 214–215
  ISO 9000 standards, 271–278
  for nuclear vessels, 231–240
  records for, 136–137
  systems, 115–140
Quick-opening closures, 111–113

Radiography:
  under alternative rules, 214
  design thickness based on degree of, 33, 38–42
  interpretation of, 151–159
  penetrameters, 155–158
Rebarreling, 242, 253, 268
Recording gages, 135, 261
Records:
  for nuclear vessels, 239
  for quality control, 136–137

Records (*Cont.*):
  software for keeping, 168
  of welders/welding operators, 167
  of welding operations, 8, 168–169
  (*See also* Data reports)
Reference chart for ASME Code, 218–223
Registration statement (cargo tanks), 265
Reinforcement of openings (*see* Openings and reinforced openings)
Repairs:
  cargo tank, 253, 267–268
  Code symbol for, 247, 256
  welded (*see* Welded repairs)
  (*See also* National Board R symbol)

Safety interlocking devices, 112
Safety relief devices, 8, 106, 113–114
Sectioning, 159–161
Service restrictions, 97, 98
Shells:
  circumferences of (*see* Circumferences of cylinders)
  cylindrical (*see* Cylindrical shells)
  volume of (*see* Volume tables and diagrams)
Shop:
  inspection of (*see* Authorized Inspectors)
  organization of, 116–117, 233
Shop Reviews, 8
Slag inclusions, 154
Specification cargo tank (*see* MC-331)
Specifications:
  design, 215, 229–230, 233
  standard welding procedures, 169–170
  user's, 212, 214–215, 222, 233–234, 240
Spherical shells, calculations for, 54–58
Spot radiography, 158
Stamping:
  with Code symbol, 103–104
  repairs or altered vessels, 16–27
Standard welding procedures specifications (SWPSs), 169–170
Standard Welding Symbols, 183–189
Statement of authority, 116, 233
States requiring Code vessels, 10–13
Statistical process control, 272
Stays and staybolts:
  calculations for, 79–82, 84
  table for pressure, pitch, and thickness, 82, 84

Steam temperature tables, 301–303
Stress analysis, 211–212
Stress relieving (see Postweld heat treatment)
Stress values of material, 35–37, 43–44
Stretching, 253, 268
Synopsis of Boiler and Pressure Vessel Laws, Rules and Regulations, 9–14

Tensile test, 173
Territories (of U.S.) requiring Code vessels, 12, 14
Test equipment control, 135–136, 237
Test gages (see Gages, test)
Tests (see specific types, e.g., Bend tests)
Thermocouples (see Heat treatment)
Thickness, minimum (see Minimum thickness)
Thickness gages (see Penetrameters)
Threaded connections, 95
Tolerances, 145–146
Toriconical heads, calculations for, 78–79
Torispherical heads, 46–53, 58–64, 71–72
  calculations for, 71–73
  external pressure, 73
  internal pressure, 71–72
Total quality control (TQC), 271
Trepanning, 159–161

U symbol stamp, 102–103
Ultrasonic testing, 205, 208, 259, 262–264
UM symbol, 102–103
Unfired steam boilers, 99, 196, 197
User's design specification, 212, 214–215, 222, 233–234, 240

Valves, safety, 8, 106, 113–114
Vendors, 240
Venting, 126, 147, 254, 261
Volume tables and diagrams, 287–296

Warehouse orders, 30
Weights (of steel plates), 297
Welded joints:
  categories of, 97–99

Welded joints (*Cont.*):
  details and symbols of, 180–181, 183–189
  effects of heat on, 189–193
  efficiency of (see Efficiency of joints)
  fit-up of, 184–189
  flame cutting plates with, 184
  grain size in, 189–193
  leg size of, 134, 144
  preparation and fit-up of, 184–189
  radiographic examination of, 144, 151–159, 202, 206–207
  sectioning, 159–161
  stress relieving (see Postweld heat treatment)
  throat size of, 144, 147
Welded repairs:
  of clad vessels, 20, 22, 27
  National Board R symbol, 8, 15, 27, 242, 265
  record of, 8
  report of, 17–18
Welders and welding operators:
  identifying symbols of, 132
  manufacturer's records of, 167
  qualification of, 168, 170, 176–179
  welder's log, 132, 133
Welding:
  accessibility for, 117
  details and symbols, 180–181, 183–189
  electrical characteristics, 174
  filler metal, 171
  fillet weld size, 134
  heat affected zone, 189–193
  manufacturer's record of, 168–169
  position, 171
  postweld heat treatment, 172
  preheat procedure qualification, 171–172
  prequalified standard welding procedures specifications, 169–170
  qualification positions, 173
  quality assurance, 231–239
  technique, 174
  variables, 173–176
  welder qualifications, 176–179

X-ray radiography (see Radiography)

ABOUT THE AUTHOR

J. Phillip Ellenberger, P.E., is Vice President of Engineering at WFI International, Inc., Houston, Texas, a design and manufacturing firm specializing in machined branch connections; and a life member of ASME, for which he serves on several codes and standards committees. Mr. Ellenberger is active with the B16 F & C subcommittees, the B31.3 design task group, and the B31 mechanical design and fabrication committee. He is also the MSS Chairman of the Coordinating Committee and Committee 113, and IS member of WG 10 SC67. He has taught piping stress analysis at the University of Houston, and numerous professional seminars on B31.3 and related topics.